Graph Theory

ENCYCLOPEDIA OF MATHEMATICS and Its Applications

GIAN-CARLO ROTA, Editor
Department of Mathematics
Massachusetts Institute of Technology
Cambridge, Massachusetts

Editorial Board

Janos D. Aczel, *Waterloo*
George E. Andrews, *Penn State*
Richard Askey, *Madison*
Michael F. Atiyah, *Oxford*
Donald Babbitt, *U.C.L.A.*
Lipman Bers, *Columbia*
Garrett Birkhoff, *Harvard*
Raoul Bott, *Harvard*
James K. Brooks, *Gainesville*
Felix E. Browder, *Chicago*
A. P. Calderon, *Buenos Aires*
Peter A. Carruthers, *Los Alamos*
S. Chandrasekhar, *Chicago*
S. S. Chern, *Berkeley*
Hermann Chernoff, *M.I.T.*
P. M. Cohn, *Bedford College, London*
H. S. MacDonald Coxeter, *Toronto*
George B. Dantzig, *Stanford*
Nelson Dunford, *Sarasota, Florida*
F. J. Dyson, *Inst. for Advanced Study*
Harold M. Edwards, *Courant*
Harvey Friedman, *Ohio State*
Giovanni Gallavotti, *Rome*
Andrew M. Gleason, *Harvard*
James Glimm, *Courant*
M. Gordon, *Essex*
Elias P. Gyftopoulos, *M.I.T.*
Peter Henrici, *ETH, Zurich*
Nathan Jacobson, *Yale*
Mark Kac, *U.S.C.*
Shizuo Kakutani, *Yale*
Samuel Karlin, *Stanford*
J. F. C. Kingman, *Oxford*

Donald E. Knuth, *Stanford*
Joshua Lederberg, *Rockefeller*
André Lichnerowicz, *Collège de France*
M. J. Lighthill, *London*
Chia-Chiao Lin, *M.I.T.*
Jacques-Louis Lions, *Paris*
G. G. Lorentz, *Austin*
Roger Lyndon, *Ann Arbor*
Robert J. McEliece, *Caltech*
Henry McKean, *Courant*
Marvin Marcus, *Santa Barbara*
N. Metropolis, *Los Alamos*
Frederick Mosteller, *Harvard*
Jan Mycielski, *Boulder*
L. Nachbin, *Rio de Janeiro* and *Rochester*
Steven A. Orszag, *M.I.T.*
Alexander Ostrowski, *Basel*
Roger Penrose, *Oxford*
Carlo Pucci, *Florence*
Fred S. Roberts, *Rutgers*
Abdus Salam, *Trieste*
M. P. Schützenberger, *Paris*
Jacob T. Schwartz, *Courant*
Irving Segal, *M.I.T.*
Oved Shisha, *Univ. of Rhode Island*
I. M. Singer, *Berkeley*
Olga Taussky, *Caltech*
René Thom, *Bures-sur-Yvette*
John Todd, *Caltech*
John W. Tukey, *Princeton*
Stanislaw Ulam, *Santa Fe, New Mexico*
Veeravalli S. Varadarajan, *U.C.L.A.*
Antoni Zygmund, *Chicago*

GIAN-CARLO ROTA, *Editor*
ENCYCLOPEDIA OF MATHEMATICS AND ITS APPLICATIONS

Volume		Section
1	LUIS A. SANTALÓ **Integral Geometry and Geometric Probability,** 1976 (2nd printing, with revisions, 1979)	Probability
2	GEORGE E. ANDREWS **The Theory of Partitions,** 1976 (2nd printing, 1981)	Number Theory
3	ROBERT J. McELIECE **The Theory of Information and Coding** A Mathematical Framework for Communication, 1977 (2nd printing, with revisions, 1979)	Probability
4	WILLARD MILLER, Jr. **Symmetry and Separation of Variables,** 1977	Special Functions
5	DAVID RUELLE **Thermodynamic Formalism** The Mathematical Structures of Classical Equilibrium Statistical Mechanics, 1978	Statistical Mechanics
6	HENRYK MINC **Permanents,** 1978	Linear Algebra
7	FRED S. ROBERTS **Measurement Theory** with Applications to Decisionmaking, Utility, and the Social Sciences, 1979	Mathematics and the Social Sciences
8	L. C. BIEDENHARN and J. D. LOUCK **Angular Momentum in Quantum Physics:** Theory and Application, 1981	Mathematics of Physics
9	L. C. BIEDENHARN and J. D. LOUCK **The Racah-Wigner Algebra in Quantum** **Theory,** 1981	Mathematics of Physics

GIAN-CARLO ROTA, *Editor*
ENCYCLOPEDIA OF MATHEMATICS AND ITS APPLICATIONS

Volume		Section
10	JOHN D. DOLLARD and CHARLES N. FRIEDMAN **Product Integration** with Application to Differential Equations, 1979	Analysis
11	WILLIAM B. JONES and W. J. THRON **Continued Fractions:** Analytic Theory and Applications, 1980	Analysis
12	NATHANIEL F. G. MARTIN and JAMES W. ENGLAND **Mathematical Theory of Entropy,** 1981	Real Variables
13	GEORGE A. BAKER, Jr. and PETER R. GRAVES-MORRIS **Padé Approximants, Part I:** **Basic Theory,** 1981	Mathematics of Physics
14	GEORGE A. BAKER, Jr. and PETER R. GRAVES-MORRIS **Padé Approximants, Part II:** **Extensions and Applications,** 1981	Mathematics of Physics
15	E. C. BELTRAMETTI and G. CASSINELLI **The Logic of Quantum Mechanics,** 1981	Mathematics of Physics
16	G. D. JAMES and A. KERBER **The Representation Theory of the Symmetric Group,** 1981	Algebra
17	M. LOTHAIRE **Combinatorics on Words,** 1982	Algebra

GIAN-CARLO ROTA, *Editor*
ENCYCLOPEDIA OF MATHEMATICS AND ITS APPLICATIONS

Volume		Section
18	HECTOR FATTORINI **The Cauchy Problem**, 1983	Analysis
19	G. G. LORENTZ, K. JETTER, and S. D. RIEMENSCHNEIDER **Birkhoff Interpolation**, 1983	Interpolation and Approximation
20	RUDOLF LIDL and HARALD NIEDERREITER **Finite Fields**, 1983	Algebra
21	W. T. TUTTE **Graph Theory**, 1984	Combinatorics
22	JULIO R. BASTIDA **Field Extensions and Galois Theory**, 1984	Algebra
23	JOHN R. CANNON **The One-Dimensional Heat Equation**, 1984	Analysis

Other volumes in preparation

GIAN-CARLO ROTA, *Editor*
ENCYCLOPEDIA OF MATHEMATICS AND ITS APPLICATIONS
Volume 21

Section: Combinatorics
Gian-Carlo Rota, *Section Editor*

Graph Theory

W. T. Tutte
Faculty of Mathematics
University of Waterloo
Waterloo, Ontario, Canada

Foreword by
Crispin St. J. A. Nash-Williams
The University of Reading

PUBLISHED BY THE PRESS SYNDICATE OF THE UNIVERSITY OF CAMBRIDGE
The Pitt Building, Trumpington Street, Cambridge, United Kingdom

CAMBRIDGE UNIVERSITY PRESS
The Edinburgh Building, Cambridge CB2 2RU, UK
40 West 20th Street, New York, NY 10011-4211, USA
10 Stamford Road, Oakleigh, Melbourne 3166, Australia
Ruiz de Alarcón 13, 28014 Madrid, Spain
Dock House, The Waterfront, Cape Town 8001, South Africa

www.cambridge.org
Information on this title: www.cambridge.org/9780521794893

© Cambridge University Press 2001

Hardback edition published by Addison-Wesley 1984

This book is in copyright. Subject to statutory exception and
to the provisions of relevant collective licensing agreements,
no reproduction of any part may take place without
the written permission of Cambridge University Press.

A catalog record for this book is available from the British Library.

Library of Congress Cataloging in Publication data

Tutte, W. T.
　Graph theory.

　(Encyclopedia of mathematics and its applications; v. 21)
　Bibliography: p.
　Includes index.
　1.Graph theory. I. Title. II. Series.
QA166.T815　1984　511'.5　83-12210

ISBN-13　978-0-521-79489-3 paperback
ISBN-10　0-521-79489-7 paperback

Transferred to digital printing 2005

Contents

Editor's Statement xiii
Foreword xv
Introduction xix

Chapter I Graphs and Subgraphs 1

    I.1 Definitions 1
    I.2 Isomorphism 5
    I.3 Subgraphs 9
    I.4 Vertices of attachment 11
    I.5 Components and connection 14
    I.6 Deletion of an edge 18
    I.7 Lists of nonisomorphic connected graphs 22
    I.8 Bridges 27
    I.9 Notes 30
          Exercises 31
          References................................. 31

Chapter II Contractions and the Theorem of Menger 32

    II.1 Contractions 32
    II.2 Contraction of an edge 37
    II.3 Vertices of attachment 41

II.4	Separation numbers	43
II.5	Menger's Theorem	46
II.6	Hall's Theorem	50
II.7	Notes	52
	Exercises	53
	References	53

Chapter III 2-Connection ... 54

III.1	Separable and 2-connected graphs	54
III.2	Constructions for 2-connected graphs	56
III.3	Blocks	60
III.4	Arms	64
III.5	Deletion and contraction of an edge	66
III.6	Notes	68
	Exercises	68
	References	69

Chapter IV 3-Connection ... 70

IV.1	Multiple connection	70
IV.2	Some constructions for 3-connected graphs	74
IV.3	3-blocks	83
IV.4	Cleavages	95
IV.5	Deletions and contractions of edges	104
IV.6	The Wheel Theorem	111
IV.7	Notes	113
	Exercises	114
	References	114

Chapter V Reconstruction ... 115

V.1	The Reconstruction Problem	115
V.2	Theory and practice	118
V.3	Kelly's Lemma	119
V.4	Edge-reconstruction	122
V.5	Notes	123
	Exercises	124
	References	124

Chapter VI Digraphs and Paths ... 125

VI.1	Digraphs	125
VI.2	Paths	129
VI.3	The BEST Theorem	133
VI.4	The Matrix-Tree Theorem	138

VI.5	The Laws of Kirchhoff	142
VI.6	Identification of vertices	149
VI.7	Transportation Theory	152
VI.8	Notes	158
	Exercises	159
	References	159

Chapter VII Alternating Paths 161

VII.1	Cursality	161
VII.2	The bicursal subgraph	163
VII.3	Bicursal units	167
VII.4	Alternating barriers	168
VII.5	f-factors and f-barriers	170
VII.6	The f-factor theorem	174
VII.7	Subgraphs of minimum deficiency	178
VII.8	The bipartite case	180
VII.9	A theorem of Erdös and Gallai	181
VII.10	Notes	183
	Exercises	183
	References	184

Chapter VIII Algebraic Duality 185

VIII.1	Chain-groups	185
VIII.2	Primitive chains	188
VIII.3	Regular chain-groups	194
VIII.4	Cycles	197
VIII.5	Coboundaries	200
VIII.6	Reductions and contractions	204
VIII.7	Algebraic duality	206
VIII.8	Connectivity	209
VIII.9	On transportation theory	215
VIII.10	Incidence matrices	217
VIII.11	Matroids	218
VIII.12	Notes	219
	Exercises	219
	References	220

Chapter IX Polynomials Associated with Graphs 221

IX.1	V-functions	221
IX.2	The chromatic polynomial	226
IX.3	Colorings of graphs	233
IX.4	The flow polynomial	237

	IX.5	Tait colorings	240
	IX.6	The dichromate of a graph	243
	IX.7	Some remarks on reconstruction	248
	IX.8	Notes	250
		Exercises	251
		References	251

Chapter X Combinatorial Maps **253**

	X.1	Definitions and preliminary theorems	253
	X.2	Orientability	257
	X.3	Duality	259
	X.4	Isomorphism	261
	X.5	Drawings of maps	263
	X.6	Angles	268
	X.7	Operations on maps	268
	X.8	Combinatorial surfaces	275
	X.9	Cycles and coboundaries	281
	X.10	Notes	283
		Exercises	284
		References	284

Chapter XI Planarity **285**

	XI.1	Planar graphs	285
	XI.2	Spanning subgraphs	288
	XI.3	Jordan's Theorem	290
	XI.4	Connectivity in planar maps	296
	XI.5	The cross-cut Theorem	306
	XI.6	Bridges	311
	XI.7	An algorithm for planarity	314
	XI.8	Peripheral circuits in 3-connected graphs	317
	XI.9	Kuratowski's Theorem	320
	XI.10	Notes	325
		Exercises	325
		References	326
		Index	327

Editor's Statement

A large body of mathematics consists of facts that can be presented and described much like any other natural phenomenon. These facts, at times explicitly brought out as theorems, at other times concealed within a proof, make up most of the applications of mathematics, and are the most likely to survive change of style and of interest.

This ENCYCLOPEDIA will attempt to present the factual body of all mathematics. Clarity of exposition, accessibility to the non-specialist, and a thorough bibliography are required of each author. Volumes will appear in no particular order, but will be organized into sections, each one comprising a recognizable branch of present-day mathematics. Numbers of volumes and sections will be reconsidered as times and needs change.

It is hoped that this enterprise will make mathematics more widely used where it is needed, and more accessible in fields in which it can be applied but where it has not yet penetrated because of insufficient information.

GIAN-CARLO ROTA

We are happy to inaugurate the Combinatorics Section of the Encyclopedia. Professor Tutte, one of the founders of graph theory, is presenting here the jewels of a subject rich in deep results, in a volume that will long remain definitive.

Foreword

It is both fitting and fortunate that the volume on graph theory in the *Encyclopedia of Mathematics and Its Applications* has an author whose contributions to graph theory are—in the opinion of many—unequaled. Indeed, the style and content of the book betray throughout the influence of Professor Tutte's own work and the distinctive flavor of his personal approach to the subject—a flavor familiar to those of use who have heard his beautifully constructed lectures on many occasions and are delighted to see so much of this exposition now recorded in permanent form for a wider audience.

The book deals with many of the central themes that one might expect to find in books on graph theory, such as Menger's Theorem and network flows, the Reconstruction Problem, the Matrix-Tree Theorem, the theory (largely created by Professor Tutte) of factors (or matchings) in graphs, chromatic polynomials, Brooks' Theorem, Grinberg's Theorem, planar graphs, and Kuratowski's Theorem. However, this is by no means "just another book on graph theory," since the treatment of all these topics is unified into a coherent whole by Professor Tutte's highly individual approach. Moreover, the more customary topics are leavened with some "pleasant surprises," such as the author's attractive theory of decomposition of graphs into 3-connected "3-blocks" (not found in other books except [5]), an interesting and remarkable approach to electrical networks, and—perhaps particularly—the classification theorem for closed surfaces. This theorem is usually considered as part of topology, but Professor Tutte shows us that it fits admirably into a work on graph theory, where indeed the essentially combinatorial nature of the arguments may be more at home.

From these remarks, it will be clear that the book has much to offer to any reader interested in graph theory. It draws together some important themes from the rapidly growing literature of the subject, and by no means duplicates any other expository writing at present available. It will also provide an excellent preparation for some slightly more specialized topics, including much of Professor Tutte's own work—for example, his extensive theory of planar enumeration and chromatic polynomials of maps, his theorem on Hamiltonian circuits in 4-connected planar graphs, the Five-Flow Conjecture (see Chapter IX, Sec. IX.4), and the interactions of matroids with graphs (see Chapter VIII, Sec. VIII.11). The treatment of minors provides background for (among many other things) the conjecture about well-quasi-ordering of graphs by the relationship of one graph being a minor of another (see the second half of Note 1 at the end of Chapter II). Current work of G. N. Robertson (a former student of Professor Tutte) and P. D. Seymour seems to promise exciting developments related to this problem. Some of these—and other—topics might provide ample material for a sequel volume, if Professor Tutte, or someone close to his thinking, is impelled to write one.

In the Introduction, Professor Tutte describes his early encounters with graph theory, particularly as a student at Cambridge. This cannot but provoke one's own early recollections, a favorite topic of conversation among graph-theorists being what first drew one's attention to graph theory. My own answer is "Nothing: I just invented it." In other words, when starting work as a research student, also at Cambridge, I felt that there ought to be a branch of mathematics dealing with this kind of thing, and, if there was not, I would create it. (References to binary relations in algebra courses might have helped to foster this idea.) It is a measure of the little known state of graph theory at that time that it took me some weeks to discover that I was not its first inventor and to hear of the one existing textbook on the subject, König's *Theorie der endlichen und unendlichen Graphen*, published eighteen years earlier in 1936. After König, I turned to the few research papers on graph theory then available, and I recall many hours in the library studying Professor Tutte's work on factors of graphs, liberally armed with colored pencils (especially red and bue ones, of course) to draw the diagrams needed to aid one's imagination. The reader is advised to equip himself similarly when he reaches Chapter VII, if indeed he has not already found this to be essential at a much earlier stage in the study of this or any other book on graph theory.

During the early part of my career, working in graph theory was a rarity and an idiosyncrasy. A graph-theorist could not expect to have others among his colleagues and might be hard pressed to find them in the same country: one simply did not expect to interact with other mathematicians except through published literature. Undergraduate and even graduate courses in the subject were virtually unknown. To some mathematicians, it even seemed questionable whether graph theory was worthwhile mathematics at all. Doubts seemed to center on its lack of elaborate techniques and its lack of unification and relevance, in the sense that it seemed to consist largely of solutions to isolated problems, not interacting closely with one another or with the rest of mathematics.

However, in fairness I must say that, for every mathematician who may have harbored such doubts, there were others who seemed sympathetic and interested. Not least among these, when I became a University lecturer, was my first Head of Department, Professor E. M. Wright (now Sir Edward Wright), whose own work has taken a distinctly graph-theoretic turn in more recent years.

In those early years, I would have thought a person almost demented if he had predicted the subsequent explosive growth in graph theory and other areas of combinatorial mathematics, which must be among the most remarkable experiences of those of us who have lived through it. Writers of futuristic novels do well to remember that truth is often stranger than fiction. It would have seemed absurd to predict that names like G. A. Dirac, F. Harary, and W. T. Tutte (then just meaningless cryptograms attached to research papers) would quite soon become those of some of the mathematicians whom I would know best personally, that graph theory would take me across the Atlantic to join a "Department of Combinatorics and Optimization," where for several years Professor Tutte and I would occupy adjacent offices, that there would be more combinatorial conferences than any one person could attend, and that the combinatorial literature would expand to its present size, including at least six journals (and perhaps more, depending on the method of counting) devoted entirely to combinatorics.

The first of these to emerge, the *Journal of Combinatorial Theory*, has for many years had Professor Tutte as its editor-in-chief (or in recent times, since proliferation of material forced it to subdivide, as editor-in-chief of half of it). The editor of this *Encyclopedia*, Professor G.-C. Rota, was one of the organizers responsible for getting the *Journal* started, and I was reliably informed by Blanche Descartes at that time that she had selected its title as an anagram of OUR FOE JIAN-CARLO ROTA BIMONTHLY.

Despite all this increased recognition of combinatorics (including graph theory) as part of mathematics, traces of the controversy about its value still persist. In the introduction to [1], L. Lovász rightly responds by pointing to the increasing body of techniques and unifying theory that the subject is acquiring. Nevertheless, some further remarks may be relevant. As another former colleague of mine, J. Sheehan, has suggested [4], there may still be important parts of combinatorics that do not fit well into any such unifying framework. There is surely *some* satisfaction in solving an obvious, natural and tough problem, even if the solution seems, at least for the present, to stand on its own. If (as sometimes seems to be suggested) a theorem can be significant only by virtue of its helpfulness in proving or illuminating other theorems, then one must go on to ask why those other theorems are significant, and a *reductio ad absurdum* is eventually reached. To prove that a branch of mathematics is interesting, must one necessarily demonstrate that it is exactly like other branches of mathematics, or might part of the appeal of mathematics lie in its diversity?

On the other hand, it is unquestionable that interplay between ideas from different sources, and elaborate techniques successfully applied, are *among* the features that make much of mathematics fascinating. Moreover, mathematics does often display a tendency to unify itself and to build up a body of technique. Therefore one may well guess (despite my caution against predicting the future) that graph theory, as it matures, will continue to develop its own characteristic techniques and that many of its results will become increasingly unified, both among themselves and with the rest of mathematics. The present book may be expected to play a considerable part in placing graph theory on a sound theoretical and technical footing.

I have written elsewhere what little I can about some aspects of the present state of graph theory [3] and Professor Tutte's immense influence on it [2]. This, and an early publication deadline, must be my excuse for a comparatively short Foreword consisting largely of random reminiscences and reflections that may interest nobody, with the possible exception of myself. Perhaps some future historian of mathematics may glean a crumb or two of something or other from them; but I will detain the reader no longer from the rich harvest in store.

C. St. J. A. Nash-Williams

REFERENCES

[1] L. Lovász, *Combinatorial Problems and Exercises*, North-Holland Publishing Company, Amsterdam, 1979.
[2] C. St. J. A. Nash-Williams, "A note on some of Professor Tutte's mathematical work," in *Graph Theory and Related Topics*, Proceedings of the conference held in honor of Professor W. T. Tutte on the occasion of his sixtieth birthday, University of Waterloo, July 5-9, 1977 (edited by J. A. Bondy and U. S. R. Murty), Academic Press, New York, 1979, pp. xxv-xxviii.
[3] C. St. J. A. Nash-Williams, "A glance at graph theory—Part I" and "A glance at graph theory—Part II," *Bulletin of the London Mathematical Society*, **14** (1982), 177-212 and 294-328.
[4] J. Sheehan, review of *Graph Theory: An Introductory Course* by B. Bollobás, *Bulletin of the London Mathematical Society*, **12** (1980), 388-390.
[5] W. T. Tutte, *Connectivity in Graphs*, University of Toronto Press, Toronto, 1966.

Introduction

The author encountered graph theory in high school, in the early thirties, while reading Rouse Ball's book *Mathematical Essays and Recreations*. He then learned of Euler paths (Sec. VI.3), map-colorings (dualized in Sec. IX.3), factors of graphs (Sec. VII.6), and Tait colorings (Sec. IX.5) [5].
    As an undergraduate at Cambridge he joined with R. L. Brooks, C. A. B. Smith, and A. H. Stone in the study of their hobby-problem of dissecting a square into unequal squares [3]. This soon called for much graph theory. It was linked, through a "Smith diagram," with the study of 3-connected planar graphs (Sec. XI.7), and with Kirchhoff's Laws for electrical circuits (Sec. VI.5). It was linked through rotor theory (Sec. VI, Notes) with graph symmetry (Sec. I.2). It was linked through the tree-number (Sec. II.2) with the theory of graph functions satisfying simple recursion formulae (Sec. IX.1).
    All this is explained in the Commentaries of [4]. That is one reason why I do not discuss squared rectangles and the analogous triangulated triangles in the present work. Another is that I visualize the book as a work on pure graph theory, making no appeal either to point-set topology or to elementary geometry.
    I became acquainted with some graph-theoretical literature at Cambridge. I read Sainte-Laguë's description of the proof of Petersen's Theorem (Sec. VII.6). I found the classical papers of Hassler Whitney, published in 1931-3, and the famous book of Dénes König, the first textbook devoted

entirely to graph theory. I was there at Cambridge at the time of the births of Smith's Theorem (Sec. IX.5) [7] and Brooks' Theorem (Sec. IX.3) [2]. Stone's discovery of flexagons came a little later.

Having meditated upon these things for 45 years I now present some of them in the present work. It is an attempt at the reference book I would have liked to have in 1936-40. In electrical theory it is important to know whether you have a connection between the two terminals, and what happens when you remove a wire. Chapter I deals graph-theoretically with these matters. Chapter II deals with the effect of contracting an edge, or shall we say making a short circuit. The theory of 3-connection is discussed in Chapter IV, and the halfway stage of 2-connection in Chapter III. Chapter V, on reconstruction, is less directly related to squared squares and rectangles. I came to it by way of reconstruction formulae for some of the above-mentioned recursive graph functions [11].

Chapter VI concerns digraphs and a generalized theory of Kirchhoff's Laws. It arose out of a study of triangulated triangles by the four undergraduates. We were sometimes reproached for basing our mathematical theory on physical laws. We protested, of course, that for us Kirchhoff's Laws were axioms of a purely mathematical system, but we were glad to be able to emphasize this by introducing generalized Laws, describing a kind of electricity that never was on land or sea.

Chapter VII derives from Sainte-Laguë's paper, with some gaps filled and some extensions made. Chapter VIII is about cycles and coboundaries, generalizations of Kirchhoff flows. It attempts to describe some parts of graph theory algebraically, and most of it derives from my doctoral thesis of 1948 [9].

Chapter VIII is about the recursive graph functions. It derives from a paper of 1947 [8]. It discusses the dichromatic polynomial, the dichromate, the chromatic polynomial, and the flow-polynomial, all of which can be referred to the theory of map-colorings and to the dual theory of vertex-colorings.

So far there is one important omission, that of a theory of planarity. The graphs of interest in connection with squared rectangles and triangulated triangles are all planar, so Chapter X prepares for the introduction of planarity by giving a general theory of maps on surfaces. But this is to be a purely graph-theoretical work, and so the maps of Chapter X are structures defined by purely combinatorial axioms. Surfaces are defined as classes of maps. The discussion is an adaptation of the classical theory of H. R. Brahana [1]. Planar maps can now be defined as maps of Euler characteristic 2.

Chapter XI gives a theory of planarity. It gives duality theorems for the tree-number and the dichromate, and it gives a combinatorial version of Jordan's Theorem. It goes on to some tests for the planarity or nonplanarity of a given graph, MacLane's and Kuratowski's among them. This part derives from my paper "How to draw a graph," of 1964, but it skips the actual drawing, that being a matter of elementary geometry rather than graph theory.

I take this opportunity to express my indebtedness to Brooks, Smith, and Stone, without whose missionary zeal I might now be writing on some other subject.

REFERENCES

[1] Brahana, H. R., Systems of circuits on two-dimensional manifolds, *Ann. Math.* (**2**) 23 (1921), 144–168.
[2] Brooks, R. L., On colouring the nodes of a network, *Proc. Cambridge Phil. Soc.* **37** (1941), 194–197.
[3] Brooks, R. L., C. A. B. Smith, A. H. Stone, and W. T. Tutte, The dissection of rectangles into squares, *Duke Math. J.* **7** (1940), 312–340.
[4] McCarthy, D., and R. G. Stanton (eds.), *Selected papers of W. T. Tutte*, The Charles Babbage Research Centre, St. Pierre, Manitoba, Canada (1979).
[5] Rouse Ball, W. W., *Mathematical recreations and essays*. The 12th edition (U. of Toronto Press, 1974) is co-authored by H. S. M. Coxeter.
[6] Sainte-Laguë, M. A., Les réseaux (ou graphes). *Memorial des Sciences Mathématiques*, Fasc. XVIII, Paris 1926.
[7] Tutte, W. T., On Hamiltonian circuits, *J. London Math. Soc.* **21** (1946), 98–101.
[8] _____, A ring in graph theory, *Proc. Cambridge Phil. Soc.* **43** (1947), 26–40.
[9] _____, An algebraic theory of graphs, Thesis, Cambridge 1948.
[10] _____, How to draw a graph, *Proc. London Math. Soc.* **52** (1963), 743–767.
[11] _____, On dichromatic polynomials, *J. Comb. Theory* **2** (1967), 301–320.

Chapter I

Graphs and Subgraphs

I.1. DEFINITIONS

In this section we give a formal definition of a graph and introduce some of the basic terminology of graph theory. We give examples of graphs and some sample theorems.

A *graph* G is defined by a set $V(G)$ of elements called *vertices*, a set $E(G)$ of elements called *edges*, and a relation of *incidence*, which associates with each edge either one or two vertices called its *ends*.

The present work is concerned only with *finite* graphs, those in which the sets $V(G)$ and $E(G)$ are both finite. Much interesting work has been done on the other graphs, the *infinite* ones. But even the theory of finite graphs is too big to be adequately covered in one volume. Let us therefore make the rule that from here on the word "graph" is to mean a finite graph unless the contrary is stated explicitly.

The terminology of graph theory is not yet standardized. Some authors prefer to use the terms "point" and "line" rather than "vertex" and "edge." This usage may be found inconvenient in problems involving both graphs and geometrical or topological structures. In some of the older papers we may find "branch" used for "edge," and "node" for "vertex."

An edge is called a *link* or a *loop* according as the number of its ends is two or one. We shall however get into the habit of saying that each edge has two ends, with the explanation that in the case of a loop the two ends are coincident. The two ends of an edge are said to be *joined* by that edge, and to be *adjacent*. Accordingly we say that a vertex is joined to itself, or is adjacent to itself, if and only if it is incident with a loop. Two or more links with the same pair of ends are said to constitute a *multiple join*. A graph without loops or multiple joins is called a *strict* graph.

There are many problems of graph theory in which only strict graphs are of interest. Accordingly some authors restrict the term "graph" to mean what we have called a strict graph. When they have occasion to add loops or multiple joins to their structures they speak of "multigraphs."

Examples of graphs are not difficult to find. For one, the edges and vertices of a convex polyhedron are the edges and vertices, respectively, of a graph G. The ends in G of an edge are its ends in the geometric sense. We call G the graph of the polyhedron.

A roadmap can be interpreted as a graph. The vertices are the junctions, and an edge is the stretch of road from one junction to the next, or from a junction back to itself. Similarly an electrical circuit may give us a graph in which the vertices are terminals and the edges wires.

It is not difficult to see graphs in genealogical tables and computer programs. All through mathematics they are visible to the graph-theoretical eye of faith, for much of mathematics can be described in terms of binary relations, and what is a binary relation but a graph?

It is customary to represent a graph G by a drawing on paper. The vertices are drawn as dots. A link with ends x and y is represented by a straight or curved line joining the dots of x and y, and not meeting any other vertex-dot. A loop with end x is drawn as a curve leaving the dot of x and returning to it again, without meeting any other vertex-dot on the way.

In such a drawing it may happen that two edge-curves intersect at some point away from all the vertex-dots. Normally such edge-crossings are ignored as not representing anything in the structure of G. But when we ask what is the least possible number of edge-crossings in a drawing of G, deep and difficult problems arise. (See [8], 122–3.)

Some examples follow. Fig. I.1.1 is a drawing of the graph of a cube, and Fig. I.1.2 shows a graph with loops and multiple joins.

Some graphs with simple structures are thought to deserve special names. For some purposes it is convenient to recognize a *null* graph, having no edges and no vertices. A *vertex-graph* is an edgeless graph having exactly one vertex (Fig. I.1.3(i)). A *loop-graph* consists of a single loop with its one end (Fig. I.1.3(ii)), and a *link-graph* consists of a single link with its two ends (Fig. I.1.3(iii)).

Let n be a nonnegative integer. Then an *n-clique* is defined as a loopless graph with exactly n vertices and $\frac{1}{2}n(n-1)$ edges, each pair of vertices being joined by a single link. Thus the n-cliques are strict graphs. Evidently the null graphs are the 0-cliques, the vertex-graphs are the 1-cliques, and the link-graphs are the 2-cliques. Figure I.1.4 shows a

Definitions

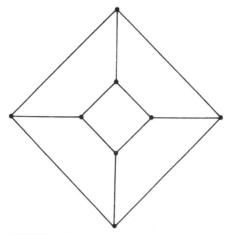

FIGURE I.1.1

3-clique, a 4-clique and a 5-clique, labeled respectively (i), (ii) and (iii). Note the edge-crossing in the diagram of the 5-clique.

Now let n be a positive integer. We define an *n-arc* as a graph G with n edges and $(n + 1)$ vertices, having the following property: The edges can be enumerated as $A_1, A_2, \ldots, A_n$, and the vertices as $a_0, a_1, a_2, \ldots, a_n$, in such a way that the ends of A_j are a_{j-1} and a_j, for each relevant suffix j.

The 1-arcs are the link-graphs. Fig. I.1.5 shows a 2-arc a 3-arc and a 4-arc, labelled (i), (ii) and (iii), respectively.

For any positive integer n we define an *n-circuit* as a graph G with n vertices and n edges, having the following property. The vertices can be enumerated as $a_1, a_2, \ldots, a_n$, and the edges as $A_1, A_2, \ldots, A_n$, in such a way

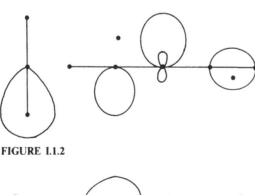

FIGURE I.1.2

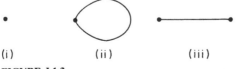

(i)   (ii)   (iii)

FIGURE I.1.3

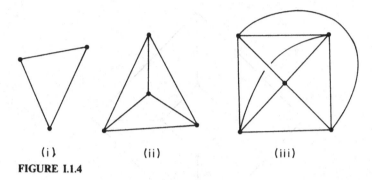

FIGURE I.1.4

that the ends of A_n are a_{j-1} and a_j (where $a_0 = a_n$) for each edge A_j. The 1-circuits are the loop-graphs and the 3-circuits are the 3-cliques. Figure I.1.6 shows a 2-circuit, a 3-circuit, and a 4-circuit, labelled (i), (ii) and (iii), respectively.

When it is unnecessary to assert or reassert the value of the integer n, an n-clique, n-arc, or n-circuit may be called simply a *clique*, *arc*, or *circuit*, respectively. In the two latter cases n is called the *length* of the arc or circuit concerned.

Let us now define the *valency* $\text{val}(G, x)$ of a vertex x in a graph G. It is the number of edges incident with x, loops being counted twice. It is clear that

$$\sum_{x \in V(G)} \text{val}(G, x) = 2|E(G)|. \qquad (\text{I}.1.1)$$

Here and in the following pages we denote the cardinality of a finite set S by $|S|$. From Eq. (I.1.1) we can deduce the following theorem.

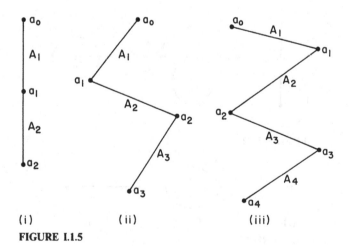

FIGURE I.1.5

Isomorphism

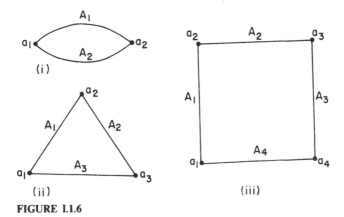

FIGURE I.1.6

Theorem I.1. *In any graph the number of vertices of odd valency is even.* (This is one of the oldest and best-known results in graph theory.)

The term "valency" or "valence" is suggested by a chemical analogy. But many workers in graph theory say "degree" instead.

A vertex of zero valency is said to be *isolated*. For example, the single vertex of a vertex-graph is isolated. It is convenient to describe a vertex of valency 1 as *monovalent*, one of valency 2 as *divalent*, and so on. Thus the single vertex of a loop-graph is divalent and the two vertices of the link-graph are monovalent. Each vertex of a circuit is divalent, and each vertex of an n-clique has valency $(n-1)$.

Consider an n-arc. Let its vertices and edges be enumerated in the manner specified in its definition. Then it is clear that a_0 and a_n are monovalent, and that every other vertex is divalent. It is usual to describe the two monovalent vertices of an arc as its *ends*. The remaining vertices, all divalent, are called the *internal vertices* of the arc.

A graph in which each vertex has the same valency n is called a *regular* graph of that valency. Thus every circuit is a regular graph of valency 2. If the integer m is positive, then each m-clique is a regular graph of valency $(m-1)$. Regular graphs of valency 3 are of special interest in graph theory. They are called the *cubic* graphs.

I.2. ISOMORPHISM

Given two graphs, such as the graphs of a large and a small cube, we may decide that they are not essentially different. We may say that they have the same structure, or that each is a copy of the other. What is meant is that the two graphs can be represented by the same diagram. To make our meaning precise we may appeal to the following notion of *isomorphism*.

Let G and H be graphs. Let f be a 1–1 mapping of $V(G)$ onto $V(H)$, and g a 1–1 mapping of $E(G)$ onto $E(H)$. Let θ denote the ordered pair (f, g). We say that θ is an *isomorphism* of G onto H if the following condition holds: The vertex x is incident with the edge A in G if and only if the vertex fx is incident with the edge gA in H.

If such an isomorphism θ exists, we say that the graphs G and H are *isomorphic*. Clearly we then have

$$|V(G)| = |V(H)| \quad \text{and} \quad |E(G)| = |E(H)|.$$

We may think of θ as an operation transforming G into H, and we accordingly write $\theta G = H$. It is also convenient to write $\theta v = fv$ and $\theta A = gA$ for each vertex v and edge A of G. Clearly G and H can be represented by the same diagram. The representative of an edge or vertex x of G can be reinterpreted as the representative of θx in H.

It is possible for G and H to be the same graph. An isomorphism of G onto itself is called an *automorphism* of G. Any graph G has the *identical* or *trivial* automorphism I such that $Ix = x$ for each edge or vertex x of G.

The relation of isomorphism between graphs is reflexive, because of the trivial automorphisms. It is symmetrical; if $\theta = (f, g)$ is an isomorphism of G onto H, then there is an inverse isomorphism $\theta^{-1} = (f^{-1}, g^{-1})$ of H onto G. Finally, it is transitive; if $\theta = (f, g)$ is an isomorphism of G onto H and $\phi = (f_1, g_1)$ is an isomorphism of H onto K, then there is an isomorphism $\phi\theta = (f_1 f, g_1 g)$ of G onto K. Here $f_1 f$ is the mapping obtained by applying first f and then f_1. Similarly $\phi\theta$ is the isomorphism obtained by applying first θ and then ϕ. It is easy to verify that the multiplication of isomorphisms thus defined is associative.

We have now verified that isomorphism is an equivalence relation. It therefore partitions the class of all graphs into disjoint nonnull subclasses, called *isomorphism classes*, such that two graphs belong to the same isomorphism class if and only if they are isomorphic.

Pure graph theory is concerned with those properties of graphs that are invariant under isomorphism, for example the number of vertices, the number of loops, the number of links, and the number of vertices of a given valency. It is therefore natural for a graph theorist to identify two graphs that are isomorphic. For example, all link-graphs are isomorphic, and therefore he speaks of the "link-graph" as though there were only one. Similarly one hears of "the null graph," "the vertex-graph," and "the graph of the cube." When this language is used, it is really an isomorphism class (also called an *abstract graph*) that is under discussion.

It is a convention rather than a theorem that all null graphs are isomorphic. We can justify it by postulating that if S and T are null sets, there is a unique 1–1 mapping of S onto T. The author remembers being taught that there is only one null set. Those who adhere to this mystical doctrine must hold that there is only one null graph, the unique member of its isomorphism class. The null graph often fits rather oddly into combinatorial definitions.

Theorem I.2. *Let G and H be strict graphs. Let f be a 1-1 mapping of $V(G)$ onto $V(H)$ having the following property: Two distinct vertices x and y of G are adjacent in G if and only if the corresponding vertices fx and fy of H are adjacent in H. Then there is a uniquely determined 1-1 mapping g of $E(G)$ onto $E(H)$ such that (f, g) is an isomorphism of G onto H.*

Proof. Let A be any edge of G. It has distinct ends x and y, since G is strict. By hypothesis there is a uniquely determined edge A' of H whose ends are fx and fy. We define a 1-1 mapping g by the rule that $gA = A'$ for each edge A of G. It is then clear that (f, g) is an isomorphism of G onto H. Conversely, if there is a g such that (f, g) is an isomorphism of G onto H, then g must satisfy the above rule. □

In a theory of strict graphs an isomorphism of G onto H is often defined as a 1-1 mapping of $V(G)$ onto $V(H)$ that preserves adjacency. This specialization can be regarded as an application of Theorem I.2.

Let n be a non-negative integer, and let G and H be n-cliques. They are strict graphs. There exists a 1-1 mapping f of $V(G)$ onto $V(H)$, and any such mapping preserves adjacency by the definition of an n-clique. Accordingly G and H are isomorphic, by Theorem I.2.

Now let n be a positive integer, and let G and H be n-arcs. In their defining enumerations let the vertices of G be enumerated as $a_0, a_1, \ldots, a_n$ and those of H as $b_0, b_1, \ldots, b_n$. Let f be the 1-1 mapping of $V(G)$ onto $V(H)$ such that $fa_j = b_j$ for each suffix j. Then f has the property specified in the enunciation of Theorem I.2, and therefore G and H are isomorphic.

A similar argument shows that if n is any integer exceeding 2, then any two n-circuits are isomorphic. It is a trivial matter to extend this result to the cases $n = 1$ and $n = 2$. But the 1-circuit and 2-circuit are not strict graphs, and so Theorem I.2 does not apply. If G and H are 2-circuits, as shown in Fig. I.1.6(i), then any 1-1 mapping of $V(G)$ onto $V(H)$ can be combined with any 1-1 mapping of $E(G)$ onto $E(H)$ to yield an isomorphism of G onto H.

The automorphisms of a graph G are the elements of a group $A(G)$ with respect to the associative multiplication of isomorphisms defined above. The unit element of $A(G)$ is the trivial automorphism. The inverse of an automorphism θ in $A(G)$ is the inverse isomorphism θ^{-1} defined above. A product $\phi\theta$ of isomorphisms is defined only when θ is a mapping onto, and ϕ is a mapping from, the same graph G. But in the case of automorphisms this condition is always satisfied. We call $A(G)$ the *automorphism group* of G.

Mathematical structures other than graphs have their theories of isomorphism. For example, two groups P and Q are said to be isomorphic if there is a 1–1 mapping f of P onto Q such that f and f^{-1} preserve products. We should note the following theorem.

Theorem I.3. *Isomorphic graphs have isomorphic automorphism groups.*

Proof. Let θ be an isomorphism of a graph G onto a graph H. Then if ξ is in $A(H)$, the isomorphism $\theta^{-1}\xi\theta$ is in $A(G)$. Similarly, if η is in $A(G)$, then $\theta\eta\theta^{-1}$ is in $A(H)$. So we have a 1–1 correspondence $\eta \to \theta\eta\theta^{-1}$ of $A(G)$ onto $A(H)$, and this clearly preserves products. □

We can now assert that the automorphism group of a graph, regarded as an abstract group, is invariant under graph isomorphism. It is thus a legitimate concern of the pure graph theorist.

Let us investigate the automorphism groups of the graphs so far defined. For the null graph, the vertex-graph, and the loop-graph, there is only the trivial automorphism. Hence the automorphism group of each of these graphs is "trivial"; that is, it consists of a unit element only.

Now consider an n-clique G, with $n > 0$. Referring to Theorem I.2 we see that any permutation of the n vertices defines a unique corresponding automorphism of G. We deduce that $A(G)$ is isomorphic to the group of permutations of n objects.

Consider next an n-arc g with ends x and y. We have a defining enumeration in which the vertices are enumerated as $a_0, a_1, \ldots, a_n$ and the edges as $A_1, A_2, \ldots, A_n$. We may suppose that $a_0 = x$ and $a_n = y$. We can get a second defining enumeration, called the *reverse* of the first, by reversing the order of edges or vertices in each of the above sequences. Adjusting the notation we can say that in the reverse enumeration $a_0 = y$ and $a_n = x$. However, a defining sequence is uniquely determined by valency considerations as soon as a_0 is given: A_1 must be the single edge incident with a_0, a_1 must be the other end of A_1, A_2 must be the other edge incident with a_1 (if $n > 1$), and so on. Evidently an automorphism of G is determined by an ordered pair of defining enumerations: It maps the first member onto the second. We deduce that $A(G)$ has exactly two elements, say I and θ. The nontrivial automorphism θ interchanges the two ends of G; it changes each defining enumeration into its reverse.

Consider next an n-circuit G, with $n > 1$. We have a defining enumeration in which the vertices are enumerated as $a_1, a_2, \ldots, a_n$ and the edges as $A_1, A_2, \ldots, A_n$. Evidently there is an automorphism θ that increases each suffix by 1. (Addition and subtraction in the suffixes is to be modulo n.) There is also an automorphism ϕ that reverses the original sequence of vertices, and that reverses the original cyclic sequence of edges. Using the powers of θ and their combinations with ϕ, we find that it is possible to obtain a defining enumeration of the circuit in which a_1 is an arbitrary vertex and A_1 is any edge incident with a_1. But it follows from valency considerations that a defining sequence is fixed uniquely when a_1 and A_1 are given. As with the arcs, an automorphism is uniquely determined by an ordered pair of defining enumerations. We deduce that $A(G)$ has exactly $2n$ elements, which we can write as

$$I, \theta, \theta^2, \ldots, \theta^{n-1}, \phi, \theta\phi, \theta^2\phi, \ldots, \theta^{n-1}\phi.$$

Indeed $A(G)$ is a dihedral group of order $2n$; its generators θ and ϕ satisfy the generating relations $\theta^n = I$, $\phi^2 = I$, and $\theta\phi = \phi\theta^{-1}$.

I.3. SUBGRAPHS

In constructing a theory of graphs, we can start with the observation that some graphs contain other graphs. A graph H contained in a graph G is called a *subgraph* of G. We shall use the following precise definition.

A graph H is said to be a *subgraph* of a graph G if

$$V(H) \subseteq V(G), \quad E(H) \subseteq E(G),$$

and each edge of H has the same ends in H as in G. Under these conditions we say also that H is *contained* in G.

According to the above definition, G is itself a subgraph of G. The other subgraphs of G are called its *proper* subgraphs. They are said to be *properly* contained in G. We write $H \subseteq G$ to indicate that H is a subgraph of G, and $H \subset G$ to indicate that H is a proper subgraph of G. We use a corresponding notation for finite sets. Thus $S \subseteq T$ means that S is a subset of T, and $S \subset T$ that S is a proper subset of T.

The above formal definition leads to the following rule for constructing subgraphs.

Rule I.4. *Let G be a graph, P a subset of $V(G)$ and Q a subset of $E(G)$. Then the necessary and sufficient condition for the existence of a subgraph H of G such that $V(H) = P$ and $E(H) = Q$ is that P shall contain both ends of each member of Q.*

Condition I.4 is always satisfied when $P = V(G)$. We then call H a *spanning* subgraph of G. We shall denote the spanning subgraph of G with edge-set S by $G:S$. Those spanning subgraphs of G that are circuits are called its *Hamiltonian circuits*; those that are regular graphs of valency 1 are called its 1-*factors*. Both Hamiltonian circuits and 1-factors are prominent in the graph-theoretical literature.

If S is any set of edges of G, let $J(S)$ denote the set of all vertices v of G such that v is incident with a member of S. By Rule I.4 there is a subgraph H of G such that $V(H) = J(S)$ and $E(H) = S$. We call H the *reduction* of G to S, and we denote it by $G \cdot S$. We note that a reduction can have no isolated vertex. In particular $G \cdot E(G)$ is the graph obtained from G by deleting all its isolated vertices.

If U is any set of vertices of G, let $J(U)$ denote the set of all edges of G having both ends in U. By Rule I.4 there is a subgraph H of G such that

$$V(H) = U \quad \text{and} \quad E(H) = J(U).$$

We call such a subgraph an *induced* subgraph of G. In particular, H is the subgraph of G induced by U; we denote it by $G[U]$.

Each graph G has a *null subgraph*, which we denote by $\varnothing$. This null subgraph is both a reduction and an induced subgraph of G, but it is not a spanning subgraph of G unless G itself is null.

Let H and K be subgraphs of G, not necessarily distinct. By I.4 there is a subgraph L of G such that

$$V(L) = V(H) \cup V(K) \quad \text{and} \quad E(L) = E(H) \cup E(K).$$

We call L the *union* of H and K, and we denote it by $H \cup K$. By I.4 there is also a subgraph M of G such that

$$V(M) = V(H) \cap V(K) \quad \text{and} \quad E(M) = E(H) \cap E(K).$$

We call M the *intersection* of H and K, and we denote it by $H \cap K$. For obvious reasons, unions and intersections of subgraphs obey algebraic rules like those for unions and intersections of subsets of a finite set. For example,

$$(H \cup K) \cap L = (H \cap L) \cup (K \cap L). \tag{I.3.1}$$

Two subgraphs H and K of G are said to be *disjoint* if they have no common edge and no common vertex. One way of asserting the disjointness is to write

$$H \cap K = \varnothing.$$

There is a similar device for asserting that a graph G is nonnull: We write $\varnothing \subset G$.

The following theorem is an easy consequence of the definition of a subgraph.

Theorem I.5. *Any subgraph of a subgraph of G is a subgraph of G.*

We can go further. Any spanning subgraph of a spanning subgraph of G is a spanning subgraph of G, any reduction of a reduction of G is a reduction of G, and any induced subgraph of an induced subgraph of G is an induced subgraph of G.

We conclude the section with a notation for relating the subgraphs of two isomorphic graphs G and H. Let $\theta = (f, g)$ be an isomorphism of G onto H.

If S is any subset of $V(G)$, then f maps S onto a subset fS of $V(H)$. This is the set of images under f of the members of S. We can denote it also by θS. The *restriction* $f \cdot S$ of f to S is the 1–1 mapping of S onto fS that agrees with f for every member of S. Similarly, if T is any subset of $E(G)$, it is mapped by g onto a subset gT of $E(H)$, also denoted by θT. We then have the restriction $g \cdot T$ of g to T, a 1-1 mapping of T onto gT.

Let L be any subgraph of G. Then there is a subgraph L' of H such that

$$V(L') = fV(L) \quad \text{and} \quad E(L') = gE(L).$$

We write $L' = \theta L$ and say that θ maps L onto L'. In this way θ sets up a 1-1 correspondence $L \to \theta L$ between the subgraphs of G and the subgraphs of H. Moreover for each L there is an isomorphism $(f \cdot V(L), g \cdot E(L))$ of L onto θL. We denote this isomorphism by $\theta \cdot L$ and call it the *restriction of θ to L*.

Let us define a *structural* property of a graph as one invariant under isomorphisms. The above discussion shows that if H is a fixed graph, then the number of subgraphs of G isomorphic to H is a structural property of G. We shall make use of this particular structural property in Chapter V on the Reconstruction Problem.

I.4. VERTICES OF ATTACHMENT

A subgraph is joined onto the rest of its graph at certain vertices called its "vertices of attachment." In this section we give a formal definition of these vertices, adopt a notation for discussing them, and derive a few theorems.

Let H be a subgraph of a graph G. A *vertex of attachment* of H in G is a vertex of H that is incident with some edge of G that is not an edge of H. We write $W(G, H)$ for the set of vertices of attachment of H in G, and $w(G, H)$ for the number $|W(G, H)|$ of such vertices. Thus $w(G, G) = w(G, \emptyset) = 0$.

In this work we shall need theorems relating the vertices of attachment of two subgraphs H and K to the vertices of attachment of their union and intersection. Some simple rules are given in the following theorem.

Theorem I.6. *Let H and K be subgraphs of a graph G. Then each vertex of attachment of $H \cup K$ or $H \cap K$ is a vertex of attachment of H or K. On the other hand, each vertex of attachment of H is a vertex of attachment of $H \cup K$ or $H \cap K$.*

Proof. Let v be a vertex of attachment of $H \cup K$. Then v is incident with an edge A of G belonging neither to H nor to K. Accordingly, v is a vertex of attachment of whichever subgraph H or K it belongs to.

Next let v be a vertex of attachment of $H \cap K$. By definition it is incident with an edge A of G that may belong to one of H and K but cannot belong to both. Accordingly, v is a vertex of attachment of whichever subgraph H or K does not have A as an edge.

Finally let v be a vertex of attachment of H or K, let us say of H. It is incident with an edge A of G not in $E(H)$. If A is not in $E(K)$, then v is a vertex of attachment of $H \cup K$. If A is in $E(K)$, then v is a vertex of $H \cap K$, and it is a vertex of attachment of that subgraph. □

If H and K are subgraphs of G, let us write $Q(G; H, K)$ for the set of all vertices x of G such that x belongs to $W(G, H)$ and $W(G, K)$ but not to $W(G, H \cup K)$. Alternatively we may characterize x as incident with an edge of H not belonging to K and with an edge of K not belonging to H, but not incident with any edge of G outside both $E(H)$ and $E(K)$. We write also

$$q(G; H, K) = |Q(G; H, K)|.$$

In later work we shall make use of the following theorem.

Theorem I.7. *Let H and K be subgraphs of a graph G. Then*

$$w(G, H \cup K) + w(G, H \cap K) = w(G, H) + w(G, K) - q(G; H, K).$$
(I.4.1)

Proof. Consider any vertex x of G. We may suppose it to belong to $W(G, H \cup K)$ or $W(G, H \cap K)$ since otherwise, by Theorem I.6, it makes no contribution to either side of Eq. (I.4.1).

Suppose first that x belongs to $W(G, H \cup K)$. Then it is incident with an edge A of G not belonging to H or K. If x belongs to $H \cap K$, then it belongs also to H and to K, and it is a vertex of attachment of each of the subgraphs H, K, $H \cup K$ and $H \cap K$. Since it does not belong to $Q(G; H, K)$, it makes the same contribution 2 to each side of Eq. (I.4.1). If x does not belong to $H \cap K$, we can assume without loss of generality that x belongs to H but not to K. It then belongs to $W(G, H)$. Since it does not belong to $Q(G; H, K)$, it makes the same contribution 1 to each side of Eq. (I.4.1).

In the remaining case x belongs to $W(G; H \cap K)$ but not to $W(G, H \cup K)$. By Theorem I.6 it belongs to one or both of $W(G, H)$ and $W(G, K)$. But if it belongs to both these sets, it belongs also to $Q(G; H, K)$. Hence x makes the same contribution 1 to each side of Eq. (I.4.1). Our proof is now complete. □

We take note of one more theorem, an obvious one.

Theorem I.8. *Let K be a subgraph of a subgraph H of a graph G. Then each vertex of attachment of K in H is a vertex of attachment of K in G. Moreover, each vertex of attachment of K in G is either a vertex of attachment of K in H or a vertex of attachment of H in G.*

Further information about vertices of attachment can be found in [9].

Let H be a subgraph of a graph G. It follows from Rule I.4 that there is a subgraph H^c of G such that

$$E(H^c) = E(G) - E(H) \quad \text{and} \quad V(H^c) = (V(G) - V(H)) \cup W(G, H).$$

We call H^c the *complementary subgraph to H in G*. Thus $G^c = \emptyset$ and $\emptyset^c = G$. For another example, let H be the vertex-graph defined by a vertex x in G. If x is an isolated vertex of G, then H^c is the graph obtained from G by deleting x. But if x is not isolated, then $H^c = G$.

Theorem I.9. *The vertices of attachment of H^c are those vertices of attachment of H that are not isolated vertices of H.*

Proof. Let x be any vertex of attachment of H^c. It is incident with an edge of H and is therefore a vertex of H. Hence, by the definition of H^c, it is a vertex of attachment of H. We note that it is not isolated in H.

On the other hand let x be a vertex of attachment of H. Then it is a vertex of H^c. It is a vertex of attachment of H^c if and only if it is not isolated in H. □

Corollary I.10. *No vertex of attachment of H^c is isolated in H^c.*

Let us now consider the complementary subgraph H^{cc} to H^c in G. It is clear that $E(H^{cc}) = E(H)$. We have also

$$V(H^{cc}) = \{V(G) - \{(V(G) - V(H)) \cup W(G, H)\}\} \cup W(G, H^c)$$
$$= \{(V(G) - W(G, H)) - (V(G) - V(H))\} \cup W(G, H^c)$$
$$= (V(H) - W(G, H)) \cup W(G, H^c).$$

Combining these results with Theorem I.9, we obtain the following theorem.

Theorem I.11. H^{cc} *is the graph obtained from H by deleting those of its vertices of attachment that are isolated in H.*

Corollary I.12. *If no vertex of attachment of H is isolated in H, then $H^{cc} = H$.*

Let us define a *complementary pair* of subgraphs of G as a pair $\{H, K\}$ such that each of H and K is the complementary subgraph to the

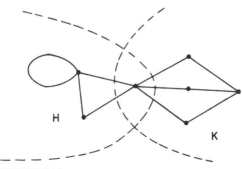

FIGURE I.4.1

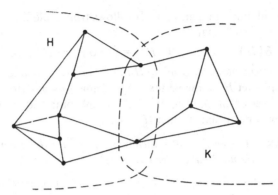

FIGURE 1.4.2

other in G. By Corollaries I.10 and I.12, such a complementary pair can be any pair of subgraphs of G of the form $\{H, H^c\}$ such that no vertices of attachment of H are isolated in H. In particular, for any subgraph H of G the pair $\{H^c, H^{cc}\}$ is complementary.

If $\{H, K\}$ is any complementary pair of subgraphs of G, we have $W(G, H) = W(G, K)$, by Theorem I.9. We call $W(G, H)$ the *binding set* and $w(G, H)$ the *binding number* of the complementary pair. Figures I.4.1 and I.4.2 show complementary pairs of subgraphs of G with binding numbers 1 and 2, respectively.

I.5. COMPONENTS AND CONNECTION

A subgraph H of a graph G is said to be *detached* in G if it has no vertices of attachment in G. Thus G itself and its null subgraph are detached in G.

Theorem I.13. *If H and K are detached subgraphs of G, then so are $H \cup K$ and $H \cap K$ (by Theorem I.6).*

Theorem I.14. *If H is a detached subgraph of G, then the complementary subgraph H^c to H in G is detached in G. Moreover, H^c is disjoint from H, and $H^{cc} = H$.*

The first part of Theorem I.14 is a consequence of Theorem I.9. Disjointness follows directly from the definition of a complementary subgraph. Theorem I.11 shows that $H^{cc} = H$.

Theorem I.15. *A detached subgraph of a detached subgraph of G is a detached subgraph of G (by Theorem I.8).*

A *component* of G is a minimal nonnull detached subgraph of G, that is, a nonnull detached subgraph of G that contains no other nonnull detached subgraph of G. From this definition and the finiteness of G, we derive the following theorem.

Theorem I.16. *If H is a nonnull detached subgraph of G, then there is a component C of G such that $C \subseteq H$.*

Theorem I.17. *Distinct components of G are disjoint.*

Proof. Let H and K be distinct components of G. Assume that they are not disjoint. Then $H \cap K$ is a nonnull detached subgraph of G, by Theorem I.13. By the minimality of H and K, we have $H = H \cap K = K$, a contradiction. □

Theorem I.18. *Let x be any edge or vertex of G. Then it belongs to exactly one component of G.*

Proof. We note that x belongs to the nonnull detached subgraph G of G. By finiteness it belongs to a nonnull detached subgraph H of G such that no proper subgraph of H is a nonnull detached subgraph of G including x.

Assume that H is not a component of G. Then it contains a nonnull detached subgraph K of G to which x does not belong. Let K' be the intersection with H of the complementary subgraph K^c to K in G. Then K' is a proper subgraph of H including x. Moreover, K' is detached in G, by Theorems I.13 and I.14. But this contradicts the definition of H. □

We have now proved that x belongs to some component H of G. It can belong to no other component of G, by Theorem I.17.

Theorem I.19. *Let x be an isolated vertex of G. Then the corresponding vertex-graph is a component of G.*

To prove this we observe that the vertex-graph is a detached nonnull subgraph of G and that it has no nonnull proper subgraph.

We denote the number of components of a graph G by $p_0(G)$. In the terminology of algebraic topology, it is "the Betti number of G of dimension zero."

If G is null, we have $p_0(G) = 0$, because components are nonnull. but for a nonnull graph G, $p_0(G)$ is always positive, by Theorem I.18. If $p_0(G) = 1$, we say that G is *connected* and if $p_0(G) > 1$, we say that G is *disconnected*.

According to these definitions the null graph is the only graph that is neither connected nor disconnected.

Theorem I.20. *Let G be an edgeless graph. Then $p_0(G) = |V(G)|$.*

This result follows from Theorems I.17 and I.19.

Let us note that $p_0(G)$ is a structural property of G. This follows from the remarks on isomorphism at the end of Sec. I.3, with the further observation that θ maps $W(G, L)$ onto $W(H, \theta L)$ for each subgraph L of G. Actually most of the properties of graphs discussed in this work are structural, being defined ultimately in terms of the incidence relations only. We need no longer emphasize structurality. It will be enough to point out the occasional nonstructural property as it comes under discussion.

We now give some examples of connected graphs. Any nonnull clique is connected, for by its definition no nonnull proper subgraph of the clique can be detached.

Consider an n-arc G. In some defining enumeration of G, let the vertices be enumerated as $a_0, a_1, \ldots, a_n$. Let C be the component of G including a_0. (See Theorem I.18.) If there are vertices of G not belonging to C, let a_j be the one with least suffix. But then a_{j-1}, being incident with A_j, is a vertex of attachment of C in G, which is impossible. Hence C includes all the vertices of G. Accordingly $p_0(G) = 1$, by Theorem I.17.

Now we consider an n-circuit G. In some defining enumeration, we suppose the sequence of vertices to be $a_1, a_2, \ldots, a_n$. We let C be the component of G including a_1, and we continue the argument as in the case of the n-arc.

We conclude that each arc and circuit is connected.

We continue with some general theorems about connected graphs and subgraphs.

Theorem I.21. *A graph G is connected if and only if no nonnull proper subgraph of G is detached in G, provided that G itself is nonnull.*

Proof. If G has a detached nonnull proper subgraph, then G has at least two components, by Theorems I.16 and I.18. In the remaining case G is a minimal nonnull detached subgraph of G, that is, a component of G. By Theorem I.17, it is the only component of G. □

Theorem I.22. *Each component of a graph G is a connected graph.*

Proof. Let C be a component of G. It has itself a component D, by Theorem I.18. But D is a detached subgraph of G, by Theorem I.15. Hence $D = C$ and $p_0(C) = 1$, by the minimality of C. □

Theorem I.23. *Let H be a subgraph of G. Then H is a component of G if and only if it is both detached in G and connected.*

Proof. If H is a component of G, it satisfies the stated condition, by Theorem I.22. Conversely, if H is both detached in G and connected, it is a minimal nonnull detached subgraph of G, by Theorems I.8 and I.21. (Theorem I.8 shows that if a subgraph of H is detached in G, it is also detached in H.) □

Theorem I.24. *Let H be a connected subgraph of G. Then there is a unique component C of G such that $H \subseteq C$.*

Proof. Let x be any vertex of H. There is a component C of G including x, by Theorem I.18. The intersection $C \cap H$ is nonnull.

Assume that $C \cap H$ is a proper subgraph of H. It is not detached in H, by Theorem I.21. Hence some vertex u of $C \cap H$ is incident with an edge A of H not belonging to C. But then u is a vertex of attachment of C in G, which is impossible.

We deduce that $C \cap H$ is the whole of H, that is, that $H \subseteq C$. □

Theorem I.25. *Let H and K be connected subgraphs of G whose intersection is nonnull. Then $H \cup K$ is connected.*

Proof. H and K are contained in components C and D, respectively, of $H \cup K$, by Theorem I.24. We must have $C = D$, by Theorem I.17. Accordingly,

$$C = H \cup K \quad \text{and} \quad p_0(H \cup K) = 1.$$ □

We conclude the section with two characterization theorems.

Theorem I.26. *A graph G is an arc if and only if it is connected, has two vertices of valency 1, and has all its other vertices of valency 2.*

Proof. Suppose first that G is an arc. We have observed in this section that G is connected, and, in Sec. I.1, that it has the other two required properties.

Conversely, suppose G to be a graph having the three stated properties. We denote one of the monovalent vertices by a_0. We then write A_1 for the single edge incident with a_0, a_1 for the other end of A_1, A_2 for the second edge incident with a_1 (if any), a_2 for the other end of A_2, and so on. Evidently no edge can be repeated in this sequence until some vertex has been repeated, and the first repetition of a vertex has been repeated, and the first repetition of a vertex would violate the valency conditions. So the sequence continues without repetition until it terminates. It can terminate only at the second monovalent vertex, a_n, say. In the sequences $a_0, a_1, \ldots, a_n$ and $A_1, A_2, \ldots, A_n$, we see the defining enumeration of an n-arc L that is a subgraph of G. But L is detached in G, for each of its vertices has the same valency in L as in G. Hence $G = L$, by Theorem I.21. □

Theorem I.27. *A graph G is a circuit if and only if it is a connected regular graph of valency 2.*

Proof. Suppose first that G is a circuit. We have observed in this section that G is connected, and, in Sec. I.1, that G is a regular graph of valency 2.

Conversely, suppose G to be a connected regular graph of valency 2. Let a_0 be one of its vertices and A_1 an edge incident with a_0. We write a_1 for the other end of A_1, A_2 for the second edge incident with a_1, a_2 for the other end of A_2, and so on, until some edge or vertex is repeated. Evidently the first repetition must be that of a vertex, and by the valency conditions it must be of the form $a_n = a_0$. (If A_1 is a loop, then $n = 1$.) In the sequences $a_1, a_2, \ldots, a_n$ and $A_1, A_2, \ldots, A_n$, we see a defining enumeration of an n-circuit J contained in G. But J is detached in G, since each of its vertices is divalent in both J and G. Hence, $G = J$, by Theorem I.21. □

I.6. DELETION OF AN EDGE

In this section we consider the effect of deleting an edge A from a graph G. The operation replaces G by its spanning subgraph $G:(E(G)-\{A\})$, which we denote also by the simpler symbol G'_A. The theory is based on the following observation.

Proposition I.28. *Let A be an edge of a graph G, and let C be the component of G including A. Then the components of G'_A are the components of G other than C, together with the components of C'_A. Moreover, each component of C'_A includes at least one end of A.*

Proof. The components of G other than C and the components of C'_A are clearly detached subgraphs of G'_A, together including all the vertices of G'_A. Hence they are the components of G'_A, by Theorems I.23 and I.7. Moreover, if C'_A had a component D not including an end of A, then D would be a detached subgraph of G. This would contradict the minimality of C. □

There are two possibilities. Either C'_A has exactly two components, each including a single end of A in G, or C'_A is connected. In the first case we say that A is an *isthmus* of G, and that the components of C'_A are the two *end-graphs* of A in G.

As an immediate consequence of Proposition I.28, we have Theorem I.29.

Theorem I.29. *Let A be an edge of a graph G. If A is not an isthmus of G, then $p_0(G'_A) = p_0(G)$. But if A is an isthmus of G, then $p_0(G'_A) = p_0(G) + 1$.*

The single edge of a link-graph is an isthmus. It can further be verified that each edge of an arc is an isthmus of the arc. Let us note also the following rules.

Theorem I.30. *Let x be a monovalent vertex of a graph G, and let A be its incident edge. Let C be the component of G including x and A. Then A is an isthmus of G. One of its end-graphs is the vertex-graph X defined by x. The other is the graph obtained from C by deleting x and A.*

Proof. X is a component of C'_A, by Theorem I.19. Since x is monovalent, A is a link, with a second end y not belonging to X. Hence A is an isthmus of G and X is one of its end-graphs. The second must be the complementary subgraph to X in C_A'. □

Theorem I.31. *A circuit has no isthmus.*

Proof. This result can be verified directly from the definitions of arcs and circuits. We observe that the deletion of an edge from a circuit changes it into a vertex-graph or an arc, that is, into a connected graph.

Another proof uses Theorem I.1. Suppose A to be an isthmus of a circuit C. Let H be an end-graph of A in C, and let x be the single end of A belonging to H. Now in C each vertex has valency 2. Hence in H the vertex x has valency 1 and every other vertex has valency 2. But this is impossible, by Theorem I.1. □

By definition, an isthmus is necessarily a link. Isthmuses satisfy the following persistence theorem.

Theorem I.32. *Let A be an isthmus of a graph G. Let H be a subgraph of G having A as an edge. Then A is an isthmus of H.*

Proof. Suppose A is not an isthmus of H. Then both ends of A are in the same component D of H_A'. But D is a connected subgraph of G_A'. It is contained in a component C of G_A', by Theorem I.24. Hence both ends of A are in the same component of G_A', contrary to the definition of A as an isthmus of G. □

A graph in which every edge is an isthmus is called a *forest*. A connected forest is a *tree*. Thus, by Theorem I.32, each component of a forest is a tree. We note that the null graph is a forest but not a tree. It is the only forest that contains no tree. As a more general application of Theorem I.32 we have:

Theorem I.33. *Any subgraph of a forest is a forest.*

We define the *cyclomatic number* $p_1(G)$ of a graph G, also known as the Betti number of dimension 1, by the following equation:

$$p_1(G) = |E(G)| - |V(G)| + p_0(G). \qquad (1.6.1)$$

Theorem I.34. *Let A be an edge of a graph G. If A is an isthmus of G, then $p_1(G_A') = p_1(G)$. But if A is not an isthmus of G, then $p_1(G_A') = p_1(G) - 1$ (by Eq. (I.6.1) and Theorem I.29).*

Theorem I.35. *The cyclomatic number of a graph G is nonnegative. It is zero if and only if G is a forest.*

Proof. Suppose first that G is edgeless. Then $p_1(G) = 0$, by Eq. (I.6.1) and Theorem I.20. An edgeless graph is, of course, a forest.

Suppose next that G is a forest having at least one edge. We delete its edges one by one until we have an edgeless graph H. Each deletion leaves the cyclomatic number unchanged, by Theorems I.32 and I.34. Hence $p_1(G) = p_1(H) = 0$.

In the remaining case G is not a forest; it has some edge A that is not an isthmus. Deleting A we reduce the cyclomatic number by 1, by Theorem 1.34. If the resulting graph is not a forest, we repeat the process, and so on. After some positive number n of such deletions of nonisthmus edges, we arrive at a forest F. We then have $p_1(G) = n + p_1(F) = n > 0$. □

A *spanning tree* of a graph G is a spanning subgraph of G that is a tree. We shall have much to say about spanning trees in later chapters. Here we prove a single theorem about them.

Theorem I.36. *A graph G has a spanning tree if and only if it is connected.*

Proof. If G has a spanning tree it is connected, by Theorem I.24.

Conversely, suppose G to be a connected graph. Let us delete nonisthmus edges one by one until no more remain. We then have a spanning subgraph H of G that is a forest. But H is connected, by Theorem I.29. It is thus a spanning tree of G. □

As examples of trees we can take the vertex-graph and the link-graph. Once we have verified that each edge of an arc is an isthmus, we can assert that each arc is a tree, by Theorem I.26.

We now take note of two rather obvious properties of trees.

Theorem I.37. *If T is any tree, then $|V(T)| = |E(T)| + 1$.*

Proof. We have $p_0(T) = 1$. But $p_1(T) = 0$, by Theorem I.35. Our theorem is thus a consequence of Eq. (I.6.1). □

Theorem I.38. *Let A be an edge of a tree T. Then the end-graphs of A in T are trees* (by Theorem I.33).

By Theorem I.37, the vertex-graph is the only tree with a single vertex, and the only edgeless tree. It is also the only tree with an isolated vertex, by Theorem I.19.

Further properties of trees can be deduced from a modification of Eq. (I.1.1). We subtract $2|V(G)|$ from each side of Eq. (I.1.1). We then use Eq. (I.6.1) to obtain the following theorem.

Theorem I.39. *If G is any graph, then*

$$\sum_{x \in V(G)} (\mathrm{val}(G, x) - 2) = 2(p_1(G) - p_0(G)). \qquad (1.6.2)$$

Theorem I.40. *If T is any tree other than the vertex-graph, it has at least two monovalent vertices. If it has only two monovalent vertices, it is an arc.*

Proof. For T the right side of Eq. (I.6.2) takes the value -2. Since T has no isolated vertex it follows from Eq. (I.6.2) that T has at least two monovalent vertices; and if it has only two, then every other vertex is divalent. The theorem follows, by Theorem I.26. □

The next theorem is easily derived directly from the definition of an arc, but as an exercise let us deduce it from Theorem I.39.

Theorem I.41. *Every arc is a tree. Let A be an edge of an arc L and let the ends of A be x and y. Let H be the end-graph of A in L including x. If x is an end of L, then H is the vertex-graph defined by x. In the remaining case H is an arc with one end at x and the other at an end of L.*

Proof. For an arc L, the left side of Eq. (I.6.2) takes the value -2 and, moreover, $p_0(L) = 1$, by Theorem I.26. Hence $p_1(L) = 0$ by Eq. (I.6.2), and L is a forest, by Theorem I.35. By connection, L is a tree.

If x is an end of L, we complete the proof by using Theorem I.30. In the remaining case, x is a monovalent vertex of H, and each other vertex of H has the same valency in H as in L. But H is a tree, by Theorem I.38, and so it has at least one other monovalent vertex, by Theorem I.40. Such an other monovalent vertex can only be an end of L. But H can include only one end of L; the other must belong to the second end-graph of A by similar reasoning. Hence H is a tree with only two monovalent vertices. The theorem now follows from Theorem I.40. □

Corollary I.42. *Let A be an edge of an arc L. Then the two ends of L belong to distinct end-graphs of A in L.*

We conclude this section with some theorems concerning the occurrence of arcs and circuits as subgraphs.

Theorem I.43. *Let x and y be distinct vertices of a graph G. In order that some subgraph of G shall be an arc with ends x and y, it is necessary and sufficient that x and y shall belong to the same component C of G. The arc is then a subgraph of C.*

Proof. Suppose first that such an arc L exists. It is contained in a component C of G, by Theorems I.24 and I.26. Then x and y both belong to C.

Conversely, suppose x and y to belong to the same component C of G. Then C has a spanning tree, by Theorem I.36. We can therefore assert the existence of a tree T that is a subgraph of C, that includes both x and y, and that has the least number of edges consistent with these conditions.

Suppose T to have a monovalent vertex z distinct from x and y. Let A be the edge of T incident with z. Then one end-graph of A in T is a tree including both x and y, by Theorems I.30 and I.38. But this contradicts the definition of T. It follows from Theorem I.40 that T is an arc with x and y as its monovalent vertices. □

Theorems I.44. *Let x and y be distinct vertices of a tree T. Then T contains exactly one arc with ends x and y.*

Proof. One such arc exists, by Theorem I.43. Suppose there are two distinct arcs, L_1 and L_2. Without loss of generality we can suppose L_2 to include an edge A not belonging to L_1. Let the ends of A be u and v. We can adjust the notation so that one end-graph H of A in L_2 includes u and x, while the other K, say, includes v and y, by Corollary I.42. (We may have $u = x$ or $v = y$.) There are components B, C, and D of T'_A containing H, K and L_1, respectively, by Theorem I.24. But then $B = D = C$, by Theorem I.17. We deduce that A is not an isthmus of T, which is a contradiction. □

Theorem I.45. *Let A be an edge of a graph G. In order that some subgraph of G shall be a circuit including A, it is necessary and sufficient that A shall not be an isthmus of G.*

Proof. Suppose G to contain a circuit K that includes A. If A is an isthmus of G, it is an isthmus of K, by Theorem I.32. Hence A is not an isthmus of G, by Theorem I.31.

Conversely, suppose A is not an isthmus of G that is, the two ends x and y of A belong to the same component C of G'_A. If $x = y$, the loop A determines a 1-circuit in G, and the theorem is satisfied. In the remaining case G'_A contains an arc L with ends x and y, by Theorem I.43. Adjoining A to L we obtain a subgraph K of G that is connected, includes A, and is regular of valency 2, by Theorem I.25. K is a circuit, by Theorem I.27. □

Because of Theorem I.45 we can characterize a forest as a graph containing no circuit, and a tree as a connected graph containing no circuit. By Theorems I.34 and I.35, we can characterize $p_1(G)$ as the least number of edges whose deletion from G destroys every circuit.

I.7. LISTS OF NONISOMORPHIC CONNECTED GRAPHS

To *add a twig* to a graph H is to adjoin a single new vertex v and a single new link A joining v to some vertex x of H. Let G be the resulting graph. The *twig* is the link-graph $G \cdot \{A\}$, and we say it is added to H at x.

If H is connected, then so is G, by Theorem I.25. In the general case, we have $p_0(G) = p_0(H)$, by Proposition I.28 and Theorem I.30, and therefore $p_1(G) = p_1(H)$ by Eq. (I.6.1). It follows that if H is a tree, then so is G.

On the other hand if T is a tree other than a vertex-graph, it has a monovalent vertex, by Theorem I.40. It can therefore be obtained from a smaller tree by adding a twig at some vertex, by Theorems I.30 and I.38.

It is clear that, to within an isomorphism, only one graph can be obtained from a given graph G by adding a twig at a given vertex x.

Let us apply these observations to the construction of a complete list L_n of nonisomorphic trees of n edges. We start with L_0 and then take L_1, L_2, L_3, etc., in turn. Evidently L_0 has the vertex-graph as its only member. For $n > 0$ we can derive L_n by listing the results of adding a twig at each vertex of each graph of L_{n-1}, and then rejecting isomorphic duplicates. It is the recognition of these duplicates that is the most difficult part of the procedure. We note one simplification: If two vertices of a member S of L_{n-1} are equivalent under some isomorphism of S, it is necessary to add a twig at only one of them.

There is only one way of adding a twig to a vertex-graph. Hence L_1 has the link-graph as its only member. Since the link-graph has an isomorphism interchanging its ends, there is essentially only one way of adding a twig to it. Accordingly L_2 has the 2-arc as its only member. A twig can be added to the 2-arc, either at an end or at an internal vertex. Accordingly, L_3 has exactly two members as shown in Fig. I.7.1. The first of these is the 3-arc; the second we call the 3-*star*. In general, for any nonnegative integer n we define an n-star as a graph defined by $(n + 1)$ vertices $a_0, a_1, \ldots, a_n$ and n edges $A_1, A_2, \ldots, A_n$, the ends of A_j being a_0 and a_j. Evidently all n-stars with the same n are isomorphic. The vertex-graph is the 0-star, the link-graph is the 1-star, and the 2-arc is the 2-star.

From L_3 we can obtain L_4, shown in Fig. I.7.2. It is made up of the 4-arc, the 4-star, and one other tree. This other tree can be obtained either by adding a twig at an internal vertex of the 3-arc or by adding one at a monovalent vertex of the 3-star. Each list L_n should be checked to see that there are no isomorphic duplicates. The checking is easy for L_4, since the three proposed members have different maximum valencies.

One more step brings us to L_5, shown in Fig. I.7.3. With each step it becomes a little more difficult to check for any remaining isomorphic duplicates. There are two members of L_5 with four monovalent vertices. But they are nonisomorphic because one has a four-valent vertex and the other does not. There are two members, each with three monovalent vertices, two

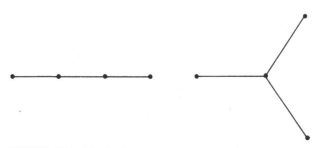

FIGURE I.7.1 The list L_3.

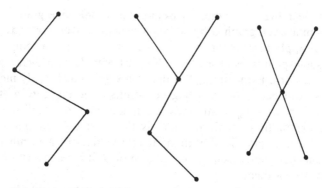

FIGURE I.7.2 The list L_4.

divalent vertices, and one trivalent one. They are nonisomorphic because one has a vertex joined to two monovalent vertices and the other does not.

Now let us write M_n for a complete list of nonisomorphic, loopless connected graphs with n edges. Clearly, M_n contains L_n. Any member of M_n that is not in L_n has a nonisthmus edge A. It can therefore be obtained from a member H of M_{n-1} by *adding a link*, that is, by adjoining a new edge A and specifying its ends as two distinct vertices of H. Apart from isomorphic duplicates, only one graph can be obtained from a given H by adding a link with given ends. On the other hand, any graph obtained from a member of M_{n-1} by adding a link is a connected loopless graph of n edges, by Theorem I.25.

We note that M_0 consists solely of the vertex-graph. Since the vertex-graph does not have two distinct vertices to be joined by a new link,

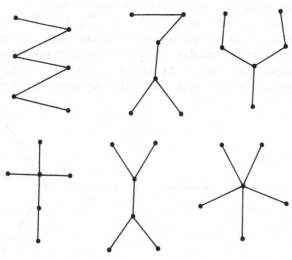

FIGURE I.7.3 The list L_5.

Lists of Nonisomorphic Connected Graphs 25

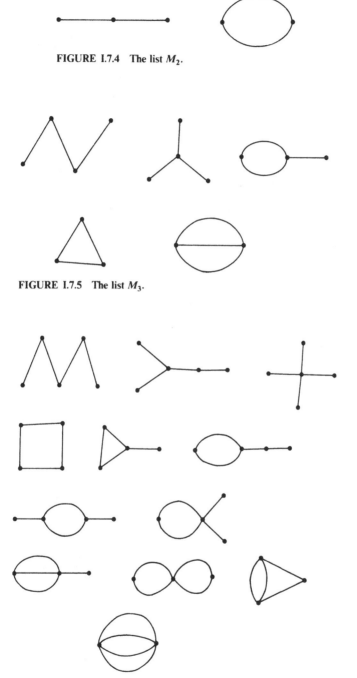

FIGURE I.7.4 The list M_2.

FIGURE I.7.5 The list M_3.

FIGURE I.7.6 The list M_4.

we have also $M_1 = L_1$. However, it is possible to join anew the two ends of the link-graph. Accordingly, M_2 has two members, the 2-arc and the 2-circuit, as shown in Fig. I.7.4.

From the 2-arc we can obtain two members of M_3 by adding a link, one of them being the 3-circuit. Another, called the 3-*linkage*, arises when a link is added to the 2-circuit. For any positive integer n we define an *n-linkage* as a graph made up of two vertices x and y and n links joining them. Thus the 1-linkage is the link-graph, and the 2-linkage is the 2-circuit. Two n-linkages with the same n are isomorphic.

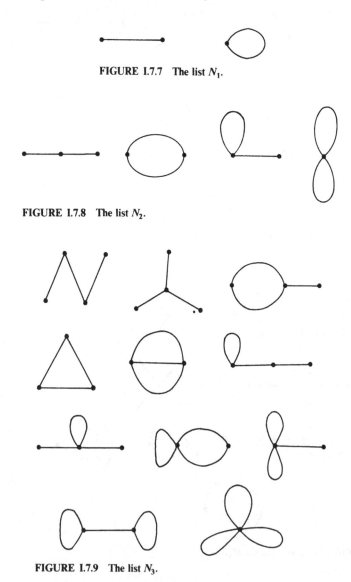

FIGURE I.7.7   The list N_1.

FIGURE I.7.8   The list N_2.

FIGURE I.7.9   The list N_3.

Let us now define N_n as a complete list of nonisomorphic connected graphs of n edges, looped as well as loopless. In constructing these lists we use the rule that the members of N_n not belonging to M_n are the graphs obtainable from the members of N_{n-1} by the operation of adding a new loop at some vertex. Thus the vertex-graph, by loop-addition, gives rise to the loop-graph. Accordingly, N_1 has two members, the link-graph and the loop-graph, as shown in Fig. I.7.7.

We obtain N_2, as shown in Fig. I.7.8, by adjoining to M_2 the graphs obtained by adding loops to the members of N_1. Another step brings us to the list N_3 of Fig. I.7.9.

Such lists lead us to some of the major enumerative problems of graph theory. How many trees are there with n edges? How many members have M_n and N_n for an arbitrary n? Such problems are now dealt with by means of a general enumerative problem of G. Pólya. But Pólya enumeration is one of many branches of graph theory not dealt with in this work. We refer interested readers to Reference [8].

I.8. BRIDGES

Here we generalize the theory of detached subgraphs set out in Sec. I.5. We discuss a graph G and a fixed subgraph J of G. A subgraph H of G is said to be J-*detached* in G if all its vertices of attachment are in J. Thus if J is the null subgraph of G, the J-detached subgraphs are simply the detached ones. Our first theorem generalizes Theorems I.13 and I.14.

Theorem I.46. *The union and intersection of any two J-detached subgraphs of G are J-detached, and the complementary subgraph in G to a J-detached subgraph of G is J-detached* (by Theorems I.6 and I.9).

A *bridge* B of J in G is a subgraph of G satisfying the three following conditions.
i) *B is not a subgraph of J.*
ii) *B is J-detached in G.*
iii) *No proper subgraph of B satisfies both (i) and (ii).*

We express Condition (iii) by saying that B is *minimal* with respect to the pair of properties (i) and (ii). Condition (i) implies that B is nonnull. Bridges generalize components. In fact, if J is null, its bridges in G are the components of G. Bridges of J are called "components mod J" in [9].

Theorem I.47. *A bridge B of J in G has no edge in J.*

Proof. Suppose A is an edge of B in J. Then B'_A satisfies (i) and (ii), contrary to the minimality of B. □

Theorem I.48. *Let B be a bridge of J in G. Let x be a common vertex of B and J. Then x is incident with at least one edge of B.*

Proof. If x is isolated in B, let B_1 be the induced subgraph of G obtained from B by deleting x. Then B_1 satisfies (i) and (ii), contrary to the minimality of B. □

A common vertex x of B and J is not necessarily a vertex of attachment of B. But if it is not, it is isolated in J, by Theorem I.47.

Consider any bridge B of J in G. We refer to the common vertices of B and J as the *outer* vertices of B. An edge of B incident with an outer vertex x of B is said to be *J-bound* to x. If it is also incident with a vertex y of B not in J, we say it is J-bound *from y to x*. By Rule I.4 there is a subgraph $N(B)$ of B obtained from B by deleting the outer vertices and the J-bound edges. We call $N(B)$ the *nucleus* of B.

Theorem I.49. *Let A be any edge of G that is not in J, but which has both its ends in J. Then the link-graph or loop-graph $G \cdot \{A\}$ is a bridge of J in G (by Conditions (i)–(iii)).*

Bridges of the kind determined by Theorem I.49 are called *degenerate*. Evidently a degenerate bridge has a null nucleus.

Theorem I.50. *Let C be a component of the induced subgraph $G[V(G) - V(J)]$ of G. Let B be the subgraph of G obtained from C by adjoining to it each link of G with one end in C and one in J, together with the end of that link in J. Then B is a nondegenerate bridge of J in G, and C is its nucleus.*

Proof. By its definition B satisfies Conditions (i) and (ii). Assume that B is not a bridge of J in G. Then it fails to satisfy Condition (iii). Hence there is a proper subgraph H of B satisfying (i) and (ii). By (i) the graph $H \cap C$ must be nonnull. Moreover, $H \cap C$ is a proper subgraph of C, for otherwise H would include every edge and vertex of B.

Applying Theorem I.21 to C, we find that some vertex x of $H \cap C$ is incident with an edge A of C not belonging to H. But then x is a vertex of attachment of H that is not in J, contrary to the definition of H. We conclude that B is in fact a bridge of J in G. By the construction of B, it is nondegenerate and C is its nucleus. □

Theorem I.51. *Any two distinct bridges of J in G have an intersection that is contained in J. The only bridges of J in G are those specified by Theorems I.49 and I.50. Any edge or vertex of G not in J belongs to just one bridge of J in G.*

Proof. Suppose H and K to be distinct bridges of J in G such that $H \cap K$ is not a subgraph of J. But then $H = H \cap K = K$ by Theorem I.46 and the minimality of H and K, and we have a contradiction.

To complete the proof we have only to observe that the bridges determined by Theorems I.49 and I.50 together include all the edges and vertices of G not in J.

By Theorem I.25 we have the following corollary. □

Corollary I.52. *Every bridge of J in G is connected.*

Figure I.8.1 shows the bridges of a circuit J in a graph G. The J-bound edges of these bridges are shown by broken lines. There are two degenerate bridges B_1 and B_2 and three nondegenerate ones B_3, B_4, and B_5.

We next state two theorems that are now rather obvious. It is convenient to be able to quote them in applications.

Theorem I.53. *Let H be any connected subgraph of G disjoint from J. Then H is contained in the nucleus of some nondegenerate bridge B of J in G* (by Theorems I.24 and I.50).

Theorem I.54. *Let B be any nondegenerate bridge of J in G. Let S be any set of outer vertices of B. Let H be the subgraph of G obtained from $N(B)$ by adjoining the members of S and those edges of B that are J-bound to members of S. Then H is connected.*

For Theorem I.54 we should perhaps remark that $N(B)$ is connected, by Theorems I.50 and I.51. This disposes of the case in which S is null. In the remaining case, each vertex of S is incident with one of the adjoined edges, by Theorem I.48 and the fact that B contains no degenerate bridge (by Theorem I.51). So the theorem follows, from Theorem I.25. It includes Corollary I.52.

We now come to two theorems that together constitute an analogue of Theorem I.43. Let us say that an arc L contained in G *avoids* J if it has no edge or internal vertex in J.

Theorem I.55. *Let L be an arc in G avoiding J. Then L is a subgraph of some bridge B of J in G.*

Proof. If L is a 1-arc its edge belongs to some B, by Theorem I.51, and $L \subseteq B$. We may therefore suppose L to be an n-arc with $n > 1$.

Delete from L its ends in J, if any, with their incident edges. There remains a connected subgraph of G disjoint from J, by Theorem I.26 and at

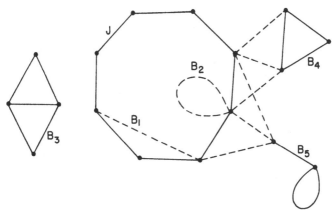

FIGURE I.8.1

most two applications of Theorem I.30. This subgraph is contained in the nucleus of some nondegenerate bridge B of J in G, by Theorem I.53. Hence $L \subseteq B$, by Condition (ii). □

Theorem I.56. *Let x and y be two distinct vertices of a bridge B of J in G. Then there exists an arc L that is a subgraph of B, which avoids J and which has x and y as its ends.*

Proof. If B is degenerate, it must be a link-graph. The theorem then holds with $L = B$.

In the remaining case, let H be the graph obtained from B by deleting each outer vertex that is not x or y together with its incident edges. Then H is connected, by Theorem I.54, and it includes both x and y. There is an arc L contained in H that has ends x and y, by Theorem I.43. But L must avoid J, by the definition of H. □

Sometimes it is necessary to relate the bridges of two different subgraphs J and K of G. The following theorem may then be helpful.

Theorem I.57. *Let J and K be subgraphs of G. Let B be a bridge of J in G having no edge in K and such that each vertex of B in K is a vertex of J. Then B is a subgraph of some bridge B' of K in G. If, in addition, each vertex of B in J is a vertex of K, then $B' = B$.*

Proof. If B is degenerate, the first part of the theorem follows from Theorem I.51, and the second part from Theorem I.49.

If B is nondegenerate, we apply Theorem I.54, with S the set of vertices of B that are in J but not in K. The corresponding subgraph H of G is contained in the nucleus of some bridge B' of K in G, by Theorems I.53 and I.54. By Condition (ii), B' includes all the J-bound edges of B, and therefore B' contains B. If B has the same vertices in K as in J, then it satisfies Conditions (i) and (ii) with respect to K. So then $B' = B$, by the minimality of B'.

I.9. NOTES

I.9.1. History

Graph theory begins in 1736 with Euler's solution of the Königsberg Bridges problem. For an account of its development during the next two centuries, reference may be made to [4].

I.9.2. Textbooks

The first textbook on graph theory is that of the Hungarian graphman Dénes König. It was preceded by an important series of papers by Hassler

Whitney, published in the years 1931-33. Collected they would have made an excellent textbook in English, had such a work been then in demand. Even before 1931 some information on graph theory was available in the semi-popular work "Mathematical recreations and essays", noted in the Introduction. Some textbooks now in common use as references are noted below as [1], [2], [5] and [7].

I.9.3. The automorphism group

Much work has been done on highly symmetrical graphs, those in which the order of the automorphism group is at least equal to the number of edges. (See [3], [6] and [9].)

EXERCISES

1. Show that all arcs of length 3 in the Thomsen graph (see Sec. IV.2) are equivalent, in the sense that any one can be transformed into any other by an automorphism of the graph. Are all the arcs of length 4 equivalent?
2. Construct a graph whose automorphism group has only three elements. Find the simplest strict graph that has only the trivial automorphism.
3. Prove the following formula for vertices of attachment:

$$W(G, H) = W(K, H \cap K) \cup \{V(H) \cap W(G, H \cup K)\}.$$

Here H and K are arbitrary subgraphs of G.

REFERENCES

[1] Berge, C., *Graphes et hypergraphes*, Dunod, Paris 1973.
[2] _____, *Graphs and hypergraphs* (translated into English by Edward Minieka), Wiley, New York, Methuen, London 1973.
[3] Biggs, N. L., *Algebraic graph theory*, Cambridge University Press, 1974.
[4] Biggs, N. L., E. K. Lloyd, and R. J. Wilson, *Graph Theory, 1736-1936*, Clarendon Press, Oxford 1976.
[5] Bondy, J. A., and U. S. R. Murty, *Graph Theory with applications*, American Elsevier, New York 1976.
[6] Coxeter, H. S. M., Self-dual configurations and regular graphs, *Bull. Amer. Math. Soc.*, **56** (1950), 413-455.
[7] Harary, F., *Graph Theory*, Addison-Wesley, Reading, Mass., 1969.
[8] Harary, F., and E. M. Palmer, *Graphical enumeration*, Academic Press, New York-London 1973.
[9] Tutte, W. T., *Connectivity in graphs*, Univ. of Toronto Press, 1966.

Chapter II

Contractions and the Theorem of Menger

II.1. CONTRACTIONS

In Chapter I we constructed a theory of graphs based on the observation that some graphs contain others. It can be said that this construction is continued in the present chapter, but with a more liberal interpretation of the term "contain."

Let G be a graph, and let S and T be complementary subsets of $E(G)$. Consider the spanning subgraph $G:S$ of G. There exists a graph H whose vertices are the components of $G:S$, whose edges are the members of T, and which has the following rule of incidence: The ends in H of an edge A of T are those components of $G:S$ that include at least one end of A in G. We call H the *contraction* of G to T, and we denote it by $G \times T$. We say also that H is formed from G by *contracting away* S.

The above rule of incidence ensures, as it should, that each edge of T has at least one but not more than two ends in $G \times T$. It should be noted, however, that a link of T may have both its ends in G in the same component of $G:S$. Hence an edge of T may be a link in G but a loop in $G \times T$. On the other hand, a loop of G in T remains a loop in $G \times T$.

Figure II.1.1 shows a contraction $G \times T$. The first diagram shows $G:S$ in G, and the edges of T are represented by broken lines. The second diagram shows $G \times T$, with its vertices as dots. The letters A to F indicate

Contractions

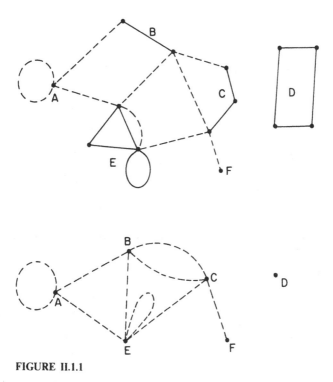

FIGURE II.1.1

components of $G:S$ in the first diagram, and corresponding vertices in the second.

Since the vertices of $G \times T$ are defined as graphs, it is possible to speak of the vertices and edges of the vertices of $G \times T$, and in this section at least it is necessary to do so. But often it is convenient to replace these vertices by others for which no internal structure is defined, perhaps by some of the vertices of G. In this connection we introduce the notion of *vertex isomorphism*. Two graphs H and K are said to be *vertex-isomorphic* if there is an isomorphism $\theta = (f, q)$ of H onto K in which g is the identical mapping of $E(H)$ onto $E(K)$. Then H and K have the same edges but may have different vertices. But vertices of H and K corresponding under f are incident with the same edges in each graph. We say that θ is a *vertex isomorphism* of H onto K. We remark that vertex isomorphism of graphs is an equivalence relation.

In such equations as (I.3.1) we use the equality sign to indicate that the graphs specified on the two sides are identical; that is, they have the same vertices, the same edges, and the same incidence relations. To indicate that two graphs H and K are vertex-isomorphic, we shall use the equality sign with the suffix v, as follows:

$$H =_v K.$$

To indicate that two graphs H and K are isomorphic, we shall write

$$H \cong K.$$

For an example of vertex isomorphism, we consider the case in which S is null. Then $G:S$ is edgeless, and its components are the vertex-graphs contained in G, by Theorem I.19. Hence there is a vertex isomorphism θ of $G \times T$ onto G. It maps each vertex-graph contained in G onto the corresponding vertex of G. Now in a formal development of graph theory, a vertex-graph is not to be regarded as the same thing as its vertex, though in the practice of graph theory the distinction between them may not be important. Thus in the present case we may say that G and $G \times T$ are essentially the same graph. But it is more precise to say that they are equivalent under a vertex isomorphism, namely θ. We note the rule that, for any graph G,

$$G \times E(G) =_v G. \qquad (\text{II}.1.1)$$

Let us now return to general S and T, and consider the relation between the subgraphs of $G \times T$ and those of G. We will say that a subgraph H of G *conforms* to S if each component of $G:S$ is either a subgraph of H or disjoint from H.

Let K be any subgraph of $G \times T$. Its vertices are components of $G:S$. By Theorem I.4 there exists a subgraph $f_S K$ of G whose vertices are the vertices of the vertices of K and whose edges are the edges of the vertices of K, together with the edges of K. Evidently $f_S K$ is uniquely determined by K, and different subgraphs K of $G \times T$ determine different subgraphs $f_S K$ of G. We can therefore regard the symbol f_S as representing a 1–1 mapping of the set X of subgraphs of $G \times T$ onto some set Y of subgraphs of G. We note further that each member of Y conforms to S.

Conversely, consider any subgraph H of G that conforms to S. The components of $H:(S \cap E(H))$ are those components of $G:S$ that are contained in H, by Theorem I.23. Hence $H \times (T \cap E(H))$ is a subgraph K of $G \times T$, and it is clear that $f_S K$ is identical with H. We can now identify Y as the set of all subgraphs of G conforming to S. We can state the rule

$$f_S K \times E(K) = K \qquad (\text{II}.1.2)$$

as the general relation between the subgraphs K of $G \times T$ and those subgraphs $f_S K$ of G that conform to S. We summarize the preceding results in the following statement.

Theorem II.1. *Let S and T be complementary subsets of $E(G)$. Then there is a 1–1 mapping f_S of the set X of subgraphs of $G \times T$ onto the set Y of subgraphs of G conforming to S, f_S being defined as follows. If $K \in X$, then*

$f_S K$ is the subgraph of G determined by the edges and vertices of the vertices of K, together with the edges of K. The mapping f_S satisfies Eq. (II.1.2).

Theorem II.2. *The mapping f_S preserves proper inclusion of subgraphs; that is, if $K \subset L \in X$, then $f_S K \subset f_S L \in Y$.*

Theorem I.2 follows immediately from the definition of f_S. In the next result we relate vertices of attachment.

Theorem II.3. *Suppose $K \subseteq L \in X$. Then a vertex x of K belongs to $W(L, K)$ if and only if some vertex x' of the component x of $G:S$ belongs to $W(f_S L, f_S K)$.*

Proof. The vertex x of K belongs to $W(L, K)$ if and only if it is incident with an edge A of $E(L) - E(K)$ in K, that is, if and only if some vertex x' of the component x of $G:S$ is incident with an edge A of $E(L) - E(K)$ in G. But an edge of $E(L) - E(K)$ incident with x' in G can equally well be described as an edge of $E(f_S L) - E(f_S K)$ incident with x' in G. For the edges of $G:S$ incident with x' in G are edges of x and therefore edges of $f_S K$. Hence x is in $W(L, K)$ if and only if some vertex x' of the component x of $G:S$ is incident with an edge A of $E(f_S L) - E(f_S K)$ in G, that is, if and only if some such x' belongs to $W(f_S L, f_S K)$. □

Let L be any member of X. We deduce from Theorem II.3 that f_S maps those subgraphs of L that are detached in L onto those subgraphs of $f_S L$ that are detached in $f_S L$ and conform to S. But any subgraph of $f_S L$ that is detached in $f_S L$ must conform to S, by Theorem I.24. Using Theorem II.2 we deduce the following:

Theorem II.4. *If L is any member of X, then f_S maps the components of L onto the components of $f_S L$.*

Putting $L = G \times T$ in Theorem II.4, we obtain:

Theorem II.5. *If T is any subset of $E(G)$, then*

$$p_0(G \times T) = p_0(G). \qquad (II.1.3)$$

Theorem II.6. *The subgraphs of the contractions of G are the contractions of the subgraphs of G.*

Proof. Any subgraph of a contraction of G can be taken as the graph K on the right of Eq. (II.1.2). The equation expresses it as a contraction of a subgraph of G.

Conversely, let H be any subgraph of G and let $H \times U$ be a contraction of H. Let Q be the complement of U in $E(H)$. The components of $H:Q$ are those components of $G:Q$ that are contained in H, by Theorem I.23. It follows that $H \times U$ is a subgraph of the contraction $G \times (E(G) - Q)$ of G. □

Theorem II.7. *Suppose $U \subseteq T \subseteq E(G)$. Then*

$$(G \times T) \times U =_v G \times U. \qquad (II.1.4)$$

Proof. Let the complementary sets of T and U in $E(G)$ be S and R, respectively.

The vertices of $(G \times T) \times U$ are the components of $(G \times T):(T-U)$. This is a spanning subgraph K of $G \times T$. It is mapped by f_S onto a spanning subgraph $f_S K$ of G defined by the edges and vertices of the components of $G:S$ and the edges of $T - U = R - S$. Thus $f_S K = G:R$. So by Theorem II.4, f_S induces a 1-1 mapping f of the vertices of $(G \times T) \times U$ onto the components of $G:R$, that is, the vertices of $G \times U$.

Let A be any edge in U, and let x and y be its ends in $(G \times T) \times U$. They are components of $(G \times T):(T-U)$, and they include the ends x' and y', respectively, of A in $G \times T$. But x' and y' are components of $G:S$, and they include the ends x'' and y'', respectively, of A in G. But by the definition of f_S, the vertices x'' and y'' belong to the subgraphs $f_S x$ and $f_X y$, respectively, components of $G:R$. Hence $f_S x$ and $f_X y$ are the ends of A in $G \times U$. Accordingly, (f, I_U) is a vertex isomorphism of $(G \times T) \times U$ onto $G \times U$, where I_U is the identical mapping of U onto itself. □

A subgraph of a contraction of G is called a *minor* of G. Any contraction of G is thus a minor of G and, to within a vertex isomorphism, any subgraph of G is a minor of G. (See Eq. (II.1.1.)) Minors obey the following simple law.

Law II.8. *A minor of a minor of G is, to within a vertex isomorphism, a minor of G.*

Proof. A minor of a minor of G is a subgraph of a contraction of a subgraph of a contraction of G. It is thus a subgraph of a subgraph of a contraction of a contraction of G, by Theorem II.6, and therefore vertex-isomorphic to a subgraph of a contraction of G, by Theorems I.5 and II.7.
□

If we ignore distinctions between vertex-isomorphic graphs, we can now recognize some symmetry between subgraphs and contractions. Under this symmetry, Theorem II.7 corresponds to Theorem I.5, and Theorem II.6 corresponds to itself. These symmetries give us our first intimations of the far-reaching theory of graph-theoretical "duality." Later on we shall attempt a rigorous definition of "duality." Meanwhile we will use it informally as a term of rather vague meaning. We will use it when we seem to have evidence of a symmetry based on or associated with that of subgraphs and contractions. Thus we will say that Theorem II.7 is the dual theorem to Theorem I.5, and that Theorem II.6 is self-dual. "Subgraph" and "contraction" are dual concepts, and "minor" is self-dual. It will be found convenient to regard "loop" and "isthmus" as dual concepts. Trouble is encountered when we try to identify the dual concept to "vertex."

II.2. CONTRACTION OF AN EDGE

Let A be an edge of a graph G. We specialize the theory of Sec. II.1 to the case in which $S = \{A\}$ and $T = E(G) - \{A\}$. We denote the contraction $G \times (E(G) - \{A\})$ by G_A'', and say it is the graph derived from G by *contracting* the edge A. By Theorem II.5 we have

$$p_0(G_A'') = p_0(G). \tag{II.2.1}$$

Let x be any vertex of G. If it is incident with A, we write x'' for the loop-graph or link-graph $G \cdot \{A\}$. But if x is not A-incident we write x'' for the vertex-graph in G with vertex x. Then the components of $G:\{A\}$ are the graphs x'', where $x \in V(G)$. These components are the vertices of G_A''. We refer to $G \cdot \{A\}$ as the A-*vertex* of G_A'', and we denote it by v_A. We say that the vertices x of G and x'' of G_A'' are *equivalent* to one another if x is not A-incident in G. When one tires of keeping up the distinction between a vertex-graph and its vertex, one says in this case that x and x'' are the same vertex.

The mapping f_S will now be written as f_A. It is a mapping of the set X of subgraphs of G_A'' onto the set Y of subgraphs of G conforming to $\{A\}$. We observe that to belong to Y a subgraph H of G must satisfy one of two conditions. Either it includes the edge A or it includes no A-incident vertex. If K is any member of X we can reformulate the definition of $f_A K$ as follows. Its vertices are the vertices x of G such that $x'' \in V(K)$. Its edges other than A are the edges of K. It includes the edge A if and only if $v_A \in V(K)$. We can add that, by the definition of G_A'', an edge of K with ends x and y in G has ends x'' and y'' in K.

Figure II.2.1 shows G and G_A'' for a case in which A is a link of G. It shows also a circuit $K = a''b''c''d''e''v_A$ of G_A'' and the corresponding graph $f_A K$ in G. Edges not in K or $f_A K$ are shown by broken lines.

We adjoin two comments, in the form of theorems, on the mapping f_A.

Theorem II.9. *Let K be a subgraph of G_A'' including v_A. Then*

$$(f_A K)_A'' = K \quad \text{(by Eq. (II.1.2))}. \tag{II.2.2}$$

Theorem II.10. *Let K be a subgraph of G_A'' not including v_A. Then the correspondence $x \to x''$ of $V(G)$ onto $V(G_A'')$ induces a vertex isomorphism of $f_A K$ onto K.*

We continue with some theorems and examples on edge-contractions.

Theorem II.11. *If A is a loop of G, then the correspondence $x \to x''$ of $V(G)$ onto $V(G_A'')$ is one-to-one, and it induces a vertex isomorphism of G_A' onto G_A''.*

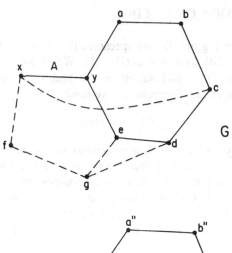
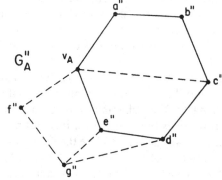

FIGURE II.2.1

This theorem follows immediately from the definitions. Because of it we usually take the contraction-edge A to be a link of G in applications. Loop-contractions can be dealt with as loop-deletions.

Theorem II.12. *If A is a link of G, then*

$$p_1(G_A'') = p_1(G), \qquad (II.2.3)$$

but if A is a loop of G, then

$$p_1(G_A'') = p_1(G) - 1. \qquad (II.2.4)$$

Proof. By Eqs. (II.2.1) and (I.6.1),

$$p_1(G) - p_1(G_A'') = 1 - \{|V(G)| - |V(G_A'')|\},$$

which is 0 or 1 according as A is a link or a loop of G. □

Theorem II.13. *Let A be a link of G with ends x and y. Then*

$$\mathrm{val}(G_A'', v_A) = \mathrm{val}(G, x) + \mathrm{val}(G, y) - 2. \qquad (II.2.5)$$

But if z is a vertex of G not incident with A, we have
$$\operatorname{val}(G_A'', z'') = \operatorname{val}(G, z). \tag{II.2.6}$$

Theorem II.13 comes directly from the definitions. We apply it in the two theorem-examples following.

Theorem II.14. *Let G be an n-circuit with $n > 1$, and let A be an edge of G. Then G_A'' is an $(n - 1)$-circuit.*

Proof. G_A'' is connected, by Eq. (II.2.1) and Theorem I.27. It is regular of valency 2, by Theorem II.13. Hence it is a circuit by Theorem I.27. □

Theorem II.15. *Let G be an n-arc with $n > 1$, and let A be an edge of G. Then G_A'' is an $(n - 1)$-arc.*

Proof. G_A'' is connected, by Eq. (II.2.1) and Theorem I.26. Its vertices are divalent except for the two corresponding to the ends of G under the mapping $x \to x''$, and these two are monovalent, by Theorem II.13. Hence G_A'' is an arc, by Theorem I.26. □

We go on to some theorems concerned with the subgraphs of G_A'' and the corresponding subgraphs of G.

Theorem II.16. *Let A be a link of G, and let K be a subgraph of G_A''. Then*
$$p_0(f_A K) = p_0(K), \tag{II.2.7}$$
$$p_1(f_A K) = p_1(K). \tag{II.2.8}$$

Proof. We may suppose k to include v_A since otherwise the theorem is a consequence of Theorem II.10. So Eq. (II.2.7) follows from Eqs. (II.2.1) and (II.2.2), while Eq. (II.2.8) is a consequence of Eqs. (II.2.3) and (II.2.2). □

Theorem II.17. *Let A be a link of G. Then f_A induces a 1-1 mapping of the set X_0 of spanning subgraphs of G_A'' onto the set Y_0 of those spanning subgraphs of G that include the edge A.*

Proof. Suppose $K \in X_0$. Then $f_A K$ is a spanning subgraph of G, and so includes both ends of A in G. Since it conforms to $\langle A \rangle$, by Theorem II.1, it must therefore include A. Thus $f_A K \in Y_0$.

Conversely, consider a member H of Y_0. It conforms to $\langle A \rangle$. Hence it is of the form $f_A K$, where K is a subgraph of G_A'', by Theorem II.1. But then $K \in X_0$, for otherwise $f_A K$ would not be a spanning subgraph of G. □

Let us define the *tree-number* $T(G)$ of a graph G as the number of spanning trees of G. By Theorem I.36, $T(G)$ is nonzero if and only if G is connected. Now a tree can be characterized as a graph H satisfying the two conditions $p_0(H) = 1$ and $p_1(H) = 0$, by Theorem I.35. It follows, from

Theorems II.16 and II.17, that the spanning trees of G_A'' are in 1-1 correspondence with those spanning trees of G that include the edge A. We can add that the spanning trees of G that do not include A are of course the spanning trees of G_A'. We thus have the following recursion theorem.

Theorem II.18. *Let A be a link of G. Then*

$$T(G) = T(G_A') + T(G_A'').\tag{II.2.9}$$

Theorem II.19. *Let us define $Q(G; i, j)$ as the number of spanning subgraphs H of G such that $p_0(H) = i$ and $p_1(H) = j$. Then if A is any link of G, we have*

$$Q(G; i, j) = Q(G_A'; i, j) + Q(G_A''; i, j)\tag{II.2.10}$$

for arbitrary nonnegative integers i and j.

The proof of Theorem II.19 is closely similar to that of Theorem II.18. We take note of a generalization of Theorem II.19, for the sake of a later application. We first agree to write

$$R(G; i_0, i_1, i_2, \ldots)$$

for the number of spanning subgraphs H of G having i_0 components of cyclomatic number 0, i_1 of cyclomatic number 1, and so on. The generalization is as follows.

Theorem II.20. *Let A be a link of G. Then*

$$R(G; i_0, i_1, i_2, \ldots) = R(G_A'; i_0, i_1, i_2, \ldots) + R(G_A''; i_0, i_1, i_2, \ldots)$$
$$\tag{II.2.11}$$

for an arbitrary sequence $(i_0, i_1, i_2, \ldots)$ of nonnegative integers having a finite sum.

Proof. If K is a spanning subgraph of G_A'', it corresponds to the same sequence $(i_0, i_1, i_2, \ldots)$ as does $f_A K$, by Theorems II.4 and II.16. Using Theorem II.17 we find that the set of spanning subgraphs of G_A'' corresponding to a given sequence is in 1-1 correspondence with the set of those spanning subgraphs of G that correspond to the same sequence and include the edge A. Moreover the spanning subgraphs of G that correspond to this sequence and do not include A are the spanning subgraphs of G_A' corresponding to the sequence. The theorem follows. □

We ought to explain why Theorem II.20 is called a generalization of Theorem II.19. To do this we need the following auxiliary theorem.

Theorem II.21. *The cyclomatic number of a graph is the sum of the cyclomatic numbers of its components.*

Proof. Let G be a graph, with components $G_1, G_2, \ldots, G_m$. For each component G_j we have

$$p_1(G_j) = |E(G_j)| - |V(G_j)| + 1 \quad \text{(by Eq. (I.6.1))},$$

$$\sum_{j=1}^{m} p_1(G_j) = |E(G)| - |V(G)| + p_0(G)$$

$$= p_1(G) \quad \text{(by Eq. (I.6.1))}.$$

From Theorem II.21 we can deduce that $Q(G; r, s)$ is the sum of those numbers $R(G; i_0, i_1, i_2, \ldots)$ for which

$$\sum i_j = r \quad \text{and} \quad \sum j i_j = s.$$

Hence Theorem II.19 is a consequence of Theorem II.20, by way of Theorem II.21. So Theorem II.20 may legitimately be called a generalization of Theorem II.19. □

There are some evidences of duality in the present section. It is fairly clear that deletion and contraction of an edge should be regarded as dual operations, and Theorems II.18, II.19, and II.20 look like self-dual theorems. The self-duality is indeed not perfect, for "link" is not a self-dual concept. To get truly self-dual theorems, we should replace it by "edge not a loop or an isthmus." The coming theory of duality will indeed recognize Theorems II.18 and II.19, *mutatis mutandis*, as self-dual. But it will not so recognize Theorem II.20, having no dual concept to "component."

II.3. VERTICES OF ATTACHMENT

Some general theorems about graphs can be proved by induction over the number of edges. Commonly the crucial step in such a proof is as follows: For some link A of G we assume that the theorem is true for G'_A and G''_A, which have fewer edges than G. From this assumption we try to deduce the theorem for G.

It is therefore desirable to make a careful study of the relations between the subgraphs of G, G'_A and G''_A. To some extent this has been done in Chapter I and the preceding sections of the present chapter. We continue the study in the present section by developing some theorems about vertices of attachment.

Theorem II.22. *Let A be any link of G, and let H be any subgraph of G'_A. Then one of the three following propositions holds.*
i) $w(G'_A, H) = w(G, H)$.
ii) $w(G, H \cup (G \cdot \{A\})) < w(G, H)$, *and both the ends of A in G are in $V(H)$.*

iii) $w(G'_A, H) = w(G, H) - 1$, and just one end x of A in G is in $V(H)$. Moreover, apart from A, x is incident in G only with edges of H.

Proof. A vertex of H is, apart from A, incident with the same edges in G as in G'_A. Hence each vertex of attachment of H in G'_A is a vertex of attachment of H in G. But consider a vertex x of H that is not a vertex of attachment of H in G'_A. It is a vertex of attachment of H in G if and only if it is incident with A.

Assume that (i) does not hold. Then, by the preceding observation, some end x of A in G is, apart from A, a vertex of H that is incident in G only with edges of H.

Suppose the other end y of A to be in H. Then, by adjoining A to H, we obtain the subgraph $H_1 = H \cup (G \cdot \{A\})$ of G. The adjunction of A to H destroys x and possibly y as vertices of attachment, retains all the other vertices of attachment, and introduces no new ones. So Proposition (ii) holds.

In the remaining case, y is not in $V(H)$. Then x is the only vertex of attachment of H in G that is not a vertex of attachment of H in G'_A. Thus Proposition (iii) is true. □

Theorem II.23. *Let A be any link of G, and let K be any subgraph of G''_A. Then one of the two following propositions holds.*
i) $w(G''_A, K) = w(G, f_A K)$.
ii) $w(G''_A, K) = w(G, f_A K) - 1$, v_A is a vertex of K and both ends of A are vertices of attachment of $f_A K$ in G.

Theorem II.23 follows from Theorem II.3, with $L = G''_A$. Our next theorem involves both G'_A and G''_A.

Theorem II.24. *Let A be any link of G, let H be any subgraph of G'_A, and let K be any subgraph of G''_A. Then one of the four following propositions holds.*
i) $w(G'_A, H) = w(G, H)$.
ii) $w(G''_A, K) = w(G, f_A K)$.
iii) $w(G, H \cup f_A K) + w(G, H \cap f_A K) < w(G, H) + w(G, f_A K)$.
iv) $w(G, H \cup (G \cdot \{A\})) < w(G, H)$, and both the ends of A in G are in $V(H)$.

Proof. Assume that (i), (ii), and (iv) do not hold. Then, as in the proof of Theorem II.22, A has just one end x in $V(H)$. Moreover, x is, apart from A, incident in G only with edges of H. But, by Theorem II.23, x is a vertex of attachment of $f_A K$ in G. We deduce that x is incident with an edge A of $f_A K$ not in $E(H)$, that x is incident with some edge of H not in $E(f_A K)$, and that x is incident with no edge of G outside both $E(H)$ and $E(f_A K)$, by Theorem II.23. So, in the notation of Sec. I.4, x belongs to the set $Q(G; H, f_A K)$ and the number $q(G: H, f_A K)$ is nonzero. Proposition (iii) of Theorem II.24 now follows as a consequence of Theorem I.7. □

II.4. SEPARATION NUMBERS

In this section we prepare for our proof of Menger's Theorem. We consider a graph G in which two disjoint subsets P and Q of $E(G)$ are specified.

Two subgraphs of G are called *edge-disjoint* if they have no common edge. We define a *cutting pair* of P and Q in G as an ordered pair (H, K) of edge-disjoint subgraphs of G such that

$$P \subseteq E(H), \quad Q \subseteq E(K), \quad \text{and} \quad H \cup K = G. \quad (II.4.1)$$

At least one such cutting pair exists. An example is $(G: P, G:(E(G)-P))$.

We define the *order* of a cutting pair (H, K) of P and Q as $|V(H \cap K)|$, the number of common vertices of H and K. It is customary to say that the member vertices of $V(H \cap K)$ *separate* P and Q in G.

Theorem II.25. *Let (H, K) be a cutting pair of P and Q in G. Then (H, H^c) and (K^c, K) are cutting pairs of P and Q in G. Moreover, $V(H \cap H^c) = W(G, H) \subseteq V(H \cap K)$ and $V(K^c \cap K) = W(G, K) \subseteq V(H \cap K)$.*

Proof. A vertex of attachment of H or K is a vertex of one of these graphs incident with an edge of the other, by Eq. (II.4.1). Hence $W(G, H)$ and $W(G, K)$ are subsets of $V(H \cap K)$. The remaining part of the theorem follows from the definitions in Sec. I.4. □

We now define $\lambda(G; P, Q)$ as the least number of vertices required to separate P and Q in G. Thus

$$\lambda(G; P, Q) = \underset{(H,K)}{\text{Min}} |V(H \cap K)|, \quad (II.4.2)$$

where the minimum is taken over all cutting pairs (H, K) of P and Q in G. By the symmetry of our definitions, we have the general rule

$$\lambda(G; P, Q) = \lambda(G; Q, P). \quad (II.4.3)$$

We call $\lambda(G; P, Q)$ the *separation number* of P and Q in G. A cutting pair of P and Q of order $\lambda(G; P, Q)$ is said to be *minimal*.

Theorem II.26. *Let m be the number of common vertices of $G \cdot P$ and $G \cdot Q$. Then*

$$m \leq \lambda(G; P, Q) \leq \text{Min}\{|V(G \cdot P)|, |V(G \cdot Q)|\}.$$

Proof. The common vertices of $G \cdot P$ and $G \cdot Q$ belong to $V(H \cap K)$ for every cutting pair (H, K) of P and Q. So the first part of the theorem follows from Eq. (II.4.2). To complete the proof we have only to observe that $(G \cdot P, (G \cdot P)^c)$ and $((G \cdot Q)^c, G \cdot Q)$ are cutting pairs of P and Q in G. □

Theorem II.27. *Let (H, K) be a minimal cutting pair of P and Q in G. Then $W(G, H) = W(G, K) = V(H \cap K)$. Accordingly, (H, K) is a complementary pair of subgraphs of G with binding number $\lambda(G; P, Q)$.*

Proof. By Theorem II.25, (H, H^c) is a cutting pair of P and Q, and $V(H \cap H^c) = W(G, H) \subseteq V(H \cap K)$. Hence $W(G, H) = V(H \cap K)$, by the minimality of (H, K). Similarly, by the symmetry of our definitions, $W(G, K) = V(H \cap K)$. The rest of the theorem follows from the definitions of Sec. I.4. □

Theorem II.28. $\lambda(G; P, Q) = \mathrm{Min}_H w(G, H)$, *where the minimum is taken over all subgraphs H of G such that $E(H)$ contains P and is disjoint from Q.*

Proof. Write μ for the expression on the right in the equation. Given an H, we note that (H, H^c) is a cutting pair of P and Q, and that its order is $w(G, H)$, by Theorem II.25. Hence $\mu \geq \lambda(G; P, Q)$, by Eq. (II.4.2).

Conversely, let (H, K) be a minimal cutting pair of P and Q. Then $w(G, H) = \lambda(G; P, Q)$, by Theorem II.27. Hence $\mu \leq \lambda(G; P, Q)$. This completes the proof. □

The remaining theorems of the section relate separation numbers in graphs G, G'_A, and G''_A.

Theorem II.29. *Let A be a link of G not in P or Q. Then*

$$\lambda(G; P, Q) - \lambda(G'_A; P, Q) = 0 \text{ or } 1.$$

Proof. Let (H, K) be a minimal cutting pair of P and Q in G. Assume that A is not in $E(H)$. Then by Theorem II.22, we have $w(G'_A, H) < w(G, H)$. For Proposition (ii) of Theorem II.22 does not hold, by Theorem II.28 and the minimality of (H, K). Hence

$$\lambda(G'_A; P, Q) < \lambda(G; P, Q),$$

by Theorem II.28. If A is in $E(H)$, we repeat the argument, interchanging P with Q and H with K, and use Eq. (II.4.3). We then get the same inequality as before.

Now let (L, M) be a minimal cutting pair of P and Q in G'_A, so that $\lambda(G'_A; P, Q) = w(G'_A, L)$, by Theorem II.27. If the edge A of G has both its ends in L, it is clear that

$$w(G'_A, L) = w\big(G, L \cup (G \cdot \{A\})\big) \geq \lambda(G; P, Q).$$

In the remaining case, we have $w(G'_A, L) > w(G, L) - 1$, by Theorem II.22. In either case, we can deduce from Theorem II.28 that

$$\lambda(G'_A; P, Q) > \lambda(G; P, Q) - 1.$$

This completes the proof. □

Theorem II.30. *Let A be any link of G not in P or Q. Then*

$$\lambda(G; P, Q) - \lambda(G''_A; P, Q) = 0 \text{ or } 1.$$

Proof. Let (H, K) be a minimal cutting pair of P and Q in G. Then $\lambda(G; P, Q) = w(G, H) = w(G, K)$, by Theorem II.27. Suppose A is in $E(H)$. Then H conforms to $\{A\}$. Hence, there is a subgraph L of G_A'' such that $f_A L = H$. We note that L, like H, includes all the edges of P but none of those of Q. By Theorem II.23 we have $w(G_A'', L) \le w(G, f_A L) = \lambda(G; P, Q)$. Hence, by Theorem II.28,

$$\lambda(G_A''; P, Q) \le \lambda(G; P, Q).$$

If A is not in $E(H)$, we have only to interchange P with Q and H with K in the above argument. Using Eq. (II.4.3), we arrive at the same inequality.

Now let (L, M) be a minimal cutting pair of P and Q in G_A''. Then $\lambda(G_A''; P, Q) = w(G_A'', L)$, by Theorem II.27. But by Theorem II.23 we have $w(G_A'', L) \ge w(G, f_A L) - 1$. Hence, by Theorem II.28,

$$\lambda(G_A''; P, Q) \ge \lambda(G; P, Q) - 1.$$

This completes the proof. □

Theorem II.31. *Let A be any link of G not in P or Q. Then either $\lambda(G_A'; P, Q)$ or $\lambda(G_A''; P, Q)$ is equal to $\lambda(G; P, Q)$.*

Proof. Assume the contrary. Let (H, H_1) be a minimal cutting pair of P and Q in G_A', and let (K, K_1) be a minimal cutting pair of P and Q in G_A''. Then by Theorems II.29 and II.30 we have

$$w(G_A', H) = w(G_A'', K) = \lambda(G_A'; P, Q)$$
$$= \lambda(G_A''; P, Q) = \lambda(G; P, Q) - 1. \quad (\text{II.4.4})$$

We note that the edge A of G does not have both its ends in H, for, if it did, we would have

$$w(G_A', H) = w\big(G, H \cup (G \cdot \{A\})\big) > \lambda(G; P, Q),$$

by Theorem II.28. But this is contrary to Eq. (II.4.4). We can therefore deduce from Theorems II.22 and II.23 that

$$w(G, H) = w(G, f_A K) = \lambda(G; P, Q). \quad (\text{II.4.5})$$

We now apply Theorem II.24. Its Propositions (i), (ii), and (iv) are ruled out by the preceding observations. Hence (iii) must hold. But then either $w(G, H \cap f_A K)$ or $w(G, H \cup f_A K)$ is less than $\lambda(G; P, Q)$, by Eq. (II.4.5), and this is impossible, by Theorem II.28. This contradiction establishes the theorem. □

We seem to smell duality again in Theorems II.29 and II.30, and self-duality in Theorem II.31.

II.5. MENGER'S THEOREM

There are many equivalent ways of stating Menger's Theorem. We start with the form that seems easiest to prove in terms of the theory of Sec. II.4.

Theorem II.32. *Let P and Q be disjoint subsets of $E(G)$. Then there exists a subgraph H of G with the following properties*:
i) *H has no edge in P or Q.*
ii) $p_0(H) = \lambda(G; P, Q)$.
iii) *Each component of H has at least one vertex in $G \cdot P$ and at least one in $G \cdot Q$.*

Proof. Let M be the set of common vertices of $G \cdot P$ and $G \cdot Q$.

We proceed by induction over the number n of links of G not in P or Q. Suppose first that n is zero. Then the cutting pair $(G \cdot P, (G \cdot P)^c)$ of P and Q in G satisfies

$$V((G \cdot P) \cap (G \cdot P)^c) = M,$$

and therefore $\lambda(G; P, Q) = |M|$, by Theorem II.26. We take H to be the edgeless subgraph of G whose vertex set is M. It satisfies (i), (ii), and (iii), by Theorem I.20.

Assume, as an inductive hypothesis, that the theorem is true whenever n is less than some positive integer q, and consider the case $n = q$. Choose a link A of G not in P or Q.

It may happen that $\lambda(G'_A; P, Q) = \lambda(G; P, Q)$. Then, by the inductive hypothesis, there is a subgraph H of G'_A satisfying the three conditions in G'_A. Clearly H satisfies them in G, too.

In the remaining case, $\lambda(G''_A; P, Q) = \lambda(G; P, Q)$, by Theorem II.31. There is a subgraph K of G''_A satisfying the three conditions in that graph. But then $f_A K$ satisfies them in G, by Theorem II.4.

Thus the induction succeeds and the theorem is established. □

Theorem II.32 states a "best possible" result, as we proceed to show.

Theorem II.33. *Let P and Q be disjoint subsets of $E(G)$. Let H be a subgraph of G having no edge in P or Q and such that each of its components meets both $G \cdot P$ and $G \cdot Q$. Then*

$$p_0(H) \leq \lambda(G; P, Q).$$

Proof. Let (K, L) be any cutting pair of P and Q in G. Let C be any component of H. Then $C \cap K$ and $C \cap L$ are nonnull subgraphs of C whose union is C. They are not disjoint, by Theorem I.21. Hence K and L have a common vertex in each component of H, and so $p_0(H) \leq |V(K \cap L)|$. Since this is true for every (K, L), the theorem follows, from Eq. (II.4.2). □

Our next version of Menger's Theorem depends on the notion of a *tie* between two disjoint subsets P and Q of $E(G)$. There are two kinds of tie. A *tie of the first kind* is a vertex-graph contained in both $G \cdot P$ and $G \cdot Q$. A *tie of the second kind* is an arc in G with one end in $G \cdot P$ but not $G \cdot Q$, with its other end in $G \cdot Q$ but not $G \cdot P$, and with no edge or internal vertex in either $G \cdot P$ or $G \cdot Q$.

Theorem II.34. *Let P and Q be disjoint subsets of $E(G)$. Then the maximum number of disjoint ties between P and Q in G is $\lambda(G; P, Q)$.*

To prove this we consider the subgraph H of Theorem II.32. Let its components be $C_1, C_2, \ldots, C_m$, where $m = \lambda(G; P, Q)$. We replace each C_j by a tie T_j contained in it.

If we can find a vertex x of C_j belonging to both $G \cdot P$ and $G \cdot Q$, we take T_j to be the corresponding vertex-graph.

In the remaining case we can, by Theorem I.43, find an arc L in C_j with one end in $G \cdot P$ and one in $G \cdot Q$. If L has an internal vertex in $G \cdot P$ or $G \cdot Q$, it contains another such arc as a proper subgraph. This is by the definition of an arc, or by an application of Theorem I.43. We deduce that L can be so chosen as to be a tie between P and Q.

We now have our $\lambda(G; P, Q)$ disjoint ties between P and Q. This is the maximum number, by Theorem II.33. □

The common form of Menger's Theorem depends on the notion of internally disjoint arcs. Let $L_1, L_2, \ldots, L_k$ be a set of arcs in G. These arcs are said to be *internally disjoint* if no edge or internal vertex of any one of them belongs to another. We note that two or more internally disjoint arcs can have the same pair of ends. The usual version of the theorem runs as follows.

Theorem II.35. *Let p and q be distinct nonadjacent vertices of G. Let λ be the least number of vertices required to separate p and q in G. Then we can find λ, but not more, internally disjoint arcs in G with ends p and q.*

In order to relate this proposition to Theorem II.34, we define P as the set of all edges of G incident with p, and Q as the set of those incident with q. We interpret, or rather define, λ as the least number of vertices required to separate P and Q in G, that is, as $\lambda(G; P, Q)$.

We appeal to Theorem II.34 for λ disjoint ties $T_1, T_2, \ldots, T_\lambda$ between P and Q. We convert each tie T_j into an arc L_j with ends p and q, by adjoining the vertices p and q, an edge A_j from p to the vertex or an end of T_j, and an edge B_j from q to the vertex or an end of T_j. (See Fig. II.5.1.) We now have λ internally disjoint arcs L_j with ends p and q. It follows from Theorem II.34 that we cannot have more. For each arc in G with ends p and q gives rise to a tie between P and Q when we delete from it its two ends and their incident edges.

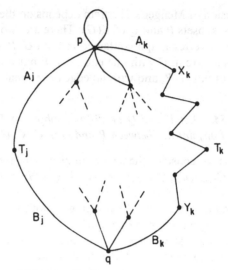

FIGURE II.5.1

The helpfulness of Menger's Theorem to the graph theorist is limited by the difficulty of determining the value of $\lambda(G; P, Q)$ in a given case. I cannot recall having used it at all in my research, except in very simple and nonessential ways. The notions involved in Menger's Theorem and its analogues are indeed important in transportation theory. But there the theorem is usually replaced by a construction that gives simultaneously a maximal set of internally disjoint arcs from p to q, and a minimal cutting pair of P and Q. Each step in this construction starts with a set X of k disjoint ties between P and Q, and either proves it maximal or derives from it a set of $(k + 1)$ such ties.

We can modify Theorem II.34 to assert this possibility of starting with an incomplete set of ties and then improving it. Given a set X of disjoint ties between P and Q, we define its P-base $X(P)$ as the set of ends in $G \cdot P$ of members of X. We define the Q-base $X(Q)$ of X analogously. The modified theorem runs as follows.

Theorem II.36. *Let P and Q be disjoint subsets of $E(G)$. Let X be a set of disjoint ties between P and Q with P-base $X(P)$. Then there exists a set Y of $\lambda = \lambda(G; P, Q)$ disjoint ties between P and Q whose P-base $Y(P)$ contains $X(P)$.*

We can call Y an improvement rather than a replacement of X because not all traces of X are lost in the change to Y. The original P-base remains as part of the new one.

We can prove Theorem II.36 by repeating the proof of Theorem II.32 with minor modifications. No change is necessary in the discussion of the case $n = 0$, since $X(P)$ is necessarily a subset of M.

Menger's Theorem

In the next step we have to choose A, assume the theorem for G'_A and G''_A, and deduce it for G. Here it is permissible to assume that the given X is *maximal*, in the sense that no tie of P and Q can be found that is disjoint from every member of X. We can also assume that X has fewer than λ members, since otherwise there is nothing to prove. The maximality of X ensures that $G \cdot P$ and $G \cdot Q$ have no common vertex outside $X(P)$, or outside $X(Q)$. Let us enumerate the members of X as $T_1, T_2, \ldots, T_\mu$, where $\mu < \lambda$.

We choose A to be incident with some vertex u of Q not in $X(Q)$. This is possible; if it were not, then the members of the cutting pair $((G \cdot Q)^c, G \cdot Q)$ of P and Q in G would have only the vertices of $X(Q)$ in common, contrary to Eq. (II.4.2).

The other end v of A may or may not be in one of the ties T_j. If v is not in T_j, we write $S_j = T_j$. If v is the vertex x_j of T_j in P we take S_j to be the corresponding vertex-graph. In the remaining case we take S_j to be the arc in T_j with ends x_j and v. Figure II.5.2 shows an example in which v is a vertex of T_1.

It may happen that $\lambda(G'_A; P, Q) = \lambda(G; P, Q)$. In that case the truth of the theorem for G'_A, with the given set X of ties, implies its truth for G.

In the remaining case, $\lambda(G''_A; P, Q) = \lambda(G; P, Q)$, by Theorem II.31. For each vertex w of G we define the corresponding vertex w'' of G''_A, as in Sec. II.2. We note that the mapping $w \to w''$ converts the set of subgraphs S_j of G into a set X'' of μ disjoint ties between P and Q in G''_A. It also induces a 1–1 mapping of $X(P)$ onto the P-base $X''(P)$ of X'' in G''_A. Since we are assuming the truth of the theorem for G''_A we can assert the existence of a set Y'' of λ disjoint ties between P and Q in G''_A with a P-base containing that of X''.

There is a subgraph K of G''_A whose components are the members of Y''. As in the proof of Theorem II.32, the subgraph $f_A K$ of G has λ

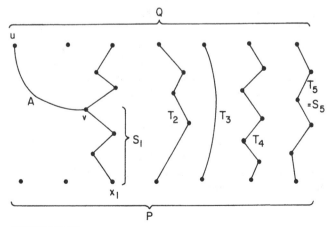

FIGURE II.5.2

components, each meeting both $G \cdot P$ and $G \cdot Q$. But now each component of $f_A K$ has only one vertex in $G \cdot P$. Moreover, the mapping $w \to w''$ induces a 1–1 correspondence betwen the set W of vertices of $f_A K$ in $G \cdot P$ and the set $Y''(P)$. Thus $X(P)$ is a subset of W.

We now reduce each component of $f_A K$ to a tie between P and Q, as in the discussion of Theorem II.34. We thus get a set Y of λ disjoint ties between P and Q whose P-base W contains $X(P)$. Thus the induction succeeds and the theorem is true. □

II.6. HALL'S THEOREM

In yet another version of Menger's Theorem, we consider two subsets U and V of $V(G)$, and we ask for the maximum number of disjoint ties between U and V. In this case, a "tie" can be a vertex-graph in G whose vertex is common to U and V. Alternatively, it is an arc in G with one end in U but not V, with the other end in V but not U, and with no internal vertex in either U or V.

We define a *cutting pair* of U and V in G as an ordered pair (H, K) of edge-disjoint subgraphs of G such that

$$U \subseteq V(H), \quad V \subseteq V(K), \quad \text{and} \quad H \cup K = G. \quad (\text{II}.6.1)$$

The *separation number* $\mu(G; U, V)$ of U and V in G is defined as follows:

$$\mu(G; U, V) = \underset{H,K}{\text{Min}} |V(H \cap K)|, \quad (\text{II}.6.2)$$

where the minimum is taken over all cutting pairs (H, K) of U and V in G.

We could develop a theory like that of Sec. II.4 but with Eqs. (II.6.1) and (II.6.2) replacing (II.4.1) and (II.4.2). However, theorems on vertices of attachment would be less directly applicable.

There is a simple transformation that relates the old theory to the new. To each vertex u of U we attach a new loop $B_u(U)$, and to each vertex v of V a new loop $B_v(V)$. A vertex x common to U and V gets two distinct new loops $B_x(U)$ and $B_x(V)$. The adjunction of these loops converts G into a new graph G_1. Let P be the set of loops $B_u(U)$, where u runs through the vertices of U, and let Q be the corresponding set of loops $B_v(V)$. The construction is illustrated in Fig. II.6.1, with $U = \{u_1, u_2, u_3\}$ and $V = \{v_1, v_2, v_3\}$.

A cutting pair (H, K) of U and V in G becomes a cutting pair of P and Q in G_1 when the loops of P and Q are adjoined to H and K, respectively. Conversely, a cutting pair (H_1, K_1) of P and Q in G_1 becomes a cutting pair of U and V in G when the loops of P and Q are deleted from its member graphs. We deduce that

$$\mu(G; U, V) = \lambda(G_1; P, Q). \quad (\text{II}.6.3)$$

Hall's Theorem

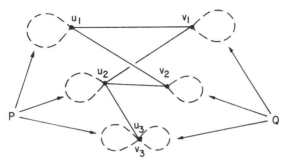

FIGURE II.6.1

A set of disjoint ties between U and V in G is a set of disjoint ties between P and Q in G_1, and conversely. So by applying Theorem II.34 to G_1, we obtain the following theorem about G.

Theorem II.37. *Let U and V be subsets of $V(G)$. Then the maximum number of disjoint ties between U and V in G is $\mu(G; U, V)$.*

A *bipartition* of a graph G is an ordered pair (U, V) of complementary subsets of $V(G)$ such that each edge of G has one end in U and one in V. A *bipartite* graph is one that has a bipartition. Thus a bipartite graph can have no loop. We proceed to apply Theorem II.37 to the theory of bipartite graphs. We shall be concerned with the 1-factors defined in Sec. I.3.

A *partial 1-factor* F of a graph G is a subgraph of G that is regular of valency 1. Evidently the components of such an F must be link-graphs. If $V(F) = V(G)$, we say simply that F is a *1-factor* of G.

Now let G have a bipartition (U, V) and let F be a partial 1-factor of G. We define the *U-deficiency* $\delta_U(F)$ of F as the number of vertices of U not incident with edges of F. The *V-deficiency* $\delta_V(F)$ of F is defined analogously.

Theorem II.38. *Let (U, V) be a bipartition of a graph G and let n be a positive integer. Then in order that G shall have no partial 1-factor whose U-deficiency is less than n, it is necessary and sufficient that the members of some subset S of U shall be adjacent to fewer than $(|S| - n)$ vertices of V.*

Proof. We deduce from Theorem II.37 that the maximum number of edges in a partial 1-factor of G is $\mu(G; U, V)$. Let (H, K) be a cutting pair of U and V in G such that $|V(H \cap K)| = \mu(G; U, V)$. Let X_U be the set of vertices of U in K, and X_V the set of vertices of V in H. Then

$$|X_U| + |X_V| = |V(H \cap K)| = \mu(G; U, V).$$

So for any partial 1-factor F of G we have

$$\delta_U(F) \geq |U| - \mu(G; U, V),$$
$$\delta_U(F) \geq |U| - |X_U| - |X_V|.$$

Now write $S = U - X_U$. The members of S can be adjacent only to members of X_V, since their incident edges are all in H. Hence the members of S are adjacent to at most $|X_V|$, and therefore at most $|S| - \delta_U(F)$, vertices of V. So if $\delta_U(F) > n$ for every F, then the vertices of S are adjacent to fewer than $(|S| - n)$ vertices of V, and the theorem is satisfied.

Conversely, suppose U to have some subset S whose members are adjacent to exactly $k < |S| - n$ vertices of V. Then for each partial 1-factor F of G, at most k edges incident with members of S can belong to F. It follows that $\delta_U(F) > n$. □

If G is to have a 1-factor, it is clear that U must have the same number of elements as V. We can regard this as a consequence of Theorem II.38 with $n = 1$, applied to the two bipartitions (U, V) and (V, U). A further application of Theorem II.38 leads to the following theorem.

Theorem II.39. *Let (U, V) be a bipartition of G. In order that G may have a 1-factor, it is necessary that $|U| = |V|$. Subject to this condition, G has a 1-factor if and only if there is no subset S of U whose members are adjacent to fewer than $|S|$ members of V.*

Theorem II.39 is known as Hall's Theorem. It was first published by P. Hall in 1935 [1]. K. Menger published his theorem in 1927 [2].

The problem dealt with in Theorem II.38 is often called the *Assignment Problem*. The members of U are called "applicants" and those of V "jobs." An edge of G indicates that the corresponding applicant is qualified for the corresponding job. A partial 1-factor of G with the maximum number of edges is an arrangement whereby as many applicants as possible are assigned to jobs for which they are qualified.

II.7. NOTES

II.7.1. Minors

For each graph G there is a least integer $n = n(G)$ such that some minor of G is an n-clique. This number is of interest in connection with a famous conjecture of H. Hadwiger, that G, if loopless, can be vertex-colored in not more than $n(G)$ colors. (See Sec. IX.3.)

Another well-known conjecture concerned with minors can be stated as follows: Let S be an infinite set of graphs. Then S has a finite subset T such that each member of S has a minor isomorphic to a member of T. For cubic graphs, this is "Kruskal's Conjecture."

II.7.2. Recursion formulae

There are many interesting functions of graphs satisfying formulae like Eqs. (II.2.9), (II.2.10), and (II.2.11). Some of them are discussed in Chapter IX.

II.7.3. Menger's Theorem

This theorem can be proved by the methods of transportation theory (Sec. VI.7). Many proofs have been given; there has been something of a competition to find the shortest one. (See [3].)

EXERCISES

1. Show that the cube graph has a minor that is a 4-clique. There is a theorem asserting that the graph of every convex polyhedron has such a minor. Test this theorem in a few simple cases.
2. Use Eq. II.2.9 in conjunction with Theorem I.36 to find the tree-number of the 4-clique.
3. For the graph of a convex polyhedron show that any two nonadjacent vertices are separated by a least three others.
4. Show that every bipartite cubic graph has a 1-factor.
5. Write a formal proof that any two distinct vertices of a circuit are joined by two and only two arcs, these being internally disjoint.
6. What graphs can be transformed into (i) the cube graph, (ii) the octahedron graph, by contracting a single edge?

REFERENCES

[1] Hall, P., On representatives of subsets, *J. London Math. Soc.* **10** (1935), 26–30.
[2] Menger, K., Zur allgemeinen Kurventheorie, *Fund. Math.* **10** (1927), 96–115.
[3] Nash-Williams, C. St. J. A., and W. T. Tutte, More proofs of Menger's Theorem, *J. Graph Theory* **1** (1977), 13–17.

Chapter III

2-Connection

III.1. SEPARABLE AND 2-CONNECTED GRAPHS

A nonnull proper subgraph of a connected graph G has at least one vertex of attachment, by Theorem I.21. The present chapter is based on a theory of those subgraphs of G that have exactly one vertex of attachment.

A *1-separation* of a connected graph G is an ordered pair (H, K) of subgraphs of G, each having at least one edge, such that $H \cup K = G$ and $H \cap K$ is a vertex-graph. The vertex of $H \cap K$ is called the *cut-vertex* of the 1-separation.

A connected graph is said to be *separable* or *nonseparable* according as it does or does not have a 1-separation. In addition, it is customary to count the disconnected graphs as separable. We shall not describe the null graph as either separable or nonseparable. The term "2-connected" is used as a synonym of "nonseparable."

According to the above definition, a connected graph with at most one edge must be 2-connected. This covers the cases of the vertex-graph, the loop-graph, and the link-graph.

Theorem III.1. *Let G be a connected graph, and let (H, K) be a 1-separation of G with cut-vertex v. Then H and K are connected graphs, and each of them has v as its only vertex of attachment.*

Proof. If H is not connected, it has a component C not including v. But then C is a detached proper subgraph of G, contrary to Theorem I.21. We conclude that H, and similarly K, is connected.

It follows that v is not isolated in either H or K. Since H and K have no common edge, v is therefore a vertex of attachment of both subgraphs.

A second vertex of attachment of H or K would be a second vertex of $H \cap K$, which is impossible. □

Consider any 1-separation (H, K) of a connected graph G. In the terminology of Sec. I.4, H and K are the members of a complementary pair of subgraphs of G with binding number 1, by Theorem III.1.

Theorem III.2. *Let G be a connected graph. Let H be a proper subgraph of G with at most one vertex of attachment. Let H have either an edge or a vertex that is not one of attachment. Then (H, H^c) is a 1-separation of G.*

Proof. By the definition of H^c, the graphs H and H^c have G as their union. But they have no common edge, and their only common vertices are the vertices of attachment of H. But H, as a nonnull proper subgraph of a connected graph, has at least one vertex of attachment. Hence, by hypothesis, it has exactly one, v, say. The intersection of H and H^c is thus the corresponding vertex-graph.

It remains to show that H and H^c each have an edge. But if H is edgeless, it has, by hypothesis, a vertex x distinct from v and so not in H^c. Moreover, if H^c is edgeless, it has a vertex y not in H, since otherwise H would not be a proper subgraph of G. In either case, x or y must be isolated in G, contrary to the connection of that graph. □

Theorem III.3. *The only 2-connected graph with a loop is the loop-graph, and the only 2-connected graph with an isthmus is the link-graph.*

Proof. Let A be an edge of a connected graph G. Suppose first that A is a loop. If $G \cdot \{A\}$ is a proper subgraph of G, then G is separable, by Theorem III.2 with $H = G \cdot \{A\}$. So if G is 2-connected, it must be the loop-graph $G \cdot \{A\}$.

Next let A be an isthmus of G. If G is not a link-graph, then A has some end-graph H that is not a vertex-graph. But then (H, H^c) is a 1-separation of G, by Theorem III.2. □

Theorem III.4. *Let G be a connected graph having a 1-separation (H, K) with cut-vertex v. Let J be a subgraph of G that is either a circuit with an edge in $E(H)$ or an arc with both its ends in $V(H)$. Then $J \subseteq H$.*

Proof. We recall that circuits and arcs are connected graphs. We note also that if $J \cap K$ has a vertex x distinct from v, then x must have the same valency 2 in $J \cap K$ as in J.

Suppose first that v is not a vertex of $J \cap K$. Then, by the preceding observations, $J \cap K$ is a proper detached subgraph of the connected graph J. Hence $J \cap K$ is null and $J \subseteq H$.

In the remaining case, v belongs to $J \cap K$. Its valency in $J \cap K$ is at most 2. If this valency is 2, then $J \cap K$ is a detached nonnull proper subgraph of the connected graph J which is impossible. The valency of v in $J \cap K$ is not 1, by Theorem I.1. But if this valency is zero, then $J \cap H$ is a detached nonnull subgraph of the connected graph J, and it is therefore the whole of J. Thus $J \subseteq H$. □

Theorem III.5. *Every circuit is 2-connected.*

Proof. Suppose some circuit G to have a 1-separation (H, K). Then G has an edge in H, and therefore $G \subseteq H$ by Theorem III.4. This is impossible, since G must have also an edge in K. □

The next two theorems are important for applications of the results of Secs. I.8 and II.5.

Theorem III.6. *Let G be a 2-connected graph. Let H be a nonnull subgraph of G that is not a vertex-graph. Let B be a bridge of H in G. Then B has at least two vertices in H.*

Proof. Assume that B has fewer than two vertices in H. Referring to the definition of a bridge in Sec. I.8, we see that all the vertices of attachment of B are in H. Hence B has at most one vertex of attachment. Moreover, B is not a subgraph of H, and so it has either an edge or a vertex that is not one of attachment. It follows by Theorem III.2 that G is separable, and this is contrary to hypothesis. □

Theorem III.7. *Let G be a 2-connected graph. Let P and Q be disjoint nonnull subsets of $E(G)$. Then $\lambda(G; P, Q) \geq 2$.*

Proof. Suppose $\lambda(G; P, Q) < 2$. By Theorem II.28, there is a subgraph H of G such that P is contained in $E(H)$ and Q is disjoint from $E(H)$, and which has at most one vertex of attachment. This implies that G is separable, by Theorem III.2, and this is contrary to hypothesis. □

III.2. CONSTRUCTIONS FOR 2-CONNECTED GRAPHS

In this section we continue the work of Sec. I.7 by showing how to make a complete list of nonisomorphic 2-connected graphs with a given number of edges. We begin with some theorems showing how some 2-connected graphs can be derived from smaller ones.

Constructions for 2-Connected Graphs

Theorem III.8. *Let G be a connected graph having a 1-separation (H, K) with cut-vertex v. Let J be a 2-connected subgraph of G having an edge, or a vertex other than v, in H. Then $J \subseteq H$.*

Proof. As a proper subgraph of J, the graph $J \cap K$ can have no vertex of attachment in J other than v. Hence, by Theorem III.2, $J \cap K$ has no edge and no vertex other than v. The theorem follows. □

Theorem III.9. *Let H and K be 2-connected subgraphs of a graph G, and let them have either an edge or two distinct vertices in common. Then $H \cup K$ is 2-connected.*

Proof. $H \cup K$ is connected, by Theorem I.25. Assume that it is not 2-connected. Then it has a 1-separation (L, M) with a cut-vertex v. We can adjust the notation so that L includes either a common edge of H and K or a common vertex of those subgraphs distinct from v. But then $H \cup K \subseteq L$, by Theorem III.8. This is impossible, since M includes an edge of $H \cup K$ not in L. □

Theorem III.10. *Let H be a 2-connected subgraph of a graph G, and let N be an arc in G whose ends are both in H. Then $H \cup N$ is 2-connected.*

Proof. $H \cup N$ is connected, by Theorem I.25. Assume it is not 2-connected. Then it has a 1-separation (L, M). We may suppose $H \subseteq L$, by Theorem III.8. Then $N \subseteq L$, by Theorem III.4. Hence $H \cup N \subseteq L$, and we have a contradiction. □

Theorem III.11. *Let G be a 2-connected graph having at least two edges. Then G can be represented as a union of a 2-connected subgraph H of G and an arc L in G that avoids H but has both its ends in H.*

Proof. Each edge A of G defines a 2-connected proper subgraph $G \cdot \{A\}$ of G. We can therefore assert the existence of a 2-connected proper subgraph H of G having the greatest possible number, n say, of edges. Clearly $n > 0$.

Since H is proper in G, there exists a bridge B of H in G, by Theorem I.51. This bridge has two distinct vertices x and y in H, by Theorem III.6. These are the ends of an arc L in B that avoids H, by Theorem I.56.

The subgraph $H \cup L$ of G is 2-connected, by Theorem III.10. It is not a proper subgraph of G since it has more than n edges. Hence it is G itself. □

We now use a construction in which a graph G is formed by uniting a graph H with an arc L. The ends of L are to be vertices of H, but no edge or internal vertex of L is to belong to H. In the final graph G, H appears as a subgraph, and L as an arc in G avoiding H. We say that G is formed from H by *adjoining* the arc L.

Theorem III.10 tells us that the result of adjoining an arc to a 2-connected graph H is always 2-connected. We learn from Theorem III.11 that every 2-connected graph with at least two edges can be derived from a smaller 2-connected graph in this way.

Let L_n denote a complete list of nonisomorphic 2-connected graphs of n edges. Thus L_0 has the vertex-graph as its only member, and L_1 consists only of the loop-graph and the link-graph. If $m > 1$ and the lists $L_0, L_1, \ldots, L_{m-1}$ are known, then we can find L_m. We have only to adjoin suitable arcs in all isomorphically distinct ways to the members of L_0 to L_{m-1}, and then reject any duplicates. Actually we can ignore L_0 here since there is no way of adjoining an arc to a graph of only one vertex.

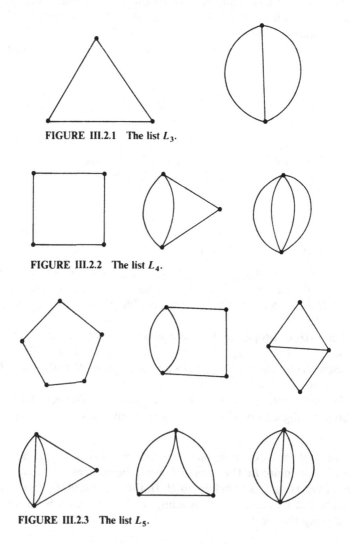

FIGURE III.2.1 The list L_3.

FIGURE III.2.2 The list L_4.

FIGURE III.2.3 The list L_5.

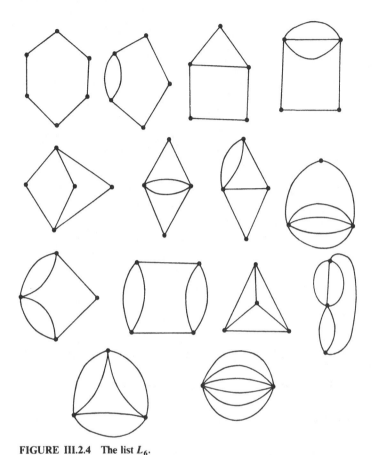

FIGURE III.2.4 The list L_6.

As an example we find the members of L_2 by adjoining a 1-arc in all possible ways to each member of L_1. In practice this means by adjoining a 1-arc in the only possible way to a link-graph. Thus L_2 has only one member, the 2-circuit.

To get L_3 we have to adjoin 2-arcs to members of L_1 and 1-arcs to members of L_2. We deduce that L_3 has just two members, the 3-circuit and the 3-linkage, as shown in Fig. III.2.1.

The rule for L_4 is to adjoin 3-arcs to members of L_1, 2-arcs to members of L_2, and 1-arcs to members of L_1. The resulting list is shown in Fig. III.2.2. The lists L_5 and L_6 are shown in Fig. III.2.3 and III.2.4, respectively.

A list of the connected separable graphs of n edges can be obtained from the list M_n of Sec. I.7 by deleting the members of our present L_n. For example, the connected separable graphs of four edges are those shown in Fig. III.2.5.

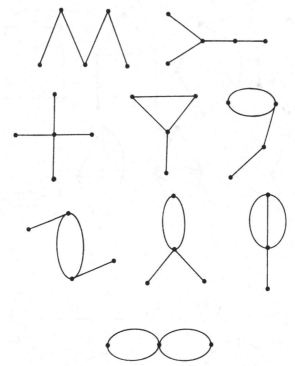

FIGURE III.2.5 Connected separable graphs of four edges.

III.3. BLOCKS

A *block* in a graph G is a maximal 2-connected subgraph of G, that is, a 2-connected subgraph of G that is not contained in any other.

Theorem III.12. *Let H be a 2-connected subgraph of a graph G. Then H is contained in some block of G. Moreover, if H is not a vertex-graph, it is contained in exactly one block of G.*

Proof. The first part of the theorem follows from the finiteness of G. To prove the second, suppose H to be contained in two blocks B_1 and B_2. Since H is not a vertex-graph, the union $B_1 \cup B_2$ is 2-connected, by Theorem III.9. Hence $B_1 = B_1 \cup B_2 = B_2$, by the maximality of B_1 and B_2. □

Theorem III.13. *Let G be any graph. Then each edge or vertex of G belongs to some block of G. Moreover, each edge belongs to exactly one block of G.*

Proof. Since the vertex-graph, loop-graph, and link-graph are 2-connected, this is a consequence of Theorem III.12. □

Theorem III.14. *The blocks of a graph G are the blocks of its components.*

Proof. A block of G, being connected, is contained in some component C of G, by Theorem I.24, and is clearly a block of C. Conversely, a block of a component C of G must be a block of G, since any 2-connected subgraph of G containing it would be a subgraph of C, by Theorem I.24. □

Because of this theorem there is no real loss of generality in restricting the further theory of blocks to the case in which G is connected.

Theorem III.15. *A 2-connected graph has exactly one block, itself. A separable graph has two or more blocks.*

Proof. The first part is true by the definition of a block. The second part follows from Theorem III.13, since each block of a separable graph G must be a proper subgraph of G. □

Theorem III.16. *Let G be a connected graph that is not a vertex-graph. Then each block of G has at least one edge.*

Proof. Let B be a block of G. If it is edgeless it must, by its connection, be a vertex-graph. Then let its vertex be x. By hypothesis, x is incident with an edge A of G. By Theorem III.13 there is a block B' of G that includes A. But then B is a proper subgraph of B', contrary to the definition of a block. □

Theorem III.17. *If A is a loop or isthmus of a graph G, then $G \cdot \{A\}$ is a block of G.*

Proof. By Theorem III.13, A belongs to a block B of G. Then A is a loop or isthmus of B, by Theorem I.32. Hence B is a loop-graph or a link-graph, by Theorem III.3. Hence $B = G \cdot \{A\}$. □

Theorem III.18. *Let A_1 and A_2 be distinct edges of a graph G. Then they belong to the same block of G if and only if some circuit in G includes them both.* [2].

Proof. Suppose A_1 and A_2 to be edges of some circuit Q in G. Then Q is a subgraph of some block B of G, by Theorems III.5 and III.12. The theorem is satisfied.

Conversely, suppose A_1 and A_2 to be edges of the same block B of G. By Theorem III.7, $\lambda(B;\{A_1\},\{A_2\}) \geq 2$. By Theorem II.34, there are two disjoint ties T_1 and T_2 between $\{A_1\}$ and $\{A_2\}$ in B. We must suppose A_1 to have one end x_1 in T_1 and the other end x_2 in T_2, while A_2 has one end y_1 in T_1 and the other end y_2 in T_2. We may have $x_1 = y_1$ or $x_2 = y_2$. Taking the union of T_1 and T_2 and adjoining to it the edges A_1 and A_2, we obtain a regular graph Q of valency 2. It is connected, by Theorem I.25. It is therefore a circuit in B and G, by Theorem I.27. □

In the remainder of this section, we study the relation between the blocks of a connected graph G and the cut-vertices of G, that is, the cut-vertices of the 1-separations of G.

Theorem III.19. *Let G be a connected graph. Let v be a cut-vertex of G, corresponding to a 1-separation (H, K). Then v is a vertex of at least two distinct blocks of G. Of these blocks, at least one is contained in H, and at least one in K.*

Proof. v is incident with an edge A_H of H and an edge A_K of K, by Theorem III.1. These edges belong to blocks B_H and B_K, respectively, of G, by Theorem III.13. But $B_H \subseteq H$ and $B_K \subseteq K$, by Theorem III.8. □

Theorem III.20. *Let G be a connected graph. Let B be a block of G, and let x be a vertex of B. Let H be the component of $G:(E(G) - E(B))$ that includes the vertex x. Then either H is a vertex-graph or (H, H^c) is a 1-separation of G. In the latter case, x is the cut-vertex of (H, H^c).*

Proof. We note first that H has no vertex other than x in B. For if y were another, there would be an arc L in H with ends x and y, by Theorem I.43. The subgraph $B \cup L$ of G would be 2-connected, by Theorem III.10, and this is impossible, by the maximality of B.

Suppose H is not a vertex-graph. Then by Theorem III.16, the connected graphs H and B each have an edge incident with x. Hence x is a vertex of attachment of H. By the result of the preceding paragraph, it is the only vertex of attachment of H. It follows from Theorem III.2 that (H, H^c) is a 1-separation of G, and from Theorem III.1, that x is the corresponding cut-vertex. □

Theorem III.21. *Let G be a connected graph, and let B be a block of G. Then the cut-vertices of G included in B are the vertices of attachment of B. They can also be characterized as the vertices at which B meets other blocks of G.*

Proof. Let x be a cut-vertex of G included in B. Then B meets another block at x, by Theorem III.19.

Now let y be any vertex at which B meets another block. Since G has more than one block, it is not a vertex-graph. So the second block must have an edge incident with y, by Theorem III.16 and the connection of a block. Hence y is a vertex of attachment of B.

Finally, let z be any vertex of attachment of B. Then the component H of $G:(E(G) - E(B))$ that includes z is not a vertex-graph. So z is a cut-vertex of G, by Theorem III.20.

The theorem follows from these three results. □

Theorem III.22. *Let G be a connected graph. Let (H, K) be a 1-separation of G with cut-vertex x. Then the blocks of G are the blocks of H*

Blocks

and K. Moreover, the cut-vertices of G are the cut-vertices of H and K, together with x.

Proof. A block of G is contained in H or K, by Theorem III.8, and must therefore be a block of H or K. But H and K can have no other blocks, by Theorems III.13 and III.16. It now follows from Theorem III.21 that the cut-vertices of G other than v are the cut-vertices of H and K other than v. □

We can indicate the relations between the blocks and cut-vertices of a connected graph G by means of its *block-graph* Blk(G). This is a strict graph with a bipartition (U,V). There is a 1–1 mapping f of the blocks of G onto the vertices of U, and a 1–1 mapping g of the cut-vertices of G onto the vertices of V. A vertex fB of U is adjacent to a vertex gx of V if and only if the block B includes the cut-vertex x in G.

If G is a vertex-graph, then Blk(G) is also a vertex-graph, its single vertex being in U.

Figure III.3.1 shows a connected separable graph G together with its block-graph Blk(G). Blocks of G are denoted by capitals and cut-vertices by small letters.

In Sec. I.3 we defined the union and intersection of two subgraphs H and K of a graph G. It should be noted that these definitions extend to sets of three or more subgraphs. Thus the *union* of subgraphs $H_1, H_2,\ldots,H_k$ of G is the subgraph of G made up of the edges and vertices of the k graphs H_j. The *intersection* of the k subgraphs is the subgraph of G made up of the edges and vertices common to them all. The k subgraphs H_j are not necessarily all distinct. We shall make use of these extended definitions from here on.

Theorem III.23. *Let G be a connected graph. Then* Blk(G) *is a tree.*

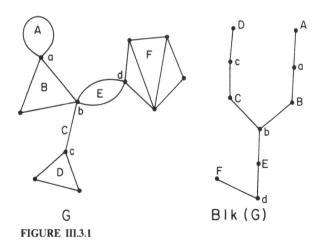

FIGURE III.3.1

Proof. Suppose first that Blk(G) is not connected. Let C be one of its components. Let H be the union of the blocks of G represented by the vertices of U in C, and K the union of all the other blocks of G. Now each component of Blk(G) includes a vertex of U, by Theorem III.19. Hence the subgraphs H and K of G have each at least one edge, by Theorem III.16. Moreover, each edge of G belongs to exactly one of them, by Theorem III.13. By the connection of G, the subgraph H has a vertex of attachment x, and this must be a common vertex of a block contained in H and a block contained in K. But then the vertex gx of Blk(G) is adjacent to a vertex of U in C and a vertex of U not in C. This contradicts the definition of C as detached in Blk(G).

Suppose next that not every edge of Blk(G) is an isthmus. Then Blk(G) contains a circuit, by Theorem I.45. Thus, since Blk(G) is strict, we can find $n \geq 2$ distinct blocks $B_1, B_2,\ldots,B_n$ of G and n distinct cut-vertices $x_1, x_2,\ldots,x_n$ of G satisfying the following rule. For each j, x_j is a common vertex of B_{j-1} and B_j, where $B_0 = B_n$. Let H be the union of all the B_j other than B_1. It is a connected subgraph of G, by repeated application of Theorem I.25. It contains an arc L with ends x_1 and x_n, by Theorem I.43. But then $B_1 \cup L$ is a 2-connected subgraph of G having B_1 as a proper subgraph, by Theorems III.13 and III.4. This contradicts the maximality of B_1.

We can now assert that Blk(G) is a connected forest, that is, a tree. □

We can now apply the theory of trees to that of blocks. For example, let us define an *extremal* block of a connected graph G as a block of G that includes exactly one cut-vertex of G. Then the extremal blocks of G correspond to the monovalent vertices of U in Blk(G). Since each vertex of V in Blk(G) is at least divalent, by Theorem III.19, we can apply Theorem I.40 to deduce the following theorem.

Theorem III.24. *Let G be a connected separable graph. Then it has at least two extremal blocks.*

III.4. ARMS

Let G be a connected graph. Let x be a vertex of G and let X be the corresponding vertex-graph. We define an *arm* of G at x as a bridge of X in G.

Theorem III.25. *An arm of G at x is a connected graph that is not a vertex-graph. It includes the vertex x.*

Proof. Let U be an arm of G at x. If U is disconnected, or if it does not include x, then it has a component C that does not include x. Since x is the only possible vertex of attachment of U in G, by the definition of a bridge, it follows that C is a detached subgraph of G. But this is impossible, by the connection of G.

We can now assert that U is connected and that it includes x. It is not a subgraph of X, by the definition of a bridge. Hence it is not a vertex-graph. □

If G is a vertex-graph, there is no arm of G at x, by Theorem III.25. In the remaining case there is at least one such arm, by Theorem I.51. If there is just one arm U of G at x, then $G = U$ by Theorem I.51. For two or more arms, we have the following theorem.

Theorem III.26. *Let there be at least two arms of G at x, and let U be one of them. Then (U, U^c) is a 1-separation of G with cut-vertex x (by Theorems III.2 and III.25).*

We go on to relate the arms to the blocks of G.

Theorem III.27. *Let U be an arm of G at x. Then there is exactly one block B of G such that B includes x and is contained in U.*

Proof. U has an edge A incident with x, by Theorem III.25. This belongs to a block B of G that includes x, by Theorem III.13. But then B is contained in U. If U is the only arm at x, this follows from the fact that $U = G$. In the remaining case, it is a consequence of Theorems III.26 and III.8.

If B is the only block of G that includes x the theorem is satisfied. Suppose therefore that G has a second such block B'. Then x is incident with some edge E of G that is in B' but not in B, by Theorem III.13. Hence the component H of $G:(E(G) - E(B))$ that includes x is not a vertex-graph. So, by Theorem III.20, (H, H^c) is a 1-separation of G with cut-vertex x. By the definition of H, all the edges of H^c incident with x are in $E(B)$.

Now H^c is X-detached in G. Hence $H^c \cap U$ is X-detached in G, by Theorems III.1 and I.46. But this intersection is not a subgraph of X, since it includes all the edges of B incident with x. It follows, by the bridge-minimality of U, that $U \subseteq H^c$. Hence all the edges of U incident with x are in B; E is not one of them and therefore B' is not contained in U. □

Theorem III.28. *Let the connected graph G be not a vertex-graph. Then the blocks of G are the blocks of the arms of G at x. The cut-vertices of G other than x are the cut-vertices of the arms of G at x. The vertex x is not a cut-vertex of any arm of G at x.*

Proof. Let B be any block of G. It has an edge A and this belongs to some arm U of G at x, by Theorem I.51. But then $B \subseteq U$, by Theorem III.8 and the fact that $U = G$ if U is the only arm at x. By its maximality in G, B must be a block of U. All the blocks of all the arms of G at x are thus accounted for, by Theorem III.13. The statements about cut-vertices now follow from Theorems III.21 and III.27. □

From Theorems III.27 and III.28, we see that the arms of G at x are in 1-1 correspondence with the blocks of G having x as a vertex. Each such arm contains exactly one such block, and each such block is contained in exactly one such arm. There are two or more arms at x if and only if x is a cut-vertex of G, by Theorem III.21.

The preceding theory is of special interest when G is a tree. Then the blocks of G are the link-graphs defined by its edges, by Theorem III.17. The number of arms at a vertex x is $\mathrm{val}(G, x)$, and each edge incident with x belongs to exactly one of these arms. Each arm of G at x is itself a tree, being a connected forest, by Theorems III.25 and I.33. An arm U of G at x has x as one of its monovalent vertices, and as its only vertex of attachment. Each other vertex of U is incident with the same edges in U as in G, and has the same valency in U as in G. By Theorem I.40, the arm U must include at least one monovalent vertex of G distinct from x.

In particular, G may be an arc L with ends y and z, and x may be an internal vertex of L. Then there are just two arms at x. By the preceding observations and by Theorem I.40, one of them is an arc L_{xy} with ends x and y, and the other is an arc L_{xz} with ends x and z. Their intersection is the vertex-graph X and their union is G, by Theorem I.51. We say that x *decomposes* L into the two arcs L_{xy} and L_{xz}. This decomposition of L is, of course, easy to describe in terms of its defining enumeration.

For an example of a connected separable graph G that is not a tree, we can take the first diagram of Fig. III.3.1. Here there are three arms at b. One of them is made up of the blocks A and B; another, of the blocks C and D; and the third, of the blocks E and F. Of the two arms at a, one is the loop-graph $G \cdot \{A\}$, and the second is the union of all the other blocks.

III.5. DELETION AND CONTRACTION OF AN EDGE

Let G be a connected graph. We describe it as a *string of blocks* if, for some integer $k \geq 2$, we can enumerate its blocks as $B_1, B_2, \ldots, B_k$ and its cut-vertices as $v_1, v_2, \ldots, v_{k-1}$ so that the following condition holds. The cut-vertex v_j belongs to B_j and B_{j+1} but to no other block, for each relevant suffix j. Clearly a connected graph is a string of blocks if and only if its block-graph is an arc.

In the above string of blocks just two blocks are extremal, B_1 and B_k. Let us adjoin a new link A with one end x in B_1 but distinct from v_1, and the other end y in B_k but distinct from v_{k-1}. We say that A *closes* the string of blocks.

Theorem III.29. *Let a graph K be formed by closing a string G of blocks with a new link A. Then K is 2-connected.*

Proof. We use the above notation but write $x = v_0$ and $y = v_k$. In each block B_j there is an arc L_j with ends v_{j-1} and v_j, by Theorem I.43. The union of the arcs L_j is connected, by repeated application of Theorem I.25. It is an arc L with ends x and y, by Theorem I.26. Adjoining the link A to it, we obtain a circuit C, by Theorems I.27 and I.43. But C has an edge in each of the blocks B_j of G. Hence K is 2-connected, by Theorems III.5 and III.9. □

Theorem III.30. *Let G be a 2-connected graph having at least two edges, and let A be an edge of G. Then either G'_A is 2-connected or it is a string of blocks that is closed by A.*

Proof. G'_A is connected and A is a link of G, by Theorem III.3. Suppose G'_A is not 2-connected, and so has at least two blocks, by Theorem III.15. Let B be an extremal block of G'_A. Then some vertex of B other than its cut-vertex must be an end of A in G, for otherwise (B, B^c) would be a 1-separation of G, by Theorem III.2. It follows, by Theorem III.24, that G'_A has exactly two extremal blocks, each including an end of A that is not a cut-vertex of G'_A. Thus $\text{Blk}(G'_A)$ is an arc, being a tree with just two monovalent vertices, by Theorems III.23 and I.40. We deduce that G'_A is a string of blocks, closed by A to form G. □

Let us now seek a corresponding theorem for G''_A. We use the notation of Sec. II.2.

Theorem III.31. *Let A be an edge of a 2-connected graph G. Then either G''_A is 2-connected or it has v_A as its only cut-vertex. In the latter case A is a link of G and for each block B of G''_A the subgraph $f_A B$ of G is 2-connected.*

Proof. G''_A is connected, by Theorem II.5. Suppose G''_A is not 2-connected. Then it has at least two blocks, by Theorem III.15, and at least one cut-vertex by Theorem III.21. It has at least two edges, by Theorem III.16, and so A is a link of G, by Theorem III.3.

Suppose some vertex x of G''_A other than v_A is a cut-vertex of G''_A. Then it corresponds to a 1-separation (H, K) of G''_A with v_A in H. But then $f_A K$ is a subgraph of G having only one vertex of attachment, by Theorem II.3. But this is impossible, by Theorem III.2 and the 2-connection of G. Since G''_A has at least one cut-vertex, we deduce that v_A is the only cut-vertex of G''_A.

Now consider a block B of G''_A. Its only vertex of attachment is v_A. Hence the only possible vertices of attachment of $f_A B$ in G are the ends x and y of A. Suppose $f_A B$ is not 2-connected. Then it has a 1-separation (H, K), with cut-vertex z, say, such that A is an edge of H. But now z is the only vertex of attachment of K in G, which is impossible by 2-connection. □

Theorem III.32. *Let A be an edge of a 2-connected graph G. Then either G''_A is 2-connected or G is the union of two or more 2-connected graphs*

$H_1, H_2, \ldots, H_k$, each with at least two edges, such that the intersection of any two of them is the link-graph $G \cdot \{A\}$.

This result follows from Theorem III.31. The H_i are the graphs $f_A B$ such that B is a block of G_A''.

Theorem III.33. *Let G be a 2-connected graph with at least two edges. Let A be an edge of G. Then either G_A' or G_A'' is 2-connected.*

Proof. A is a link of G, by Theorem III.3. Let its ends be x and y.

Assume that neither G_A' nor G_A'' is 2-connected. Then by Theorem III.30, no block of G_A' includes both x and y. Referring to Theorem III.32, we see that each of the graphs $(H_i)_A'$ is connected, by Theorem III.3. Hence x and y are the ends of an arc L_1 in $(H_1)_A'$ and an arc L_2 in $(H_2)_A'$, by Theorem I.43. These two arcs are internally disjoint and their union is a circuit, by Theorems I.27 and I.43. By Theorems III.5 and III.12, this circuit is contained in a block of G_A', and this block includes both x and y. Our assumption has led to a contradiction, and the theorem is established. □

III.6. NOTES

III.6.1. 2-connection and planarity

The property of 2-connection or non-separability is of special interest in connection with planar graphs. When a connected graph is drawn on the 2-sphere, the remainder of the surface is decomposed into a finite number of topological discs. In the 2-connected case, each of these discs is bounded by a circuit of the graph. This is a topological theorem: the corresponding combinatorial result is Theorem XI.22. One of the classical papers of Hassler Whitney is entitled "Nonseparable and planar graphs" [2].

EXERCISES

1. If a, b, and c are three distinct vertices in a 2-connected graph G, show that (i) there is an arc in G with ends b and c that does not pass through a, (ii) there is an arc in G with ends b and c that does pass through a.
2. Let the number of blocks of a connected graph G be N, and the number of cut-vertices r. Let n_i be the number of arms at the ith cut-vertex. Prove Harary's formula ([1])

$$N - \Sigma n_i + r = 1.$$

3. Let x and y be distinct cut-vertices of a connected graph G, y being in the arm U of G at x. Describe the arms of G at y in terms of those of G at x and those of U at y.

REFERENCES

[1] Harary, F., An elementary theorem on graphs, *Amer. Math. Monthly* **66** (1959), 405–407.
[2] Whitney, H., Nonseparable and planar graphs, *Trans. Amer. Math. Soc.*, **34** (1932), 339–362.

Chapter IV

3-Connection

IV.1. MULTIPLE CONNECTION

The notion of 2-connection studied in Chapter III can be generalized as follows:

Let G be a connected graph. For any positive integer n, we define an n-*separation* of G as an ordered pair (H, K) of edge-disjoint subgraphs of G satisfying the three following conditions:
i) $H \cup K = G$.
ii) H and K have exactly n common vertices.
iii) H and K have each at least n edges.

We say that G is m-*connected*, where m is a positive integer, if it has no n-separation for any $n < m$. Thus, to say that G is 1-connected is simply to repeat that it is connected. Our definition of m-connection agrees with that of 2-connection in Chapter III.

There are well-known variants on the above definitions. Suppose, for example, we replace Condition (iii) by the following:

iii)' H and K have each a vertex not belonging to the other.

We shall then call a pair (H, K) satisfying (i), (ii) and (iii)' a *vertical n-separation* of G. We shall likewise say that G is *vertically m-connected* if it has no vertical n-separation for any $n < m$.

In another variant, Condition (iii) is replaced by

(iii)″ *H and K have each a circuit.*

The pair (H, K) will then be called a *cyclic n-separation* of G, and the new kind of m-connection will be called *cyclic m-connection*.

In this work we shall usually be concerned with n-separation and m-connection as defined by Conditions (i), (ii), and (iii), but we shall sometimes have occasion to mention the other two kinds of m-connection.

It should be noted that most authors use the term "m-connection" for what is called here "vertical m-connection." Our "m-connection" has the advantage of generalizing to the theory of matroids in a self-dual way.

We begin our study with two theorem-examples.

Theorem IV.1. *Let G be a connected graph having a circuit C such that $|E(C)| \leq \frac{1}{2}|E(G)|$. Then (C, C^c) is an n-separation of G for some $n \leq |E(C)|$.*

Proof. The number n of common vertices of C and C^c is at most $|V(C)| = |E(C)|$. By hypothesis, C and C^c have each at least $|E(C)|$ edges. Hence, Conditions (i), (ii), and (iii) are satisfied with $H = C$ and $K = C^c$. □

Corollary IV.2. *A 2-connected graph with at least two edges is loopless. A 3-connected graph with at least four edges is strict.*

The first part of the Corollary is, of course, included in Theorem III.3.

Theorem IV.3. *Let G be a connected graph which is not a vertex-graph. Let x be a vertex of G such that $\mathrm{val}(G, x) \leq \frac{1}{2}|E(G)|$. Let H be the subgraph of G defined by x, its incident edges, and their other ends. Then (H, H^c) is an n-separation of G for some positive integer $n \leq \mathrm{val}(G, x)$.*

Proof. We note that $\mathrm{val}(G, x)$ is positive, by the connection of G. Let n be the number of common vertices of H and H^c. Allowing for the possibility of loops, we can write

$$1 \leq |E(H)| \leq \mathrm{val}(G, x). \qquad (IV.1.1)$$

Hence, by hypothesis, H and H^c have each at least one edge. By the connection of G, they must have at least one common vertex. Thus the integer n is positive.

Now n cannot exceed $|E(H)|$. Hence by Eq. (IV.1.1) and our hypothesis, it cannot exceed $|E(H^c)|$. Accordingly, Conditions (i), (ii), and (iii) are satisfied with $K = H^c$. □

Theorem IV.4. *If a 3-connected graph has four edges, it has at least six. Moreover, it has at least four vertices.*

Proof. Since such a graph G is strict, by Corollary IV.2, it must have more than three vertices. Each vertex is at least trivalent by Theorem IV.3. So the number of edges is at least six, by Eq. (I.1.1). □

If there is a greatest integer k such that the connected graph G is k-connected, we call k the *connectivity* of G and write it as $\kappa(G)$. If there is no such integer, we say that the connectivity $\kappa(G)$ is infinite. We note that a connected graph G is m-connected for every positive integer m not exceeding its connectivity. Let us make a list of the graphs of infinite connectivity. We can use the following rule.

Theorem IV.5. *Let G be a k-connected graph with at most $(2k-1)$ edges. Then $\kappa(G) = \infty$.*

This is because Condition (iii) cannot be satisfied for any $n \geq k$. Our list is given in the next theorem. We do not count the null graph because it is not connected, and so the definition of k-connectivity does not apply to it.

Theorem IV.6. *The graphs of infinite connectivity are the 2-connected graphs of at most three edges. Thus they are the vertex-graph, the loop-graph, the link-graph, the 2-circuit, the 3-circuit, and the 3-linkage.*

Proof. The six graphs here listed are the 2-connected graphs of at most three edges, by the results of Sec. III.2. They have infinite connectivity, by Theorem IV.5.

Suppose G is another graph of infinite connectivity. It has at least four edges. It is strict, by Corollary IV.2, and has at least four vertices, by Theorem IV.4. The valency of each vertex is at least 3.

Choose a vertex x of minimum valency b. Let H be the graph obtained from G by deleting x and its incident edges. Then H has at least three vertices, each with valency at least $(b-1)$. By Eq. (I.1.1) the number of edges of H is at least $3(b-1)/2$. Since b is at least 3, this number is not less than b. Hence G is at most b-connected, by Theorem IV.3, contrary to our assumption of infinite connectivity. □

We conclude with some theorems about deletions and contractions.

Theorem IV.7. *Let A be an edge, not an isthmus, of a connected graph G of finite connectivity κ. Then $\kappa(G'_A) \geq \kappa - 1$.*

Proof. G'_A is connected by hypothesis, so $\kappa(G'_A)$ is defined. Assume that $\kappa(G'_A) < \kappa - 1$. Then there exists an n-separation (H, K) of G'_A with $n \leq \kappa - 2$. If the ends of A are both in H or both in K, we can get an n-separation of G by adjoining A to H or K, respectively. This being impossible, we must suppose A to have one end x in H but not K, and the other end y in K but not H.

Each of H and K has at least n edges. If each has only n, then G has infinite connectivity, by Theorem IV.5. This being contrary to hypothesis we can assume without loss of generality that $|E(K)| > n$.

Form H_1 by adjoining A and y to H. Then (H_1, K) is an $(n + 1)$-separation of G, contrary to hypothesis. The theorem follows. □

Sometimes the connectivity of a graph G is increased by the deletion of an edge A. Obvious examples occur in which G has just one loop, or just one multiple join. When m-connection is defined as what we have called vertical m-connection, it is permissible to ignore loops and multiple edges and to treat each graph as strict. It is then usual to make a special convention about the connectivity of the n-clique, and so to arrange that the deletion of an edge can never increase the connectivity. But with our definition we shall have to endure the possibility of increases on deletion.

Theorem IV.8. *Let A be an edge of a connected graph G of finite connectivity κ. Then $\kappa(G_A'') \geq \kappa - 1$.*

Proof. G_A'' is connected, by Eq. (II.2.1). Accordingly $\kappa(G_A'')$ is defined. Assume that $\kappa(G_A'') < \kappa - 1$. Then there exists an n-separation (H, K) of G_A'' with $n \leq \kappa - 2$. By taking the least possible value of n, we ensure that the n common vertices of H and K are the vertices of attachment of H, and of K, in G_A''. For example, a common vertex isolated in H could be deleted from that subgraph but retained in K. If $f_A H$ has only n vertices of attachment in G, then G has the n-separation $(f_A H, (f_A H)^c)$. This being impossible, we can deduce from Theorem II.23 that $f_A H$, and similarly $f_A K$, has just $(n + 1)$ vertices of attachment in G, and that two of these are the ends x and y of A in G.

Each of H and K has at least n edges. If each has exactly n, then G has infinite connectivity, by Theorem IV.5. This being contrary to hypothesis, we can assume without loss of generality that $|E(K)| > n$.

We can now infer that $(f_A H, (f_A H)^c)$ is an $(n + 1)$-separation of G, which is contrary to hypothesis. The theorem follows. □

Theorem IV.9. *Let G be a connected graph with at least two vertices. Let x be a vertex, but not a cut-vertex of G. Let H be the graph obtained from G by deleting x and its incident edges. Then $\kappa(H) \geq \kappa(G) - 1$.*

Proof. H is connected, for otherwise x would have to be a cut-vertex of G. Accordingly, $\kappa(H)$ is defined. Assume that $\kappa(H) < \kappa(G) - 1$. Then H has an n-separation (L, M) for some $n \leq \kappa(G) - 2$. If no vertex of $V(L) - V(M)$ is joined to x by an edge of G, then (L, L^c) is an m-separation of G for some $m \leq n$, contrary to hypothesis. We deduce that some vertex of $V(L) - V(M)$, and similarly some vertex of $V(M) - V(L)$ is joined to x by an edge of G. Let L_1 be the subgraph of G formed from L by adjoining x

and the edges from x to vertices of $V(L)-V(M)$. Then (L_1, L_1^c) is an m-separation of G for some $m \leq n+1$, contrary to hypothesis. This contradiction completes the proof. □

IV.2. SOME CONSTRUCTIONS FOR 3-CONNECTED GRAPHS

From now on we shall consider the theory of 3-connected graphs. We begin by showing that for strict graphs is does not matter whether we speak of 3-connection or of vertical 3-connection.

Theorem IV.10. *Let G be a strict connected graph. Then the 1-separations of G are its vertical 1-separations. Moreover, if G is 2-connected, then the 2-separations of G are its vertical 2-separations. Hence G is 2-connected if and only if it is vertically 2-connected, and 3-connected if and only if it is vertically 3-connected.*

Proof. Let (H, K) be a 1-separation of G. Each of H and K has an edge. Hence, by strictness, each has a vertex not belonging to the other.

Conversely, let (H, K) be a vertical 1-separation of G. Each of H and K has a vertex not belonging to the other. Hence, by connection, each has an edge.

This completes the first part of the proof. We note the implication that G is 2-connected if and only if it is vertically 2-connected. In the rest of the proof we may suppose G to be 2-connected.

Let (H, K) be a 2-separation of G. H has at least two edges. By strictness they cannot both join the two common vertices of H and K. Hence H has a vertex not in K, and similarly K has a vertex not in H.

Conversely, let (H, K) be a vertical 2-separation of G. Then H has a vertex h not in K, and K has a vertex k not in H. By the 2-connection of G, h and k are at least divalent. Hence H and K have each at least two edges. This completes the proof. □

In view of Corollary IV.2, the requirement of strictness in Theorem IV.10 asks very little.

We go on to consider some constructions whereby 3-connected graphs are derived from smaller ones. First, we have the operation of *link-adjunction*, described in the following theorem.

Theorem IV.11. *Let x and y be nonadjacent vertices of a 3-connected graph H. Let a new link A be adjoined to H, with ends x and y, to form a new graph G. Then G is 3-connected.*

Proof. G is 2-connected, by Theorem III.10. Assume that it is not 3-connected. Then it has a 2-separation (L, M), where we may suppose A in L. This implies that H has at least three edges, and is therefore loopless, by

Corollary IV.2. But L must have exactly two edges. For otherwise (L'_A, M) would be a pair of complementary detached subgraphs of H, or a 1-separation or 2-separation of H, contrary to the 3-connection of H.

The two edges of L do not have the same two ends, by hypothesis. Hence L must have a vertex x that does not belong to M. But then x can be incident with only one edge of H. This is contrary to the 3-connection of H, by Theorem IV.3. □

The next operation is a kind of inverse of edge-contraction. We call it *vertex-splitting*. Let x be a vertex of a strict graph H. We replace x by two new vertices x_1 and x_2. Each edge of H joining x to another vertex y is replaced by an edge joining y and x_1, by an edge joining y and x_2, or by both. Finally we adjoin a new link A with ends x_1 and x_2. If we replace each edge incident with x by one edge only, then the operation is vertex-splitting "without duplication of edges." The next three theorems are concerned with the properties of the resulting graph G. The process is illustrated in Fig. IV.2.1.

Theorem IV.12. Let a graph G be derived from a strict connected graph H by splitting a vertex x. Then G is strict and connected.

Proof. We first observe that, to within a vertex-isomorphism, G''_A has H as a spanning subgraph. Hence G''_A is connected, and therefore G is connected, by Eq. (II.2.1). As for strictness, we note that the only new edges in G are those incident with x_1 and x_2. By the definition of vertex-splitting these are links, and no two of them have the same two ends. □

Theorem IV.13. Let a graph G be derived from a strict 2-connected graph H by splitting a vertex x. Let the new vertices x_1 and x_2 be at least divalent in G. Then G is 2-connected.

Proof. G is connected and strict, by Theorem IV.12. Assume it not 2-connected. Then by Theorem IV.10 it has a vertical 1-separation (L, M). We may suppose the new edge A, joining x_1 and x_2, to belong to L. The common vertex u of L and M may or may not be incident with A.

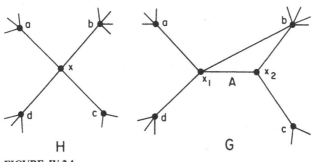

FIGURE IV.2.1

L has a vertex ℓ not in M, and M has a vertex m not in L. By connection, each of these is incident with an edge of G. Since x_1 and x_2 are at least divalent each of ℓ and m is incident in G with an edge of H. Let M'' be the subgraph of H defined by the edges and vertices of M, u being replaced by x if u is x_1 or x_2. Then M'' is a proper subgraph of H having at least one edge and at most one vertex of attachment. But this implies that H is separable, by Theorem III.2. We conclude that in fact G is 2-connected. □

Theorem IV.14. *Let a graph G be derived from a strict 3-connected graph H by splitting a vertex x. Let the new vertices x_1 and x_2 be at least trivalent in G. Then G is 3-connected.*

Proof. G is 2-connected and strict, by Theorem IV.13. Assume it not 3-connected. By Theorem IV.10 it has a vertical 2-separation (L, M), where we can suppose the new edge A joining x_1 and x_2 to be in L. Let u and w be the two common vertices of L and M.

L has a vertex ℓ not in M, and M has a vertex m not in L. By Theorem IV.3 each of these vertices is incident with at least two edges of G. Since x_1 and x_2 are at least trivalent, each of ℓ and m is incident in G with at least two edges of H. Let M'' be the subgraph of H defined by the edges and vertices of M, a vertex u or w being replaced by x if it is incident with A. In the extreme case in which A joins u and w, these two vertices of M are identified in M''. By the preceding considerations M'' has at least two edges, and so does its complementary subgraph $(M'')^c$ in H. But M'' and $(M'')^c$ have at most two common vertices, the only possibilities being u and w, or their replacements. But this means that H has a 1-separation or 2-separation $(M'', (M'')^c)$, which is contrary to hypothesis. □

So far the only 3-connected graphs to be noted are the six graphs of infinite connectivity listed in Theorem IV.6. The operation of link-adjunction is not applicable to any of these. The operation of vertex-splitting, with the requirement of trivalency, is applicable only to the 3-circuit, and it converts that graph into the 4-clique. We can now assert that the 4-clique is 3-connected. But this result is part of the following more general theorem.

Theorem IV.15. *For each positive integer n, the n-clique is 3-connected.*

Proof. Let G be an n-clique. We can assume $n \geq 4$, by Theorem IV.6. We noted in Sec. I.5 that such a G must be connected. If it is not 3-connected, it has a vertical b-separation (L, M), where b is 1 or 2, by Theorem IV.10. L has a vertex ℓ not in M, and M has a vertex m not in L. But ℓ and m are joined by an edge A in the clique G. This edge must belong either to L or to M, and therefore ℓ and m are both in L or both in M. This contradiction establishes the theorem. □

Our next construction can be called *link-adjunction with subdivision*. To *subdivide* an edge A with ends x and y, in a graph H, is to replace it by a new vertex z, a new edge A_1 with ends x and z, and a new edge A_2 with ends y and z. (See Fig. IV.2.2.) We say that the resulting graph G is obtained from H by *subdividing A at z*.

We note that when a loop is subdivided it is replaced by a 2-circuit. Hence by two applications of the operation of subdividing every edge, we can convert any graph into a strict graph. In a strict graph H any subdivision of an edge can be described as a vertex-splitting. In terms of Fig. IV.2.2 we subdivide A by splitting the vertex x, to within a vertex-isomorphism, and the two "new" vertices are x and z.

From the definitions of arc and circuit, we can infer the following rules. When an edge of an n-arc L is subdivided, the graph is converted into an $(n+1)$-arc with the same ends. When an edge of an n-circuit C is subdivided, the graph is converted into an $(n+1)$-circuit.

The operation of link-adjunction with subdivision occurs in two forms. The first kind involves a link A and a vertex t not incident with it. It is described in the following theorem.

Theorem IV.16. *Let H be a 3-connected graph with at least four edges. Let A be a link of H and let t be a vertex of H not incident with A. Let a graph G be formed from H by subdividing A at z and then adjoining a new link B with ends t and z. Then G is 3-connected.*

Proof. We first observe that each vertex of H is at least trivalent in H, by Theorem IV.3.

Let the ends of A in H be x and y. The procedure is illustrated in Fig. IV.2.3. There may or may not be a link C in H with ends x and t. The possibility of such a link is indicated in Fig. IV.2.3 by a broken line.

If the link C exists we can obtain G from H by splitting the vertex x. In this operation C is replaced by two links, one called C from t to x and one called B from t to z. The requirement of trivalency is clearly satisfied. So in this case G is 3-connected, by Theorem IV.14.

In the remaining case, H has no link from t to x. We then adjoin a new link D with these ends. The resulting graph H_1 is 3-connected, by Theorem IV.11. But G can be derived from H_1 by splitting the vertex x, the link D being replaced by a single new link with ends t and z. Hence G is 3-connected, by Theorem IV.14. □

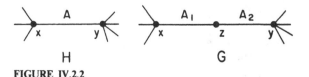

FIGURE IV.2.2

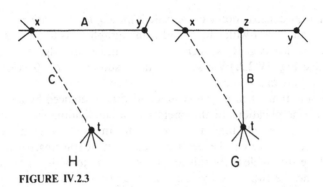

FIGURE IV.2.3

The first kind of link-adjunction with subdivision is important in connection with the theory of wheels. A *wheel* W_n, *of order* n, is formed from an n-circuit C_n by adjoining a single new vertex h, and then joining x to each vertex of C_n by a single new link. We call h the *hub* of the wheel W_n, C_n its *rim*, and the new edges its *spokes*. We note that the 4-clique can be regarded as the wheel of order 3. For it, the choice of one vertex as hub is arbitrary. The second graph of Fig. III.2.2 is the wheel of order 2. Figure IV.2.4 shows the wheels from W_1 to W_5.

Theorem IV.17. *For* $n \geq 3$ *the wheel* W_n *is 3-connected.*

Proof. We use induction over n. For $n = 3$ the wheel is the 4-clique, and it is 3-connected, by Theorem IV.15. Assume as an inductive hypothesis that the theorem holds whenever n is less than some integer $q > 3$, and consider the case $n = q$.

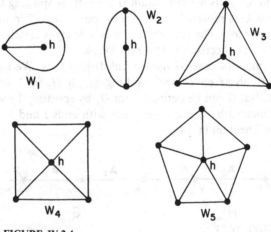

FIGURE IV.2.4

By the inductive hypothesis, the wheel W_{q-1} is a 3-connected graph of at least six edges. Let us subdivide some edge A of its rim at a new vertex z. Then the rim becomes a q-circuit. We now obtain W_q by adjoining a new link from z to the hub of W_{q-1}. Thus W_q is derived from W_{q-1} by the operation described in Theorem IV.16. Accordingly, W_q is 3-connected. The induction succeeds and the theorem is true. □

We go on to the second kind of link-adjunction with subdivision. It is the process described in the following theorem.

Theorem IV.18. *Let H be a 3-connected graph with at least four edges. Let A and B be distinct edges of H. Let them be subdivided at new vertices z_A and z_B, respectively, and then let z_A and z_B be joined by a new link C. Then the resulting graph G is 3-connected.*

Proof. The graph H is of course strict, and each of its vertices is at least trivalent, by Theorem IV.3 and Corollary IV.2.

Our procedure is illustrated in Fig. IV.2.5. We first derive from H a graph H_1 by subdividing A at z_A and then joining z_A to an end u of B not incident with A in H. The graph H_1 is 3-connected, by Theorem IV.16. But now G can be obtained from H_1 by splitting the vertex u. Hence G is 3-connected, by Theorem IV.14. □

The second kind of link-adjunction with subdivision has the interesting property of changing cubic graphs into cubic graphs. Figure IV.2.6 shows an example in which A and B are links of the 4-clique with no common end. The resulting graph is called the *Thomsen graph*.

The Thomsen graph is an example of a *complete bipartite graph*. By definition such a graph is strict and has a bipartition (U, V). Moreover, each vertex of U is joined to each vertex of V by a single link. If $|U| = r$ and $|V| = s$, then the complete bipartite graph, or rather its isomorphism class, is denoted by $K_{r,s}$. The Thomsen graph is $K_{3,3}$. In Fig. IV.2.6 the outer edges in the drawing of $G = K_{3,3}$ define a 6-circuit, and members of U and V occur alternately in this circuit. The construction illustrated in Fig. IV.2.6 shows that $K_{3,3}$ is 3-connected.

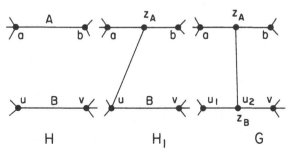

FIGURE IV.2.5

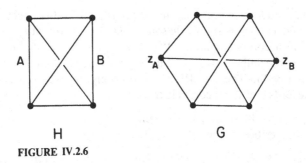

H G

FIGURE IV.2.6

Since the Thomsen graph is strict, any of its automorphisms is determined by the corresponding permutation of the vertex-set, by Theorem I.2. An automorphism leaving U and V invariant can be any permutation of U combined with any permutation of V. An automorphism interchanging U and V can be any 1-1 mapping of U onto V combined with any such mapping of V onto U. There is no other possibility. We conclude that the automorphism group of $K_{3,3}$ has 72 elements.

Clearly, $K_{3,3}$ is a cubic graph having no circuit of fewer than four edges. It can be shown that $K_{3,3}$ has the least number of edges consistent with these properties. Indeed $K_{3,3}$ is the only cubic graph of nine edges having no circuit with fewer than four.

We have one more construction to discuss in this section. It is applied to a 3-connected graph H of at least four edges. We have seen that such a graph is strict and that each of its vertices is at least trivalent. It is called *expansion of a vertex into a circuit*. Let v be a vertex of H of valency b. Let S_v be the set of the b edges of H incident with v. We replace v by an n-circuit C, made up of new edges and vertices, where $3 \le n \le b$. Each edge of S_v is made incident with some vertex of C, instead of with v. The incidences between edges of S_v and vertices of C are assigned arbitrarily, except that each vertex of C is required to be incident with at least one member of S_v. This requirement ensures that the vertices of the resulting strict graph G are at least trivalent in G. In Fig. IV.2.7 we see the operation of vertex-expansion applied to the 4-clique and the 4-wheel.

In these two examples, G can equally well be derived from H by the second kind of link-adjunction with subdivision. Figure IV.2.8 shows a more complicated example in which the hub of the 5-wheel is expanded into a 5-circuit. The graph G of this figure is known as the *Petersen graph*.

Theorem IV.19. *Let v be a vertex of a 3-connected graph H having at least four edges. Let G be obtained from H by expanding v into a circuit C of at least three edges. Then G is 3-connected.*

Proof. Let n be the number of vertices of C. We begin by considering the case in which n has its least possible value, 3. If the valency b of v in

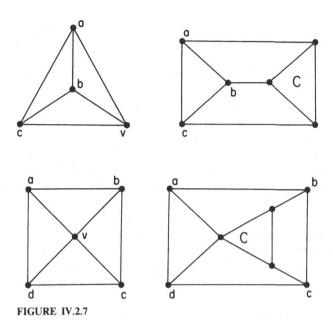

FIGURE IV.2.7

H is also 3, then the vertex-expansion can be effected by an edge-adjunction with subdivision, as we saw in the case of the first example of Fig. IV.2.7. So G is 3-connected, by Theorem IV.18.

Suppose now that b exceeds 3. Let the vertices of C be a_1, a_2, and a_3. Let T_i be the set of members of S_v that are to be made incident with a_i ($i = 1, 2, 3$). We can adjust the notation so that T_1 has at least two members. We now split the vertex v in H so that, apart from the new link A, one new vertex v_1 is incident only with the members of T_1 and the other new vertex

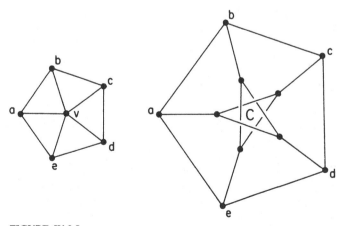

FIGURE IV.2.8

v_2 is incident only with the members of T_2 and T_3. The resulting graph H_1 is 3-connected, by Theorem IV.14, the requirement of trivalency being obviously satisfied. But now G can be derived from H_1 by splitting the vertex v_2, the link A being replaced by two edges of C. So G is 3-connected by Theorem IV.14.

We have proved that the theorem is true when $n = 3$. Assume as an inductive hypothesis that it is true whenever n is less than some integer $q > 3$, and consider the case $n = q$.

We wish to expand v into a q-circuit C_q. Let the sequence of vertices in some defining enumeration of C be $a_1, a_2, \ldots, a_q$, and the corresponding sequence of edges $A_1, A_2, \ldots, A_q$. Let T_i denote the set of those edges of S_v that are to be made incident with a_i ($1 \le i \le q$). Let us, however, first expand v into a $(q-1)$-circuit C_{q-1}. We can take the sequence of vertices in some defining enumeration of C_{q-1} to be $a_1, a_2, \ldots, a_{q-1}$, and the corresponding sequence of edges to be $A_1, A_2, \ldots, A_{q-1}$. For $i \le q-1$ we make the members of T_i incident with a_i, and we make the members of both T_{q-1} and T_q incident with a_q. By the inductive hypothesis, the resulting graph H_{q-1} is 3-connected. But the required graph G can be obtained from it by splitting the vertex a_{q-1}. The new link introduced by this operation, joining the two new vertices, is called A_q. The two new vertices are denoted by a_{q-1} and a_q. The new a_{q-1} is taken to be incident with A_{q-1}, A_q and the members of T_{q-1}. The new vertex a_q is taken to be incident with A_1, A_q and the members of T_q. Figure IV.2.9 illustrates the case $q = 5$. So G is 3-connected, by Theorem IV.14.

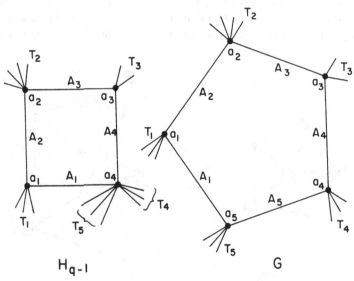

FIGURE IV.2.9

The induction has succeeded, and we can assert the truth of the theorem. □

Using this theorem we can assert the 3-connection of the Petersen graph. Instead of defining this graph by reference to a diagram, we can introduce it more formally as follows: It is formed from the union of two disjoint 5-circuits $a_1a_2a_3a_4a_5$ and $b_1b_3b_5b_2b_4$ by adjoining five new links, one from a_i to b_i for each i.

We note a similarity between the Thomsen and Petersen graphs. For any graph G that is not a forest, we define the *girth* of G as the least integer n such that G contains an n-circuit. We have remarked, though without a complete proof, that the Thomsen graph is the simplest cubic graph of girth 4. Here "simplest" means "with the least number of edges." It can be shown that the Petersen graph is the simplest cubic graph of girth 5.

The Petersen graph is related to the regular dodecahedron. If diametrically opposite points of the surface of that solid are identified, its graph of geometrical edges and vertices condenses into a Petersen graph. This observation shows that the Petersen graph has high symmetry. It is found, indeed, that the automorphism group has 120 elements.

IV.3. 3-BLOCKS

In Sect. III.3, we discussed the breakdown of a connected graph into its maximal 2-connected subgraphs or "blocks." Here we describe a somewhat similar decomposition of a 2-connected graph G into its "3-blocks." These are not exactly subgraphs of G. A 3-block must be either a 3-connected graph, an n-circuit, or an n-linkage, with $n \geq 3$ in the last two cases.

The given graph G may be 3-connected. We then say that it has exactly one 3-block, which is G itself.

In the remaining case, G has a 2-separation (H, K). It therefore has at least four edges, and is loopless, by Theorem III.3. We refer to the two common vertices of H and K as the *hinges* of the 2-separation. Some basic properties of the 2-separations of G are recorded in the following theorems.

Theorem IV.20. *Let a 2-connected graph G be the union of two edge-disjoint subgraphs H and K with just two common vertices, b and c. Let H and K have each at least one edge. Then H and K are connected graphs.*

Proof. Suppose H is disconnected. Then either some component C of H includes neither b nor c, or some component D of H has at least one edge and includes one, but not both, of b and c. In the first case, C is a detached nonnull proper subgraph of G, contrary to Theorem I.21. In the second case, G is separable, by Theorem III.2. This being contrary to hypothesis, we conclude that H, and similarly K, is connected. □

Theorem IV.21. *Let (H, K) be a 2-separation, with hinges b and c, of a 2-connected graph G. Let graphs H_1 and K_1 be formed from H and K, respectively, by adjoining a new edge A with ends b and c. Then H_1 and K_1 are 2-connected.*

Proof. H_1 and K_1 are connected, by Theorems IV.20 and I.25. Suppose H_1 separable, having a 1-separation (L, M) with cut-vertex v. We may suppose A to be an edge of M. But then L is a nonnull proper subgraph of G with v as its only vertex of attachment. Accordingly, G is separable, contrary to hypothesis, by Theorem III.2. We conclude that H_1, and similarly K_1, is 2-connected. □

It is convenient to say that the graph H_1 of Theorem IV.21 is derived from G by replacing K by an *equivalent* edge A. "Equivalent" in this sense means that the ends of A are the hinges of (H, K).

Considering the graphs G that are 2-connected but not 3-connected, we classify them into three types with respect to a distinguished edge A. Type I consists of those graphs G for which the subgraph $G \cdot \{A\}$ has two or more bridges in G. Type II consists of those for which $G \cdot \{A\}$ has just one bridge B, and B is separable. Type III represents the remaining case, in which $G \cdot \{A\}$ has just one bridge B, and B is 2-connected.

In all these cases let us denote the ends of A by b and c. A bridge B of $G \cdot \{A\}$ in G must have b and c as its two vertices of attachment, by Theorem III.6, and it must be connected and so have have at least one edge, by Corollary I.52. If B has only one edge it cannot be the only bridge of $G \cdot \{A\}$ in G, by Theorem I.51. Hence such a bridge can occur only for Type I. The bridge B of Types II and III is identical with G'_A.

A 2-separation (H, K) of G is said to be *A-maximal* if A is an edge of H and there is no other 2-separation (H', K') of G such that A is in H' and K is a subgraph of K'. Now it is clear that no two distinct 2-separations (H, K) of G can have the same second member K, for K determines the edge-set of H and therefore it determines H, by connection. (See Theorem IV.20.) Hence the requirement "K is a subgraph of K'" in the above definition can be replaced by "K is a proper subgraph of K'". It follows that if (L, M) is any 2-separation of G with A in L, there exists an A-maximal 2-separation (H, K) of G such that M is a subgraph of K. We need the following theorem about A-maximal 2-separations.

Theorem IV.22. *Let A be an edge of a 2-connected graph G of Type III. Let (H_1, K_1) and (H_2, K_2) be distinct A-maximal 2-separations of G. Then K_1 and K_2 are edge-disjoint, and each of their common vertices is a common hinge of the two 2-separations. Moreover, the two 2-separations do not have both hinges in common.*

Proof. Consider the four intersections $H_1 \cap H_2$, $H_1 \cap K_2$, $H_2 \cap K_1$, and $K_1 \cap K_2$. No two of them have an edge in common, and the union of the four is G. The first one has at least the edge A. If the second is edgeless, then K_2 has all its edges in K_1. But then $K_2 \subseteq K_1$, since K_2 is connected, by Theorem IV.20. This is contrary to our hypothesis that the given 2-separations are A-maximal and distinct. We conclude that $H_1 \cap K_2$, and similarly $H_2 \cap K_1$, has at least one edge.

We need to prove that $K_1 \cap K_2$ is edgeless. Let us therefore assume that it has at least one edge, and study the consequences of this assumption.

Now each of the four intersections has at least two vertices of attachment in G, by Theorem III.2. Moreover, we can show that each of them has exactly two such vertices, for otherwise one of the four unions $H_i \cup K_j$ would have fewer than two vertices of attachment, by Theorem I.7, and so G would be separable, by Theorem III.2. We can now deduce from Theorem IV.20 that each of our four intersections is connected.

The intersection $H_1 \cap H_2$ must be the link-graph $G \cdot \{A\}$, for otherwise the intersection, being connected, would have at least two edges and the pair $(H_1 \cap H_2, K_1 \cup K_2)$ would be a 2-separation of G. (See Theorem I.7.) But this is not possible because of the A-maximality and distinctness of the given 2-separations.

Let us discuss the vertices of attachment of the four intersections. Any common vertex of two of them is a vertex of attachment of both, by their connection, and of course any vertex of attachment of one of the intersections is a vertex of two of them. Moreover, a common vertex of two of the intersections must be a common vertex either of H_1 and K_1 or of H_2 and K_2. It must therefore be a hinge of (H_1, K_1) or (H_2, K_2).

Let the hinges of (H_i, K_i) be denoted by x_i and y_i, $(i = 1, 2)$. We can now assert that each intersection includes exactly two distinct vertices of the collection $\{x_1, x_2, y_1, y_2\}$. Moreover, these two vertices are its vertices of attachment, and the only vertices at which it can meet another of the four intersections.

The vertices of attachment of $H_1 \cap H_2$, that is, $G \cdot \{A\}$, must be the two ends of A. We can adjust the notation so that one end of A is x_1. Then the other is not y_1, for if it were, then $G \cdot \{A\}$ would have at least two bridges, one a subgraph of K_1 and one including an edge of $H_1 \cap K_2$. Then G would be of Type I, contrary to hypothesis. By a further adjustment of notation, we can arrange that the second end of A is x_2. We have proved incidentally that x_2 is distinct from x_1 and y_1. Similarly, x_1 is distinct from y_2. The possibility that $y_1 = y_2$ is not yet ruled out.

Let us next consider the subgraph H_1 with its two vertices of attachment x_1 and y_1. The intersection $H_1 \cap H_2$ consists of A and its two ends x_1 and x_2. Moreover, x_2 is distinct from both x_1 and y_1. (See Fig. IV.3.1.) For later reference we infer that x_2 is not a vertex of K_1 and, similarly, x_1 is not a vertex of K_2. The intersection $H_1 \cap K_2$ includes y_1 since that vertex of H_1 is not in $H_1 \cap H_2$. It must also include x_2 since that vertex is not monovalent in G, by Theorem III.3. It follows that the two vertices of

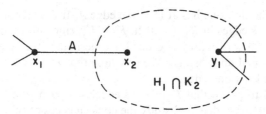

FIGURE IV.3.1 The subgraph H_1.

attachment of $H_1 \cap K_2$ in G are x_2 and y_1. Similarly, those of $H_2 \cap K_1$ are x_1 and y_2.

From the foregoing observations, neither x_1 nor x_2 can be a vertex of $K_1 \cap K_2$. The vertices of attachment of that intersection are therefore y_1 and y_2, which are now seen to be distinct. We deduce that G has the form shown in Fig. IV.3.2. The graph $B = G'_A$ is separable, having at least the two 1-separations: $(K_1, H_1 \cap K_2)$, and $(K_2, H_2 \cap K_1)$, with cut-vertices y_1 and y_2, respectively. This means that G is of Type II, contrary to hypothesis.

We can now assert that $K_1 \cap K_2$ is edgeless. By Theorem IV.20, this implies that $K_1 \subseteq H_2$ and $K_2 \subseteq H_1$. So any common vertex of K_1 and K_2 belongs also to H_1 and H_2. It is thus a common hinge of the two given 2-separations.

It only remains to prove that the two given 2-separations do not have both hinges in common. Assume that they do, the two hinges being x and y. If these are the two ends of A, then $G \cdot \{A\}$ has at least two bridges in G, one being contained in K_1 and the other in K_2. But then G is of Type I, contrary to hypothesis. In the remaining case, $H_1 \cap H_2$ has at least two edges, incident with an end of A not x or y. Accordingly, $(H_1 \cap H_2, K_1 \cup K_2)$ is a 2-separation of G, contrary to the A-maximality and distinctness of the given 2-separations. This completes the proof of the theorem. □

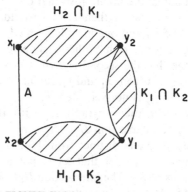

FIGURE IV.3.2

We return to the consideration of a graph G that is 2-connected but not 3-connected, and in which some edge A is distinguished. We call A the *intake* of G.

We propose to distinguish certain 2-separations (H_i, K_i), $1 \le i \le m$, of G as its *A-major* 2-separations. We require these to satisfy the following conditions.
i) A is in H_i for each i.
ii) If $i \ne j$, then K_i and K_j are edge-disjoint. Moreover, any common vertex of K_i and K_j is a common hinge of (H_i, K_i) and (H_j, K_j).

If these conditions are satisfied we denote the intersection of all the graphs H_i by J.

Theorem IV.23. *If Conditions (i) and (ii) are satisfied, then an edge of G belongs to J if and only if it belongs to none of the graphs K_i. Moreover, J includes the hinges of all the A-major 2-separations.*

Proof. The first part follows from Condition (ii) and the definition of J. By Condition (ii), H_i contains K_j for each $j \ne i$. Hence each H_i includes the hinges of all the A-major 2-separations. □

Theorem IV.24. *If Conditions (i) and (ii) are satisfied, then each edge or vertex of G that is not in J belongs to exactly one of the graphs K_i.*

Proof. For each such edge or vertex, there is an H_i to which it does not belong, and therefore a K_i to which it does belong. But it cannot belong to two of the K_i, by Condition (ii). □

With Conditions (i) and (ii) satisfied, it is convenient to introduce m new edges, called *virtual edges* of G. We enumerate them as $A_1, A_2, \ldots, A_m$ and postulate that the ends of A_i are the hinges of (H_i, K_i), for each suffix i. Adjoining these virtual edges to J we obtain a graph D called the *leading 3-block* of G, with respect to A. We refer to A as the *intake* of D, and to the virtual edges A_i as the *outlets* of D. The graph K_i is called the *outflow* from D at the outlet A_i. Adjoining A_i to K_i as an extra edge, we obtain a graph $F(A_i)$ which we call the *closed outflow* from D at A_i. The graph $F(A_i)$ is 2-connected, by Theorem IV.21. We adopt A_i as its intake.

We go on to complete the definition of A-major 2-separations for each of the three types.

For Type I we enumerate the bridges of $G \cdot \{A\}$ as $B_1, B_2, \ldots, B_k$, where $k \ge 2$. Each bridge has the two ends b and c of A as its vertices of attachment, by Theorem III.6. Each has at least one edge incident with b and at least one incident with c, by Theorem I.48, and each is connected, by Theorem IV.20. Thus a bridge B_i may be a link-graph, but otherwise it must have at least two edges.

It may happen that each B_i is a link-graph. Then G is a $(k+1)$-linkage, and we say that G is its own leading 3-block D. In this case, no A-major 2-separations are defined, and D has no outlets.

In the remaining case we may suppose the bridges that are not link-graphs to be $B_1, B_2, \ldots, B_s$, for some positive integer $s \le k$. Since each such B_i has at least two edges, it is the second member of a 2-separation (H_i, B_i) of G, with A in H_i. We adopt the 2-separations (H_i, B_i), $1 < i < s$, as the A-major ones, Conditions (i) and (ii) being then obviously satisfied. Each such pair (H_i, B_i) has a corresponding virtual edge A_i. For $j > s$ we write A_j for the actual edge of B_j. The leading 3-block D of G is a $(k+1)$-linkage with vertices b and c, and its edges other than A are $A_1, A_2, \ldots, A_k$. The first s of these are the outlets of D. The construction just described is illustrated in Fig. IV.3.3. Here virtual edges are shown by broken lines.

For Type II the single bridge $B = G'_A$ of $G \cdot \{A\}$ is separable. It follows by Theorem III.30 that B is a string of blocks, closed by A. We can enumerate the blocks of B as $B_1, B_2, \ldots, B_k$, where $k > 2$, and the cut-vertices of B as $v_1, v_2, \ldots, v_{k-1}$. By the definition of a string of blocks, we can suppose v_i to belong to B_i and B_{i+1}, but to no other block, for each i.

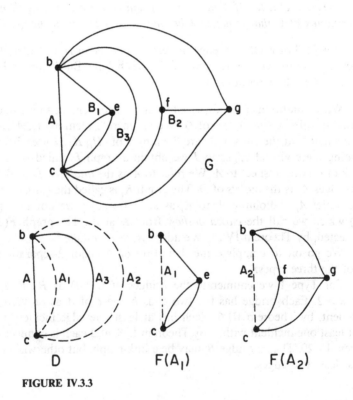

FIGURE IV.3.3

3-Blocks

We now introduce a $(k+1)$-circuit D. Its vertices are $b, c, v_1, v_2, \ldots, v_{k-1}$. One of its edges is A, with ends b and c. The other edges are enumerated as $A_1, A_2, \ldots, A_k$. The ends of A_1 are b and v_1, and the ends of A_k are v_{k-1} and c. The ends of any other A_i are v_{i-1} and v_i.

It may happen that each B_i is a link-graph. Then G can be identified with the $(k+1)$-circuit D. We say that G, that is, D, is its own leading 3-block. No A-major 2-separations are defined, and D has no outlets.

If the loopless 2-connected graph B_i is not a link-graph, it must have at least two edges. It is then the second member of a 2-separation (H_i, B_i) of G with A in H_i. We adopt the 2-separations of G of this form as the A-major ones, Conditions (i) and (ii) being obviously satisfied. Each such 2-separation (H_i, B_i) has its associated virtual edge A_i. But if B_i is a link-graph, we use the symbol A_i for its actual edge. We now take D to be the leading 3-block of G, with respect to A. The virtual edges A_i are its outlets. An example of the procedure just described is shown in Fig. IV.3.4. As before, virtual edges are indicated by broken lines.

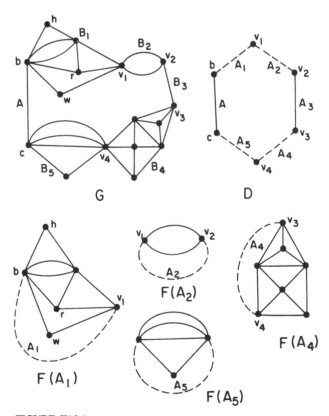

FIGURE IV.3.4

In Case III we take the A-major 2-separations to be the A-maximal ones as defined for Theorem IV.22. By that theorem, Conditions (i) and (ii) are satisfied. □

Theorem IV.25. *For Type III, the leading 3-block of G is a 3-connected graph with at least six edges.*

Proof. If there is only one A-maximal 2-separation the graph J includes two edges of G, and there is one virtual edge to be added in the formation of D. If there is more than one A-maximal 2-separation, then D includes A and at least two virtual edges. Moreover, G, not being 3-connected, has at least one 2-separation. We conclude that the leading 3-block D must have at least three edges.

Now D is 2-connected, by repeated application of Theorem IV.21. If D has only three edges, then, by the results of Sec. III.2, there are only two possibilities: D is a 3-circuit or a 3-linkage. But then it is clear that G is of Type I or Type II, contrary to hypothesis. We infer that D has at least four edges.

Suppose D is not 3-connected. Then it has a 2-separation (L, M) with A in L, and with hinges x and y, say. From L we form a subgraph L' of G by taking the union of $L \cap J$ and the subgraphs K_j corresponding to virtual edges A_j of D in L. Analogously, we form M' from M. Using Theorems IV.23 and IV.24, we find that L' and M' are edge-disjoint subgraphs of G whose union is G and which have each at least two edges. Moreover, their only common vertices are x and y. Hence (L', M') is a 2-separation of G with A in L'. But now there must be an A-maximal 2-separation (L'', M'') of G with $M' \subseteq M''$. But $M'' = K_j$ for some j. It follows that M can have only one edge, the corresponding A_j. This contradicts the definition of (L, M) as a 2-separation of D.

We can now assert that D is a 3-connected graph with at least four edges. It has at least six edges, by Theorem IV.4. □

Figure IV.3.5 shows the extraction of the leading 3-block of G for a graph of Type III.

A *complete* set of 3-blocks of G, with respect to A, is obtained by taking the leading 3-block of G, then the leading 3-blocks of the graphs $F(A_i)$, then the leading 3-blocks of their closed outflows, and so on. To justify the procedure we should make the following observation.

Theorem IV.26. *Let A be the intake of a 2-connected graph G that is not 3-connected. Let D be the leading 3-block of G with respect to A, and let $F(A_i)$ be the closed outflow of D at an outlet A_i. Then $F(A_i)$ has fewer edges than G.*

Proof. For all three types, D has at least two edges other than A_i. If one of these is a virtual edge A_j, we have a subgraph K_j of G with at least

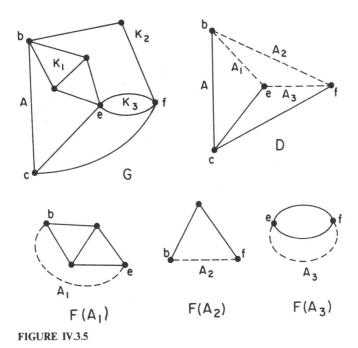

FIGURE IV.3.5

two edges, none of them in $F(A_i)$, by Theorem IV.24. So G always has at least two edges not in $F(A_i)$. On the other hand, all the edges of $F(A_i)$, except A_i, are in G. The theorem follows. □

We can describe the determination of the 3-blocks of G, with respect to A, more formally as follows: For each nonnegative integer n in turn, we make two lists L_n and M_n. Initially we have L_0 null and M_0 with the single member G. In general, L_n is to consist of recognized 3-blocks and M_n of 3-connected graphs, each member of M_n having at least three edges and having one of them distinguished as its intake. When the pair (L_s, M_s) has been found, we construct (L_{s+1}, M_{s+1}) as follows: L_{s+1} consists of the members of L_s and the leading 3-blocks of the members of M_s. M_{s+1} consists of the closed outflows of the leading 3-blocks of the members of M_s. By Theorem IV.26 the procedure must lead eventually to a null M_n. Thereafter, increases in the suffix leave the pair of lists unchanged. The 3-*blocks of* G, *with respect to* A, are the members of L_n at this final state.

At each state in the process we may have to introduce more virtual edges. These are always to be distinct new edges, that is, no two 3-blocks are to be given a common outlet. All these virtual edges can be referred to as *virtual edges of* G, *with respect to* A.

Of course, the chosen set of virtual edges of G can be replaced by another set by way of a one-to-one correspondence. When we speak of "the" 3-blocks or virtual edges of G, with respect to A, we admit the possibility of such a substitution.

We can exploit the recursive nature of our definition of 3-blocks with respect to A in the proof of theorems about them. For it implies that the 3-blocks of G with respect to A are the leading 3-block D and the 3-blocks of the closed outflows from D, each with respect to its own intake if it is not 3-connected. Since no two of these closed outflows have any edge in common, by Theorems IV.23 and IV.24, and since no virtual edge is used twice, it follows inductively that no 3-block is counted twice in the above classification.

We can now prove the following theorem by a further obvious induction.

Theorem IV.27. *Each edge of G belongs to exactly one 3-block of G with respect to A. Also, each virtual edge of G with respect to A occurs in exactly two 3-blocks, in one as an outlet and in the other as the intake.*

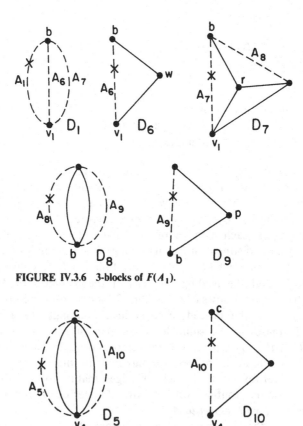

FIGURE IV.3.6 3-blocks of $F(A_1)$.

FIGURE IV.3.7 3-blocks of $F(A_5)$.

3-Blocks

Figure IV.3.6 shows the complete decomposition of the graph $F(A_1)$ of Fig. IV.3.4 into 3-blocks, with respect to A. Each intake is marked with a cross. Figure IV.3.7 shows the decomposition of $F(A_5)$ of Fig. IV.3.4.

In view of Theorem IV.27, we can regard the 3-blocks of G with respect to A as the vertices of a graph $\text{Blk}_3(G, A)$ whose edges are the virtual edges of G. In $\text{Blk}_3(G, A)$ the ends of an edge E are the two 3-blocks of G in which it appears, once as intake and once as outlet. If A is an edge of a 3-connected graph G, we define $\text{Blk}_3(G, A)$ to be a vertex-graph, with G as its vertex. This happens also when G is an n-circuit or n-linkage, with $n \geq 4$.

In the case of the graph G of Fig. IV.3.4, we can easily construct $\text{Blk}_3(G, A)$ by making use of Fig. IV.3.6 and IV.3.7. The result is shown in Fig. IV.3.8. The arrow on each edge is directed from the 3-block in which the edge is an outlet, and to the one in which it is the intake.

This diagram suggests the general theorem that $\text{Blk}_3(G, A)$ is a tree. In preparation for this theorem, we supplement the results on trees in Sec. I.6 with the following proposition.

Theorem IV.28. *Let a graph H be the union of two trees T and U whose intersection is a tree. Then H is a tree.*

Proof. H is connected, by Theorem I.25. Using the definition in Eq. (I.6.1), we can verify that

$$p_1(H) = p_1(T) + p_1(U) - p_1(T \cap U).$$

Hence H is a connected forest, that is, a tree, by Theorem I.35. □

Theorem IV.29. *Let A be an edge of a 2-connected graph G. Then $\text{Blk}_3(G, A)$ is a tree.*

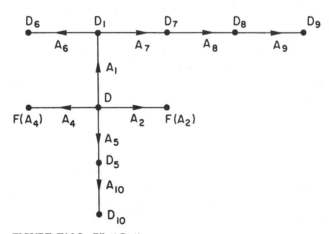

FIGURE IV.3.8 $\text{Blk}_3(G, A)$.

Proof. Let n_A be the number of 3-blocks of G with respect to A. If $n_A = 1$, the theorem is trivial, $\text{Blk}_3(G, A)$ being a vertex-graph.

We induce over n_A. Assume the theorem true for $n_A < q$, where q is an integer not less than 2, and consider the case $n_A = q$. Then the 3-block D of G has outlets $A_1, A_2, \ldots, A_s$, where $s \geq 1$. Let $F(A_i)$ denote the closed outflow at A_i. By our recursive definition of 3-blocks, and by Theorems IV.23 and IV.24, the graphs $\text{Blk}_3(F(A_i), A_i)$ occur as disjoint subgraphs of $\text{Blk}_3(G, A)$. That graph is formed from their union by adjoining the vertex D and the edges $A_1, A_2, \ldots, A_s$ joining D to the leading 3-blocks of the outflows $F(A_i)$. By the inductive hypothesis, each $\text{Blk}_3(F(A_i), A_i)$ is a tree. Hence $\text{Blk}_3(G, A)$ is a tree, by repeated application of Theorem IV.28.

The induction has succeeded, and the theorem is proved. □

We conclude this section with a supplement to Theorem IV.27.

Theorem IV.30. *Let A be an edge of a 2-connected graph G. Let x be a vertex of G that belongs to two distinct 3-blocks D_r and D_s with respect to A. Then x is an end in G of each edge of the unique arc in $\text{Blk}_3(G, A)$ joining D_r and D_s.* (See Theorem I.44.)

Proof. We use the notation of Theorem IV.29, and make an analogous induction. If $n_A = 1$, the theorem is vacuously true. If n_A is the "q" of this induction, we consider the following cases.

Case I: $D_r = D$, and D_s is a 3-block of some closed outflow $F(A_i)$.

By the definition of a leading 3-block, the only common vertices of D and $F(A_i)$ are the hinges of the corresponding A-major 2-separation, that is, the ends of A_i. (See Theorem IV.23.) So x is an end of A_i and a vertex of D and the leading 3-block D_i of $F(A_i)$, which are joined by A_i in $\text{Blk}_3(G, A)$. Now D_s may be D_i. But if not, then x is an end in G of each edge of the arc joining D_i and D_s in $\text{Blk}_3(F(A_i), A_i)$, by the inductive hypothesis. In either case the theorem is satisfied in $\text{Blk}_3(G, A)$.

Case II: D_r and D_s are 3-blocks of distinct closed outflows $F(A_i)$ and $F(A_j)$, respectively.

In this case, x is a common end in G of A_i and A_j, by Condition (ii). It is therefore a vertex of D. In $\text{Blk}_3(G, A)$ the arc joining D_r and D_s is clearly the union of the arcs joining D to D_r and D_s. So the theorem is satisfied, by the result of Case I.

Case III: The remaining case.

Now D_r and D_s are 3-blocks of the same closed outflow $F(A_i)$. The theorem is satisfied in $\text{Blk}_3(F(A_i), A_i)$ by the inductive hypothesis, and therefore it is satisfied in $\text{Blk}_3(G, A)$.

The theorem thus holds when $n_A = q$. By induction, it is true in general. □

We describe some more properties of $\text{Blk}_3(G, A)$ is the next section. The most important result perhaps is that $\text{Blk}_3(G, A)$ is independent of A.

IV.4. CLEAVAGES

The "cleavages" of this section can be defined either in terms of a given edge or in an absolute manner. They can therefore be used in a proof that $\text{Blk}_3(G, A)$ is independent of A. (See [4].)

Let (H, K) be a 2-separation of a 2-connected graph G, with hinges x and y. We say that H or K is *split* in (H, K), or by x and y, if it is the union of two edge-disjoint subgraphs of G, each with at least one edge, having x and y as their only common vertices. Let us define the *hinge-graph* X of (H, K) as the edgeless subgraph of G constituted by x and y. Then our two edge-disjoint subgraphs are unions of bridges of X in H or K.

We call (H, K) a *cleavage* of G if H and K are not both separable, and not both split in (H, K).

Theorem IV.31. *Let (H, K) be a 2-separation of a 2-connected graph G. Then if H or K is split in (H, K), it is 2-connected.*

Proof. Let H_1 be formed from H by adjoining a new edge A whose ends are the hinges of (H, K). Then H_1 is 2-connected, by Theorem IV.21. Suppose H to be split in (H, K). Then $(H_1)''_A$ is separable, with cut-vertex v_A, by Theorem II.3. Hence $(H_1)'_A$, that is, H, is 2-connected, by Theorem III.33. Since H and K can be interchanged in this argument, the theorem follows. □

We now come to some propositions about a given 2-separation (H, K) of a graph G that is 2-connected but not 3-connected. We suppose some edge A of G to be chosen as intake, and we write D for the leading 3-block of G with respect to A. We adjust the notation so that A is in H.

Theorem IV.32. *Suppose that no outflow from D contains K as a subgraph. Then D is a linkage or a circuit, and any outflow from D either is contained in K or is edge-disjoint from K.*

Proof. Let the hinges of (H, K) be x and y. Let the A-major 2-separations of G be enumerated as (H_i, K_i), $(1 \le i \le k)$. Let A_i be the outlet of D associated with the outflow K_i, and let $F(A_i)$ be the corresponding closed outflow.

Suppose first that G is of Type III with respect to A. There is an A-maximal, and therefore A-major, 2-separation (H', K') of G such that $K \subseteq K'$. But this is contrary to hypothesis. Hence G is of Type I or Type II, that is, D is a linkage or a circuit.

Let us say that K breaks K_i if it includes one edge but not all the edges of K_i. Then $K \cap K_i$ has at least two vertices of attachment in $F(A_i)$, by Theorem IV.21. These are vertices of attachment of K in G, and so can only be x and y.

Suppose G to be of Type I. We assume first that K breaks some K_i, so that x and y are vertices of K_i. Then by hypothesis, K includes an edge of some bridge B of $G \cdot \{A\}$ other than K_i. If B is an outflow K_j broken by K, then x and y must be the common vertices of K_i and K_j, the two ends of A. (See Fig. IV.3.3.) If B is not a broken K_j, there are two possibilities: It may still be a K_j, or it may be a link-graph contained in D. In either case, it is a subgraph of K, wherefore the ends of A are vertices of attachment of K in G, that is, they are again x and y. But if x and y are the ends of A, then K_i is split in (H_i, K_i), into the two nonnull subgraphs $K \cap K_i$ and $H \cap K_i$. This is inconsistent with the definitions of Type I. We conclude that, in fact, no outflow is broken by K; that is, that the theorem is satisfied.

In the remaining case, G is of Type II. Suppose first that K breaks some K_i. Then, by hypothesis, K has an edge in some block B of G'_A other than K_i. Such a B cannot be an outflow of D broken by K, for two distinct blocks of the string G'_A have at most one vertex in common. (See Fig. IV.3.4.) So B is either an unbroken outflow or a link-graph contained in D. In either case, B is contained in K. But if K contains one or more blocks of the string other than K_i, then K must have at least one vertex of attachment in G that is not in K_i, and more than two vertices of attachment in all. This being impossible, we deduce that, in fact, K breaks no outflow K_i. This completes the proof of the theorem. □

The next theorem is concerned with the same graphs but goes into further detail.

Theorem IV.33. *Let the graph K of Theorem IV.32 be the union of two edge-disjoint subgraphs L and M, each with exactly two vertices of attachment in G, and each with at least one edge. Then under the conditions of Theorem IV.32, each outflow from D is either contained in L, contained in M, or edge-disjoint from both L and M.*

Moreover, D is a linkage if L and M have the same two vertices of attachment, and a circuit if they do not.

Proof. Suppose some outflow K_i from D to have one edge in L but not all its edges in L. Then $L \subset K_i$. For if L has only one edge, this result is trivial, and in the remaining case it follows from Theorem IV.32, with L replacing K. But now K_i is contained in K, by Theorem IV.32. It follows that K_i has one edge, but not all its edges, in M. Hence, $M \subset K_i$, by the above reasoning, with M replacing L. We deduce that K is a subgraph of K_i, contrary to the hypothesis of Theorem IV.32.

We infer that each outflow K_i from D is either contained in L or is edge-disjoint from L. Similarly, it is contained in M or is edge-disjoint from M. The first part of the theorem follows.

Suppose L and M to have the same two vertices of attachment in G. Then they must be x and y, the vertices of attachment of K. Hence K is 2-connected, by Theorem IV.31. If G is of Type II, it follows that K is contained in some block of G'_A, that is, of some outflow from D, by Theorem IV.20. But this is contrary to the hypothesis of Theorem IV.32. Since G is not of Type III, by the proof of Theorem IV.32, we deduce that, in fact, G is of Type I, and hence that D is a linkage.

It remains to consider the case in which L and M do not have the same vertices of attachment. They do have one common vertex of attachment, z, say, since K is connected, by Theorem IV.20. It follows that K has the 1-separation (L, M), with cut-vertex z, by Theorem III.2. Since the vertices of attachment of K are x and y, and G is 2-connected, we can take the vertices of attachment of L to be x and z, and those of M to be y and z. (See Fig. IV.4.1.)

Let X be the hinge-graph of (H, K). Then K is a bridge of X in G, since K is not split in (H, K) by Theorem IV.31. Hence, by Theorem I.57, K is contained in some bridge of $G \cdot \{A\}$ in G. Accordingly, G is not of Type I, by the hypothesis of Theorem IV.32. So G must be of Type II, and D is a circuit. □

In connection with the last two theorems, we recall that, by the theory of Sec. IV.3, the leading 3-block D is a linkage if G is of Type I with respect to A, a circuit if G is of Type II, and a 3-connected graph with at least six edges if G is of Type III. In the last case, D is not a linkage or a circuit, by Theorems IV.1 and IV.3. In the first two cases, D has at least three edges.

We proceed to some theorems having to do with the structure of $\text{Blk}_3(G, A)$. The following definition is convenient. Let C be any 3-block of G with respect to A. Then the *overflow* of C is the union of C, with its intake and outlets deleted, and the outflows from C. Adjoining the intake of C to this overflow, we obtain the *closed overflow* of C. Thus the closed overflow of D is G. In any other case, the intake of C is an outlet of some other

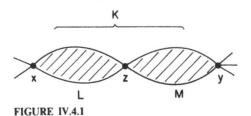

FIGURE IV.4.1

3-block C' of G with respect to A. Then the overflow of C is the outflow from C' at this outlet. In general, we denote the overflow of C by K_C, and the closed overflow by F_C.

Theorem IV.34. *Let C be any 3-block of G with respect to A. Let L be any subgraph of its overflow K_C. Then L has the same vertices of attachment in F_C as in G.*

Proof. If $C = D$, there is nothing to prove. In the remaining case, K_C is an outflow from another 3-block C' with respect to A. Any vertex of attachment of L in F_C is either a vertex of attachment of L in K_C or a vertex of attachment of K_C in $F_{C'}$. (See Theorem I.8.) But vertices of the latter kind, being incident in F_C with the intake of C, are still vertices of attachment of L in F_C. We conclude that L has the same vertices of attachment in $F_{C'}$ as in F_C. By repetition of this argument we obtain the required result. □

We return to the given 2-separation (H, K) of G.

Theorem IV.35. *Let K be a proper subgraph of the overflow K_C of some 3-block C of G with respect to A. Then K is the second member of a 2-separation (H', K) of the closed overflow F_C of C, with hinges x and y.*

Proof. By Theorem IV.34, K is a subgraph of F_C with x and y as its vertices of attachment. Its complementary subgraph H' in F_C has at least two edges, the intake of C being one. The theorem follows. □

We define an *A-carrier* of (H, K) as follows. It is a 3-block C of G with respect to A such that K is a proper subgraph of the overflow of C, and such that each outflow from C is either contained in K or is edge-disjoint from K.

Theorem IV.36. *The 2-separation (H, K) has exactly one A-carrier.*

Proof. We return to the lists L_n and M_n defined in Sec. IV.3, remarking that the members of the lists M_n are the closed overflows of the 3-blocks of G with respect to A. For a given n there can be at most one member of M_n having K as a subgraph, by repeated application of Theorem IV.24. On the other hand K is a proper subgraph of the single member G of M_0. Hence there is a greatest integer q such that K is a proper subgraph of some member F_q of M_q. For each integer i such that $0 \le i \le q$ there is a unique member F_i of M_i having K as a proper subgraph. Moreover, if $i < q$, then F_{i+1} must be a closed outflow from the leading 3-block C_i of F_i.

By the definition of an A-carrier, any A-carrier of (H, K) must be the leading 3-block C_i of some F_i, and by the preceding observation F_q is the only possibility. But F_q is indeed an A-carrier of (H, K) by Theorems IV.32 and IV.35, the former being applied to F_q. □

Cleavages 99

In the next two theorems we record some properties of A-carriers.

Theorem IV.37. *If H is split in (H, K), then the A-carrier of (H, K) is a linkage.*

Proof. Let X be the hinge-graph of (H, K). We can find two bridges L and M of X in H such that A is in M. It may happen that M has only the one edge A. Then the ends of A are x and y, G is of Type I, and D is a linkage. Moreover, each outflow from D is a bridge of X in H or K, by Theorem I.46. Accordingly, D is the A-carrier of (H, K), and the theorem holds.

In the remaining case, M has at least two edges and there is a 2-separation (H', K') of G such that $K' = K \cup L$. Its hinges are x and y. Let C be the A-carrier of (H', K').

It may happen that K' is an outflow from C at some outlet E. Then the ends of E are x and y. The leading 3-block C' of the closed outflow at E, like D in the above argument, must be a linkage and the A-carrier of K. In the remaining case, C is a linkage and the A-carrier of (H, K), by Theorem IV.33. □

Theorem IV.38. *If H is separable, then the A-carrier of (H, K) is a circuit.*

Proof. By Theorems IV.21 and III.30, H is a string of blocks that would be closed by an edge joining x and y. From this string we select one block L including x or y but not A, and the block M containing A. We adjust the notation so that L includes x. It will then not include y.

It may happen that M has only the one edge A, so that A is an isthmus of H, by Theorems I.45 and III.5. Then A has end-graphs H_x and H_y in H including x and y, respectively, since G is 2-connected. Of these, H_x is not a vertex-graph; it contains L, by Theorem III.8.

We now observe that G'_A is separable, having the 1-separation $(H_x, H_y \cup K)$, with cut-vertex x. If H_y is not a vertex-graph, then G'_A has also the 1-separation $(H_y, H_x \cup K)$, with cut-vertex y. (See Fig. IV.4.2.)

Now G is of Type II and D is a circuit. Moreover, each outflow from D is a block of H_x, H_y, or K, by Theorem III.22. Accordingly, D is the A-carrier of K.

In the remaining case, M has at least two edges and there is a 2-separation (H', K') of G such that $K' = K \subseteq L$. We note that K' has the 1-separation (K, L) with cut-vertex x. The hinges of (H', K') are y and the cut-vertex z of H that belongs to L. Let C be the A-carrier of (H', K').

It may happen that K' is an outflow from C at some outlet E. Then the leading 3-block C' of the closed outflow from E, like D in the preceding argument, must be a circuit and the A-carrier of (H, K). In the remaining case, C is a circuit and the A-carrier of (H, K), by Theorem IV.33. □

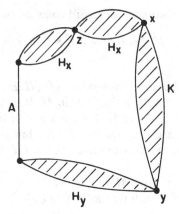

FIGURE IV.4.2

Theorem IV.39. (H, K) *is a cleavage of G if and only if K is an outflow from some 3-block C of G with respect to A.*

Proof. Suppose K is such an outflow. Then C is the A-carrier of (H, K). If H is split in (H, K), then C is a linkage, by Theorem IV.37. This means that F_C is of Type I with respect to its intake. Accordingly, K is not split by x and y. If H is separable, then C is a circuit, by Theorem IV.38. Then F_C is of Type II and its outflow K is 2-connected. We deduce that, in all cases, (H, K) is a cleavage of G.

Conversely, suppose (H, K) to be a cleavage of G. Assume that K is not an outflow from C, the A-carrier of (H, K). By Theorems IV.32 and IV.35, C must be a linkage or a circuit.

Suppose, first, that C is a linkage. Then K is the union of two or more bridges of $F_C \cdot \{E\}$, where E is the intake of C. But it is not the union of all such bridges, being a proper subgraph of K_C. The ends of E are vertices of attachment of K and are therefore x and y. It follows that each of H and K is split in (H, K), by Theorem IV.34, contrary to our supposition that (H, K) is a cleavage.

Suppose next that C is a circuit, with intake E. Then K is the union of two or more blocks of $(F_C)'_E$, but not of all such blocks. Accordingly, K is separable. The blocks belonging to K must be consecutive in the string of blocks $(F_C)'_E$, since K has only two vertices of attachment. We can therefore assert that $(F_C)'_E$ is the union of K with two other graphs K_x and K_y such that K and K_x have only the vertex x in common, such that K and K_y have only the vertex y in common, such that K_x and K_y are not both vertex-graphs, and such that E has one end b in K_x and one end c in K_y. (See Fig. IV.4.3.)

Now $(F_C)'_E$ appears as a subgraph of G with the same vertices of attachment b and c, by Theorem IV.34. It follows that H is separable,

Cleavages

having at least one of b and c as a cut-vertex. This is contrary to our supposition that (H, K) is a cleavage.

We conclude that if (H, K) is a cleavage of G, then K is an outflow from the A-carrier C of (H, K). The proof is now complete. □

We can express Theorem IV.39 by saying that there is a 1-1 correspondence between the cleavages of G and the virtual edges of G with respect to A. A virtual edge A_i is an outlet of exactly one 3-block C of G with respect to A. The outflow K from C at A_i is the second member of the cleavage (H, K) corresponding to A_i. If in what follows (H, K) is a cleavage of G, we shall refer to H and K as its *wings*. We shall write $F(K)$ for the closed outflow from the A-carrier associated with the outflow K.

The importance of the correspondence is that the definition of a cleavage of G does not depend on any intake. We are now entitled to say that, for every choice of intake, we get the same set of virtual edges, one for each cleavage. Let us go on to show that for every such choice we get the same 3-blocks and the same tree $\text{Blk}_3(G, A)$.

Theorem IV.40. *Let (H_1, K_1) and (H_2, K_2) be distinct cleavages of G. Then one wing of (H_1, K_1) is a proper subgraph of one wing of (H_2, K_2).*

Proof. In the notation of Sec. IV.3 we may suppose $F(K_1)$ to belong to M_r and $F(K_2)$ to M_s, for an arbitrarily chosen intake A, and we can adjust the notation so that $r \leq s$. Using Theorem IV.24, as in the proof of Theorem IV.36, we find that just one member of M_r, $F(K'_2)$ say, has K_2 as a subgraph. If $K'_2 = K_1$, then $r < s$ and K_2 is a proper subgraph of K_1 by Theorem IV.26.

In the remaining case $F(K_1)$ and $F(K'_2)$ are distinct members of M_r. Then, by Theorem IV.24, K'_2 is a subgraph of H_1. Moreover, K'_2 is a proper

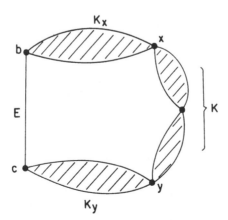

FIGURE IV.4.3

subgraph of H_1. For we must have $r > 0$, whence A belongs to H_1 but not to K_2'. Hence K_2 is a proper subgraph of H_1. □

Let us now consider three distinct cleavages (H_i, K_i) of G ($i = 1, 2, 3$). It may happen that each (H_i, K_i) has a wing W_i such that $W_1 \subset W_3 \subset W_2$. If so, we say that (H_3, K_3) *intervenes* between (H_1, K_1) and (H_2, K_2). We say that two distinct cleavages are *adjacent* if no third cleavage intervenes between them. We note that this definition of adjacency does not depend on any intake. Nevertheless, adjacency can be characterized in terms of the 3-blocks with respect to a particular intake A.

Theorem IV.41. *Let (H_1, K_1) and (H_2, K_2) be distinct cleavages of G. Let A be any edge of G. Then the two cleavages are adjacent if and only if one of the two following conditions is satisfied. (We take A to be in H_1 and H_2.)*
i) *K_1 and K_2 are distinct outflows from some 3-block C of G with respect to A.*
ii) *One of K_1 and K_2 is the overflow of some 3-block C of G with respect to A, and the other is an outflow from C.*

Proof. We use the notation of Theorem IV.40, observing that K_2' is one wing of a cleavage (H_2', K_2') of G, by Theorem IV.39.

Suppose first that $K_2' = K_1$. If $s > r + 1$, then some cleavage (H'', K''), where $F(K'') \in M_{r+1}$, intervenes between (H_1, K_1) and (H_2, K_2), and neither (i) nor (ii) can be satisfied. We assume therefore that $s = r + 1$, in which case Condition (ii) holds. If now some cleavage (H'', K'') intervenes between (H_1, K_1) and (H_2, K_2), and if we take A to be in H'', then we can have only

$$K_2 \subset K'' \subset K_1 \quad \text{or} \quad K_1 \subset K'' \subset K_2.$$

But we must assign $F(K'')$ to some M_t, and these inclusions imply $r + 1 > t > r$. We infer that the two given cleavages are in fact adjacent.

In the remaining case, K_1 and K_2' are distinct. If K_2' is not K_2, then neither (i) nor (ii) can hold. On the other hand, (H_2', K_2') intervenes between (H_1, K_1) and (H_2, K_2): $K_2 \subset K_2' \subset H_1$. We therefore assume that $K_2' = K_2$.

If now K_1 and K_2 are outflows from distinct 3-blocks of G with respect to A, they are contained in distinct members $F(K_1'')$ and $F(K_2'')$, respectively, of M_{r-1}. Then neither (i) nor (ii) holds. On the other hand there is a cleavage (H_1'', K_1'') that intervenes between (H_1, K_1) and (H_2, K_2): $K_1 \subset K_1'' \subset H_2$.

There remains only the case in which K_1 and K_2 are outflows from the same 3-block C with respect to A. They are then contained in the same member $F(K)$ of M_{r-1}, K being the overflow of C. Thus Condition (i) applies. If some cleavage (H'', K'') intervenes between (H_1, K_1) and (H_2, K_2), it must have a wing W''' such that $K_1 \subset W''' \subset H_2$ or $K_2 \subset W''' \subset H_1$.

We can replace the subgraphs involved in these formulae by their complementary subgraphs, provided that we reverse the inclusions. We can therefore adjust the notation so that $K_1 \subset K'' \subset H_2$ and A is in H''. We must assign K'' to some M_t, and by the inclusions, $t < r$. But the member of M_t that contains K_1 contains also the overflow of C, and this contains K_2. This contradicts our conclusion that $W''' \subset H_2$. So in this case, the original two cleavages must be adjacent. The proof is now complete. □

If v is a vertex of a graph N, let us define the corresponding *vertex-bond* as the set of all links of N incident with v. In a tree it is clear that no two distinct vertices can have the same vertex-bond unless they are both monovalent and the tree reduces to a link-graph. Let us consider the vertex-bonds of the tree $\mathrm{Blk}_3(G, A)$.

Theorem IV.42. *The vertex-bonds of* $\mathrm{Blk}_3(G, A)$ *are independent of the choice of* A.

Proof. Theorem IV.41 tells us that the edges of a vertex-bond S of $\mathrm{Blk}_3(G, A)$ must correspond to mutually adjacent cleavages of G. But consider a particular virtual edge A_i of G, corresponding to a cleavage (H_i, K_i). It occurs in two 3-blocks C_1 and C_2 of G with respect to A, in C_1 as an outlet and in C_2 as intake. In $\mathrm{Blk}_3(G, A)$ the two ends of A_i are C_1 and C_2. By Theorem IV.41 the cleavages adjacent to (H_i, K_i) correspond to the members, other than A_i, of the vertex-bonds of C_1 and C_2 in $\mathrm{Blk}_3(G, A)$. Moreover, two of them are adjacent if and only if they correspond to members of the same one of these two vertex-bonds.

These observations imply that we can determine the vertex-bonds of $\mathrm{Blk}_3(G, A)$ containing the edge A_i without reference to A. We need only the cleavages of G and their relations of adjacency. The theorem follows. □

Theorem IV.43. *Let S be a vertex-bond of* $\mathrm{Blk}_3(G, A)$ *corresponding to a vertex C of that graph. Let its members be enumerated as $A_1, A_2, \ldots, A_k$. Then the cleavage of G corresponding to A_i can be written as (U_i, W_i), for each i, so that no two of the wings W_i have any edge or vertex in common, apart from common hinges of their cleavages. Moreover, if $k \geq 2$, the wings W_i are uniquely determined by this rule.*

If $k \geq 2$, then C is formed from the intersection of the wings W_i by adjoining the members of S. If $k = 1$, it is formed from U_1 or W_1 by adjoining the member of S.

Proof. Let us write the cleavage corresponding to A_i also as (H_i, K_i), with A in H_i. We now write $W_i = H_i$ or K_i according as A_i is the intake or an outlet of C. Referring to the definition, in Sec. IV.3, of a 3-block with respect to A, we see that the wings W_i now satisfy the required rule. Moreover, C is as specified in the enunciation.

Suppose $k \geq 2$. Let two distinct members A_i and A_j of S correspond to cleavages (H_i, K_i) and (H_j, K_j), respectively, where we no longer assume A to be in H_i and H_j. By Theorem IV.40 we may suppose K_i to be a proper subgraph of K_j. Then H_i must have an edge in each of H_j and K_j. We are therefore forced to put $W_i = K_i$. The proof is now complete. □

Theorem IV.44. *Let S be a vertex-bond of* $\text{Blk}_3(G, A)$, *and let this tree be not a link-graph. Then the corresponding 3-block C of G with respect to A is the same for all choices of A, apart from the location of its intake.*

Proof. If $|S| \geq 2$, this follows from Theorem IV.43. Suppose therefore that $|S| = 1$. In the notation of Theorem IV.43, C is either U_1 or W_1. By hypothesis, G has a second cleavage, and one wing of this is a proper subgraph of either U_1 or W_1. Irrespective of the choice of A, C must be the other member of the pair (U_1, W_1). □

The excluded case in which $\text{Blk}_3(G, A)$ is a link-graph is trivial. There is only one cleavage (U, W) and the two 3-blocks are obtained from U and W by adjoining the virtual edge, a construction independent of the choice of A.

In view of this observation we can express Theorem IV.44 by saying that G has the same 3-blocks with respect to every edge, provided that we ignore the distinction between intake and outlets for each 3-block. So we may now speak simply of the 3-blocks of G, not of the 3-blocks with respect to some specified intake A. We likewise write $\text{Blk}_3(G, A)$ simply as $\text{Blk}_3(G)$, calling it the *tree of 3-blocks of G*. In this section we have supposed G to have a 2-separation. But if G is 3-connected, we take its only 3-block to be G itself, so that $\text{Blk}_3(G)$ is a vertex-graph.

We conclude this section with a theorem on the structure of $\text{Blk}_3(G)$.

Theorem IV.45. *Let C and C' be 3-blocks of G that are adjacent vertices of* $\text{Blk}_3(G)$. *Then they are not both linkages and not both circuits.*

Proof. With respect to some intake A, we may suppose C' to be the leading 3-block of a closed outflow from C. This closed outflow will be the closed overflow $F_{C'}$ of C'. Let E and E' be the intakes of C and C', respectively.

Suppose C to be a linkage. Then F_C is of Type I with respect to E, E and E' have the same ends, and the outflow $K_{C'}$ is not split by these ends. So $F_{C'}$ is not of Type I with respect to E', and C' is not a linkage.

Suppose next that C is a circuit. Then F_C is of Type II with respect to E, and the outflow $K_{C'}$ is 2-connected. Hence $F_{C'}$ is not of Type II with respect to E', and C' is not a circuit. □

IV.5. DELETIONS AND CONTRACTIONS OF EDGES

In this section, G is a 3-connected graph having at least six edges. (See Theorem IV.4.) A set $\{A, B, C\}$ of three edges of G is a *triangle* if it is the edge-set of a 3-circuit of G, and a *triad* if it is the vertex-bond of a trivalent vertex. By Theorems IV.1 and IV.3, we have the following theorem:

Theorem IV.46. *If A is a member of a triangle of G, then G_A'' is not 3-connected. If A is a member of a triad of G, then G_A' is not 3-connected.*

Let A be an edge of G with ends b and c. We proceed to a detailed study of the case in which G_A' is not 3-connected. We note first that G_A' is still 2-connected, by Theorem IV.7. We can decompose it into its 3-blocks by the method explained in Sec. IV.3.

Consider an *extremal* 3-block of G_A', that is, one corresponding to a monovalent vertex of $\mathrm{Blk}_3(G_A')$. It has only one virtual edge E and it must have at least two edges in G. Let J be the subgraph of G formed from it by deleting E. Since J is not to be a member of a 2-separation of G, we must suppose that one end of A is in J and is distinct from each end of E.

Now $\mathrm{Blk}_3(G_A')$ has at least two monovalent vertices, by Theorem I.40. It must therefore have exactly two, one for each end of A. It is thus an arc, again by Theorem I.40. We express these results by saying that G_A' is a *string of 3-blocks, closed* by A.

By the definition of an arc, the 3-blocks of G_A' can be enumerated as $C_1, C_2, \ldots, C_{n+1}$, and the virtual edges as $A_1, A_2, \ldots, A_n$, where $n \geq 1$, so that A_i belongs only to C_i and C_{i+1}. We can adjust the notation so that b belongs to C_1 and c to C_{n+1}. We denote the ends of A_i by $a(i,1)$ and $a(i,2)$. It is sometimes convenient to write also

$$a(0,1) = a(0,2) = b$$

and

$$a(n+1,1) = a(n+1,2) = c.$$

We write J_i for the subgraph of G obtained from C_i by deleting its virtual edge or edges. We call J_i a *segment* of G_A', the ith segment counting from b and the $(n+2-i)$th counting from c.

Figure IV.5.1 gives an example in which C_1 and C_3 are circuits and C_4 is a 3-linkage.

In the terminology of Sec. II.5, the ends of A_i separate b and c. Hence there are at most two internally disjoint arcs in G_A' with ends b and c,

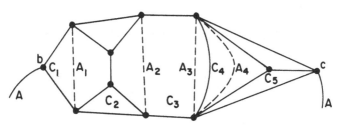

FIGURE IV.5.1

by Theorem II.35. On the other hand, it follows from Theorems II.35 and III.7 that at least one pair of such internally disjoint arcs exists.

Theorem IV.47. *If G'_A is not 3-connected, then the maximum number of internally disjoint arcs in G'_A joining b and c is 2.*

Perhaps we should make it explicit that no edge of G'_A joins b and c, by Theorem IV.1.

Consider two internally disjoint arcs F_1 and F_2 in G'_A joining b and c. Their union is a circuit F. We call F an *A-frame* of G and say that F_1 and F_2 are its *sides*. It is clear that each virtual edge A_i has one end in F_1 and one in F_2. Having chosen an A-frame F, we adjust the notation so that $a(i, j)$ is always in F_j ($j = 1, 2$).

A *cross-join* of the frame F is an arc in G'_A avoiding F, with one end in F_1 but not F_2, and with the other end in F_2 but not F_1. The A-frame F is *doubly crossed* if it has two disjoint cross-joins.

A virtual edge A_i corresponds to a cleavage (U_i, W_i) of G'_A, where U_i is the union of the segments J_k with $k \leq i$ and W_i is the union of the segments with $k > i$. Of the two arms of F_j at $a(i, j)$, one is an arc in U_i with ends b and $a(i, j)$, and the other is an arc in W_i with ends $a(i, j)$ and c. (See Sec. III.4.) Neither of them can include the other end of A_i.

Suppose $1 \leq i < n$. It may happen that $a(i, j) = a(i + 1, j)$ for some j. Then the arm of F_j at $a(i, j)$ is contained in W_{i+1}. Accordingly, the intersection of J_i and F_j is a vertex-graph, with vertex $a(i, j)$. We then say that J_i is *pinched* on F_j. In Fig. IV.5.1 we can take F to be the outer boundary of G'_A, F_1 being above and F_2 below. Then C_4 is pinched on both F_1 and F_2. No such effect can occur for either extremal 3-block C_1 and C_{n+1}, since neither b nor c is incident with any virtual edge of G'_A. Suppose, however, that $a(i, j)$ and $a(i + 1, j)$ are distinct. Then the intersection of J_i with F_j is an arc $L(i, j)$, with ends $a(i, j)$ and $a(i + 1, j)$. We can find it by taking the arm L of F_j at $a(i, j)$ that is in W_i, and then taking the arm of L at $a(i + 1, j)$ that is not in W_{i+1}. This latter rule extends to the cases $i = 0$ and $i = n$: The intersection of J_1 and F_j is the arm of F_j at $a(1, j)$ including b, and the intersection of J_{n+1} and F_j is the arm of F_j at $a(n, j)$ including c. In every case, we denote the intersection of J_i and F_j by $F(i, j)$ and call it the *border* of J_i in F_j. We note that F_j is the union in it of the borders of the $(n + 1)$ segments of G'_A, by Theorem I.26.

Theorem IV.48. *If C_1 is a circuit, it is a 3-circuit, and its two edges incident with b form with A a triad of G.*

Proof. If C_1 has a vertex other than b, $a(1, 1)$, and $a(1, 2)$, then that vertex is divalent in G, contrary to Theorem IV.3. Since b is not incident with A_1, the theorem holds. □

Deletions and Contractions of Edges

We associate Theorem IV.48 with the following triviality.

Theorem IV.49. *If F_j has just two edges, these form with A a triangle of G.*

We have seen that F_j must have an edge in J_1 and an edge in J_{n+1}. If it has only two edges, then all the other segments are pinched on F_j. Their borders in F_j coincide, the vertex of each being the single internal vertex of F_j.

Theorem IV.50. *Let there be a 2-arc L in G'_A with ends b and c. Then there is an A-frame of G having L as one of its sides.*

Proof. Consider any A-frame F of G, with sides F_1 and F_2. We can adjust the notation so that F_1 does not include the internal vertex of L. Then L and F_1 are internally disjoint; they constitute the required A-frame. □

Theorem IV.51. *If C_i is not a circuit, then J_i contains a cross-join of any A-frame F of G.*

Proof. Let H_i be the union of the borders of J_i in F_1 and F_2. It is not the whole of J_i, for otherwise C_i would be a circuit. Hence there is at least one bridge of H_i in J_i.

Some bridge B of H_i in J_i has one vertex of attachment p in $F(i,1)$ but not $F(i,2)$, and one vertex of attachment q in $F(i,2)$ but not $F(i,1)$, for otherwise the union of one of these borders with the bridges having all their vertices of attachment in it would be a member of a 1-separation or 2-separation of G. By Theorem I.56 the required cross-join exists in B, with ends p and q. □

Theorem IV.52. *Let G'_A be not 3-connected. Let F be an A-frame of G with sides F_1 and F_2, and let F_2 be a 2-arc with edges X and Y. Then either A belongs to a triad of G or G'_X is 3-connected.*

Proof. Assume that A is in no triad of G. Then neither C_1 nor C_{n+1} is a circuit, by Theorem IV.48, with the enumerations reversed in the case of C_{n+1}. Let x denote the internal vertex of F_2. We can adjust the notation, reversing enumerations if necessary so that X is incident with b. (See Fig. IV.5.2.)

Since C_1 is not a circuit or a linkage it is, by the theory of Sec. IV.3, a 3-connected graph with at least six edges. So $(C_1)'_X$, like G'_A, is 2-connected. By Theorems II.35 and III.7, there are two internally disjoint arcs N_1 and N'_2 in it, with ends b and x. But b and x are joined also by a 2-arc N_3 in G with edges Y and A. If neither N_1 nor N'_2 includes A_i, then N_1, N'_2, and N_3 are three internally disjoint arcs in G'_X with ends b and c. Accordingly, G'_X is 3-connected, by Theorem IV.47.

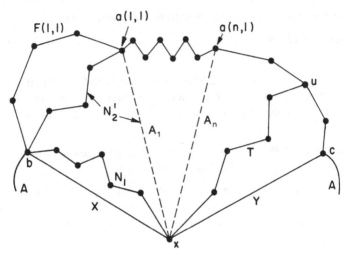

FIGURE IV.5.2

In the remaining case, we may suppose A_1 to be in N_2', as indicated in Fig. IV.5.2. There is a cross-join T of F in J_{n+1}, by Theorem IV.51. One end of T must be x. The other is a vertex u of $F(n+1,1)$, necessarily distinct from c. If u is $a(1,1)$, let S be the corresponding vertex-graph. Otherwise let L be the arm of F_1 at $a(1,1)$ including c, and let S be the arm of L at u that includes $a(1,1)$ (and so not c). In either case the union of S, T, and the graph obtained from N_2' by deleting A_1 is an arc N_2 in G_X' joining b and x. The three arcs N_1, N_2, and N_3 in G_X' are internally disjoint, and so G_X' is 3-connected, by Theorem IV.47. □

We must now pay some attention to the case in which G_A'', but not necessarily G_A', has a 2-separation.

Theorem IV.53. *Let A be an edge of G and let G_A'' be not 3-connected. Then G_A' is the union of two edge-disjoint connected subgraphs H and K with exactly three vertices in common, two of these being the ends b and c of A in G.*

Proof. G_A'' has a 2-separation (P, Q), by Theorem IV.8. Hence G is the union of the two subgraphs $f_A P$ and $f_A Q$, whose only common edge is A. Their common vertices correspond to the hinges of (P, Q). Hence there are at most three common vertices, and if three, then two of them are the ends b and c of A.

Write $H = (f_A P)_A'$ and $K = (f_A Q)_A'$. Then G_A' is the union of its two edge-disjoint subgraphs H and K, each with at least two edges. By the 3-connection of G, any component of H or K with at least one edge must have at least two vertices of attachment in G, and any such component with at least two edges must have at least three vertices of attachment in G. It follows from these observations that H and K are connected and that they have just three common vertices b, c, and d, where b and c are the ends of A in G. □

Theorem IV.54. *Suppose that neither G'_A nor G''_A is 3-connected. Suppose further than A belongs to no triangle of G. Let H and K be as in Theorem IV.53, with common vertices b, c, and d. Then no A-frame of G is contained in H or K. Moreover, there is an A-frame F of G, not including d, which has one side F_1 in H and the other side F_2 in K. This A-frame is not doubly crossed.*

Proof. By Theorem I.43, there is an arc in H with ends b and c. Suppose that all such arcs include d. Then d is the cut-vertex of a 1-separation (H_1, H_2) of H, since the deletion from H of d and its incident edges would put b and c in different components of the resulting graph. But then H_1 and H_2 have each only two vertices of attachment in G, either b and d or c and d. Hence each is a link-graph by the connection of H and the 3-connection of G. But then the two edges of H form with A a triangle of G, which is contrary to hypothesis.

We deduce that there is an arc F_1 from b to c in H that does not include d. Similarly, there is an arc F_2 in K from b to c that does not include d. The union of these arcs is the required A-frame F. By applying Theorem II.34 to the sets of edges of F_1 and F_2, we find that F is not doubly crossed. (We then have $\lambda(G'_A; P, Q) \le 3$ by virtue of the cutting pair (H, K)).

To complete the proof, it is sufficient to show that no A-frame of G is contained in H. But the two sides of such a frame, together with the above arc F_2, would make three internally disjoint arcs from b to c in G'_A, which is contrary to Theorem IV.47. □

Let us say that an edge A of a 3-connected graph G is *essential* if neither G'_A nor G''_A is 3-connected.

Theorem IV.55. *If the edge A is essential, it belongs either to a triangle or a triad of G.*

Proof. Assume that A belongs to no triangle of G. Let H, K, and F be as in Theorems IV.53 and IV.54. Then d is incident with no virtual edge of G'_A. Hence d belongs to exactly one of the segments J_i by the construction of Sec. IV.3 (or by Theorem IV.30). We can adjust the notation, reversing enumerations if necessary, so that this segment is not J_1. But then $J_1 \cap H$ and $J_1 \cap K$ have only the vertex b in common, and therefore J_1 can contain no cross-join of F. Accordingly, C_1 is a circuit, by Theorem IV.51, and A belongs to a triad of G, by Theorem IV.48. □

Theorem IV.56. *Let A be essential. Let b be trivalent in G, and incident with edges X and Y in G'_A. Then either A belongs to some triangle of G or G''_X is 3-connected.*

Proof. Let H, K, and F be as in Theorems IV.53 and IV.54. We may suppose X and Y to be in F_1 and F_2, respectively. Let the ends other than b of X and Y be x and y, respectively. Let us assume that A belongs to no triangle of G, and that G''_X is not 3-connected.

Deleting from F_1 and F_2 the end b and the edges X and Y, we obtain arcs L_1 and L_2, respectively, the first from x to c and the second from y to c. Deleting b, X, and Y from G'_A, we obtain a graph G_b containing L_1 and L_2. At least one of the segments J_i contains a cross-join of F, necessarily in G_b, by Theorems IV.45 and IV.51. On the other hand, F does not have two disjoint cross-joins, by Theorem IV.54. Applying Theorem II.34 to the edge-sets of L_1 and L_2 in G_b, we find that G_b must be the union of two edge-disjoint subgraphs H' and K' such that $L_1 \subseteq H'$, such that $L_2 \subseteq K'$, and such that H' and K' have just one common vertex t other than c. (See Fig. IV.5.3.)

Now consider G'_X. By Theorem IV.3, it is not 3-connected. By Theorem IV.54, G has an X-frame V that is not doubly crossed. Let its sides be V_1 and V_2, with A in V_1 and Y in V_2. The arm R of V_1 at c that includes x must be a subgraph of H', for H' has only two vertices of attachment t and c in G'_X, and t must be in V_2. Similarly, the arm S of V_2 at t that includes x must be a subgraph of H'.

By an argument like that in Theorem IV.51, we now show that V has a cross-join T_1 in H' that does not include c. Suppose first that $R \cup S = H'$. Then R is a 1-arc, since otherwise G would have a divalent vertex, contrary to Theorem IV.3. This contradicts our assumption that A is in no triangle of G. In the remaining case, there is at least one bridge of $R \cup S$ in H'. In

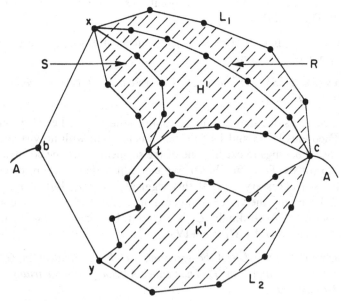

FIGURE IV.5.3

order that G may have no 2-separation, there must be a bridge B of $R \cup S$ in H' with not all its vertices of attachment in R and not all of them in S. If one of the vertices of attachment of B is an internal vertex of R, then the required cross-join T_1 exists, by Theorem I.56. Suppose, however, that no such bridge B has any vertex of attachment in R other than x and c. Then, since R cannot be a 1-arc, the union of R and those bridges of $R \cup S$ in H' that have all their vertices of attachment in R must be one member of a 2-separation of G with hinges x and c. But this is impossible.

There exists an arc in K' that is disjoint from T_1, which has one end in V_1 but not in V_2, and which has its other end in V_2 but not in V_1, for L_2 is an example. A shortest arc with these properties is a cross-join T_2 of V, for if it had an internal vertex u in V we could replace it by one of its arms at u, for u could not be one of the common vertices b and x of V_1 and V_2. But we have now established that V is doubly crossed, which is a contradiction. The theorem follows. □

In the following section we apply the results of the present one to the case in which every edge of G is essential.

IV.6. THE WHEEL THEOREM

Let G be any 3-connected graph with at least six edges.

Theorem IV.57. *In order that every edge of G shall be essential, it is necessary and sufficient that each edge of G shall belong to both a triangle and a triad of G.*

Proof. Sufficiency follows from Theorem IV.46. In proving necessity, we assume that each edge is essential. Then each edge belongs either to a triangle or a triad, by Theorem IV.55, each edge belonging to a triangle belongs also to a triad, by Theorem IV.52, and each edge belonging to a triad belongs also to a triangle, by Theorem IV.56. □

Theorem IV.58 (The Wheel Theorem). *In order that every edge of G shall be essential, it is necessary and sufficient that G shall be a wheel, of order at least* 3.

Proof. Sufficiency follows from Theorem IV.57. In proving necessity, we assume that each edge is essential, and apply Theorem IV.57. We first note that G has a 3-circuit C and that each edge of C is incident with a trivalent vertex. We denote the vertices of C by h, a_1, and a_2, so that a_1 and a_2 are trivalent in G. We write S_1 for the edge joining h and a_1, S_2 for the edge joining h and a_2, and A_1 for the edge joining a_1 and a_2. By its trivalency, a_2 is incident with a third edge A_2. Let the other end of A_2 be a_3. By strictness, a_3 is distinct from a_1, a_2, and h. Since A_2 belongs to a triangle, there is an edge S_3 joining a_3 either to h or to a_1. (See Fig. IV.6.1.)

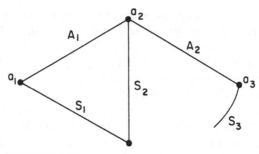

FIGURE IV.6.1

Let H be the reduction of G with edges A_1, A_2, S_1, S_2, and S_3. If S_3 is incident with a_1, then H has only the two vertices of attachment a_3 and h. The complementary subgraph H^c must be a link-graph, by the 3-connection of G. But then G is a 4-clique, that is, a wheel of order 3. We may assume from now on that S_3 joins a_3 and h. If h is trivalent in G, the same sort of argument applies, the two vertices of attachment of H being now a_1 and a_3. So we may assume from now on that $\mathrm{val}(G, h) \geq 4$.

We may now assert the existence in G of an arc L and a vertex h not belonging to L, such that the following conditions hold:
i) L has at least two edges.
ii) Each internal vertex of L is trivalent in G.
iii) Each vertex of L is adjacent to h in G.
iv) $\mathrm{val}(G, h) \geq 4$.

We have constructed such an L and h. Our arc L is a 2-arc, with edges A_1 and A_2. But it may be possible to satisfy the conditions with a longer arc L. Let us now choose L and h, satisfying the four conditions, so that L has as many edges as possible.

We write the defining sequence of vertices of L as $(a_1, a_2, \ldots, a_n)$, and that of edges as $(A_1, A_2, \ldots, A_{n-1})$, with $n \geq 3$. Let S_j be the edge of G joining h and a_j ($1 \leq j \leq n$). (See Fig. IV.6.2.)

Since S_n is in a triad, the vertex a_n must be trivalent. It is therefore joined by a new edge A_n to some vertex a_{n+1} of G that is not h or a_{n-1}. Suppose a_{n+1} is not in L. Then since A_n is in a triangle of G, there is an edge S_{n+1} of G joining a_{n+1} to a_{n-1} or h. But S_{n+1} is not incident with a_{n-1}, by the trivalency of that vertex. Hence S_{n+1} joins a_{n+1} to h. This means that we can replace L by a longer arc, adjoining A_n and a_{n+1}, and still satisfy the four conditions. This being contrary to the choice of L and h, we infer that a_{n+1} belongs to L. But A_n cannot be incident with an internal vertex of L, by Condition (ii). We must therefore identify a_{n+1} with a_1. We observe further that since S_1 belongs to a triad, the vertex a_1 is trivalent in G, that is, incident only with S_1, A_1, and A_n.

The edges A_j and S_j, $1 \leq j \leq n$, with their incident vertices, now constitute a wheel of order n, a subgraph of G that can have no vertex of

Notes 113

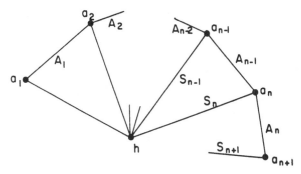

FIGURE IV.6.2

attachment other than h. Since G is 3-connected, this wheel must be the whole of G. □

In connection with the Wheel Theorem, we recall that all wheels of order 3 or more are 3-connected (Theorem IV.17).

By the Wheel Theorem, any 3-connected graph of six or more edges that is not a wheel can be obtained from a 3-connected graph of one edge fewer, either by link-adjunction or by vertex-splitting, as these operations are defined in Sec. IV.2. (See Theorems IV.11 and IV.14.) This result could be used in the construction of a catalogue of 3-connected graphs. Indeed it has been so used, quite extensively, in the case of the planar ones [2].

To start such a catalogue, we would observe that there is no 3-connected graph of five edges, by Theorem IV.4. Hence the only one with six edges is the wheel of order 3, commonly called the 4-clique. Since this does not admit the kind of link-adjunction or vertex-splitting (without duplication of edges) required by Theorems IV.11 and IV.14, we deduce that there is no 3-connected graph of seven edges, and that the only one of eight edges is the wheel of order 4.

We find three 3-connected graphs of nine edges. One is the graph of the trigonal bipyramid, obtained from the wheel of order 4 by link-adjunction. The other two are obtained by splitting the hub of W_4. They are the Thomsen graph and the graph of the triangular prism. In extending this catalogue, the main difficulty is that of recognizing isomorphic duplicates.

IV.7. NOTES

IV.7.1. 3-Connection and Planarity

In the theory of planar graphs, the connectivity $\kappa(G)$ is of interest as an invariant under duality (Theorem XI.1). The six nonnull graphs of infinite connectivity are all planar. Using Theorem XI.4, we can show that every other planar graph has connectivity not exceeding 3. Steinitz' Theorem asserts that the planar graphs of connectivity 3 are, to within isomorphism, the graphs of the convex polyhedra [1].

IV.7.2. 2-Isomorphism

It is possible for two nonisomorphic graphs to have the same set of 3-blocks. In the terminology of Hassler Whitney, such graphs are 2-*isomorphic* ([5]).

EXERCISES

1. What is the connectivity of the graph of the octahedron?
2. A graph G has a bipartition (U, V) such that each vertex of U is joined to each vertex of V by exactly one edge. Discuss the connectivity, vertex-connectivity, and cyclic connectivity of G for the case $4 \leq |U| \leq |V|$.
3. Use the operations of Sec. IV.2 to show that the graph of (i) a cube and (ii) a regular dodecahedron is 3-connected.
4. What are the 3-blocks of a theta graph? (Definition p. 307).
5. Let G be as in Exercise 2, but with $|U| = 2$ and $|V| = 3$. Determine the 3-blocks and the cleavages of G.
6. Interpret the monovalent vertices of $Blk_3(G)$ in terms of the structure of G.
7. Show how to transform the cube-graph into a wheel by a sequence of edge-deletions and edge-contractions, each of which preserves the property of 3-connection.

REFERENCES

[1] Barnette, D. W., and B. Grünbaum, "On Steinitz' Theorem concerning convex 3-polytopes and on some properties of planar graphs," In *The many facets of graph theory*, Springer, Berlin 1969, pp. 27–40.
[2] Duijvestijn, A. J. W., Thesis, Eindthoven. *Philips Res. Reports* **17** (1962), 523–613.
[3] Tutte, W. T., A theory of 3-connected graphs, *Konink. Nederl. Akad. van W.*, Proc. **64** (1961), 441–455.
[4] ———, *Connectivity in graphs*, Chap. 11, University of Toronto Press, 1966.
[5] Whitney, H., 2-isomorphic graphs, *Amer. J. Math.* **55** (1933), 245–254.

Chapter V

Reconstruction

V.1. THE RECONSTRUCTION PROBLEM

There are deep and difficult questions concerned with the isomorphism of graphs, and the best known of them is the Reconstruction Problem.

Let G be a graph, with vertices enumerated as $v_1, v_2, \ldots, v_k$. Let G_j denote the graph obtained from it by deleting the vertex v_j and its incident edges. We will refer to the k subgraphs G_j of G as its *primal* subgraphs.

C_1 (*The Reconstruction Conjecture*). *Let the isomorphism classes of the k primal subgraphs of G be given. Then if $k \geq 3$, the isomorphism class of G is uniquely determined.*

Two or more of the graphs G_j might be isomorphic. It is assumed in the conjecture that each isomorphism class of a primal subgraph is given with the appropriate multiplicity. Up to the time of writing, the conjecture has remained unproved

It is not difficult to see why k is required to be at least 3 in C_1. If $k=1$, the only primal subgraph of G is the null graph, and this tells us nothing about the number of loops of G. If $k = 2$, the two primal subgraphs, having only one vertex each, tell us nothing about the number of links of G. So we assume from now on that $|V(G)| \geq 3$.

If some property of G can be inferred from the isomorphism classes of the G_j, it is said to be *reconstructible*, and it is said also that the graphs having that property are *recognizable*. An example of a reconstructible property of G is its number k of vertices. It exceeds by 1 the vertex-number of each G_j, and it is the number of given isomorphism classes.

Theorem V.1. *The numbers $\alpha(G)$ and $\beta(G)$ of loops and links of G, respectively, are reconstructible.*

Proof. Each loop of G occurs in exactly $(k-1)$ of the graphs G_j, and each link of G in exactly $(k-2)$ of them. Hence,

$$(k-1)\alpha(G) = \sum_j \alpha(G_j), \qquad (\text{V.1.1})$$

$$(k-2)\beta(G) = \sum_j \beta(G_j). \qquad (\text{V.1.2})$$

There is no harm in speaking of the vertex v_j of G as something known. It is the vertex of G that corresponds to the jth given isomorphism class. □

Theorem V.2. *The numbers $\alpha_j(G)$ and $\beta_j(G)$ of loops and links of G, respectively, incident with the vertex v_j, are reconstructible. Hence* $\text{val}(G, v_j)$ *is reconstructible.*

Proof. The theorem follows from Theorem V.1, since $\alpha_j(G) = \alpha(G) - \alpha(G_j)$ and $\beta_j(G) = \beta(G) - \beta(G_j)$. Then $\text{val}(G, v_j) = 2\alpha_j(G) + \beta_j(G)$, by Sec. I.1. □

Corollary V.3. *Regular graphs are recognizable.*

Let $M(G)$ be the maximum valency, and $m(G)$ the minimum, to occur in G. Both these numbers are reconstructible, by Theorem V.2.

We supplement Theorem V.2 with the following obvious rule.

Rule V.4. *Let H_j be a member of the isomorphism class of G_j, and let θ be an (unknown) isomorphism of H_j onto G_j. Let x be any vertex of H_j, and let $r_j(x)$ be the number of links of G joining θx to v_j. Then*

$$M(G) - \text{val}(H_j, x) \geq r_j(x) \geq m(G) - \text{val}(H_j, x). \qquad (\text{V.1.3})$$

A graph G is said to be *reconstructible* if its isomorphism class is reconstructible, that is, if it satisfies the Reconstruction Conjecture.

Theorem V.5. *If G is regular, it is reconstructible.*

Proof. Let H_j be as in Rule V.4. By Theorem V.2 and Corollary V.3, we can recognize G as regular and determine its valency $\text{val}(G)$. This being both $M(G)$ and $m(G)$, we can determine the numbers $r_j(x)$ of Rule V.4, by Eq. (V.1.3). We can therefore construct a graph G' isomorphic with G by adjoining to H_1 a new vertex w, joining each vertex x of H_1 to w by exactly $r_1(x)$ new links, and then attaching exactly $\alpha_1(G)$ new loops on w. □

The Reconstruction Problem

Theorem V.6. *The numbers $p_0(G)$ and $p_1(G)$ are reconstructible.*

Proof. Suppose we find, using Theorems V.1 and V.2, that $\beta(G) = \beta(G_j)$ for some j. Then one component of G must consist of v_j with its incident loops, and the others must be the components of G_j. So, in this case, $p_0(G)$ is determined by the formula

$$p_0(G) = p_0(G_j) + 1. \tag{V.1.4}$$

In the remaining case, each component of G has at least one vertex in each G_j. Let C_j be the one including v_j. Let H_j be the graph obtained from C_j by deleting the loops. It is connected, for a loop cannot be an isthmus. Clearly, $p_0(G) \leq p_0(G_j)$, with equality if and only if v_j is not a cut-vertex of H_j.

If the graph H_j is separable, it has an extremal block B. This has at least two vertices, and only one of them is a cut-vertex of H_j. (See Theorems III.16 and III.24.) We conclude that, in every case, H_j has at least one vertex that is not a cut-vertex. Accordingly, $p_0(G)$ is determined as follows:

$$p_0(G) = \underset{j}{\mathrm{Min}}\{p_0(G_j)\}. \tag{V.1.5}$$

The reconstructibility of $p_1(G)$ now follows from Eq. (I.6.1) and the observations of the present section. □

Since a tree can be characterized as a graph with $p_0(G) = 1$ and $p_1(G) = 0$, we have the following corollary.

Corollary V.7. *Trees are recognizable.*

Theorem V.8. *2-connected graphs are recognizable.*

Proof. By Theorem V.6, we can determine whether G is connected. By Theorem V.1, we can find if it has a loop. If it does, it is, of course, separable. We may therefore suppose that G is connected and loopless. If it has a cut-vertex then one of the graphs G_j is disconnected. If it is 2-connected, then each of the graphs G_j is connected, by Theorem IV.9. □

This sequence of theorems could be continued. We could show, for example, that 3-connected graphs are recognizable, the condition being that G is 2-connected and that no G_j has a 2-circuit or 1-separation. We could expand the argument of Theorem V.6 to show that disconnected graphs are reconstructible. A theorem of P. J. Kelly asserts that trees are reconstructible, and there are several later variations on his result [1, 3].

In the following sections we discuss theorems applicable to general graphs. We shall note, however, that one of them implies, in a special case, the reconstructibility of disconnected graphs.

V.2. THEORY AND PRACTICE

There seem to be two main problems in the theory of reconstruction. One, of course, is the theoretical proof of the Reconstruction Conjecture. The other is that of finding a good algorithm for finding the graph or graphs G whose primal subgraphs have given isomorphism classes, or showing that no such G exists. Following J. Edmonds, we can define a good algorithm as one whose difficulty, measured by computer time required, increases only as a polynomial in the number of vertices, not exponentially or worse. Thus the instruction to test every graph with the right number of vertices, though short and simple, is bad.

The following algorithm for determining the component of a graph G including a given vertex v is acknowledged to be good. We list the vertex v, then the set S_1 of vertices adjacent to v, then the set S_2 of the remaining vertices adjacent to members of S_1, and so on, until the process terminates with a null S_n. The induced subgraph H of G determined by the listed vertices is then clearly detached in G. It is connected, by Theorem I.25. It is the required component, by Theorem I.23.

The algorithm for finding components can be used to find the bridges in G of a subgraph H, by way of Theorems I.49, I.50, and I.51. We can find the cut-vertices and 1-separations of a graph G by finding the bridges of the vertex-graphs contained in it. (See Sec. III.4.) We can thus determine if G is 2-connected. If it is, we can infer its 2-separations from the bridges of its edgeless subgraphs of 2 vertices. We thus determine whether or not G is 3-connected. If it is not, then the list of 2-separations will enable us to break it up into its 3-blocks by the methods of Sec. IV.3.

Reconstruction is concerned with isomorphism classes. Its practice is likely to lead us to the Isomorphism Problem, that of determining whether or not two given graphs are isomorphic. The author is not an expert in the algorithmic art, but he understands that no good algorithm is known for the Isomorphism Problem, and it is suspected that no such good algorithm exists.

This does not mean that the Isomorphism Problem cannot be easy in particular cases. Given two graphs G and H, it is usually possible to subdivide their vertex-sets into corresponding smaller subsets, so that any isomorphism of G onto H must map each such subset of $V(G)$ onto the corresponding subset of $V(H)$. We can assume, of course, that $|V(G)| = |V(H)|$. Initially, we can subdivide $V(G)$ and $V(H)$ according to valency. Having obtained subdivisions, we may be able to proceed to finer ones, perhaps classifying the vertices of one subset according to the numbers of adjacent vertices in the other subsets. No doubt other possibilities will occur to the reader. We can hope, in any ordinary case, to reduce the Isomorphism Problem to a managable size or even make it completely trivial by such subdivisions.

Presumably the difficult cases of the Isomorphism Problem occur when the number of different valencies is unusually small. In the extreme case, G would be regular. Yet for regular graphs, the Reconstruction Problem is trivial, by Theorem V.5. An optimist might seize upon this fact to argue that there may be a good reconstruction algorithm after all; perhaps whenever the Isomorphism Problem becomes difficult, it can be avoided.

It is at least possible to give a reconstruction algorithm that will work if the various cases of the Isomorphism Problem occurring in it can be dealt with. We shall be given graphs H_1 and H_2, isomorphic with G_1 and G_2, respectively. The algorithm requires us to find all the isomorphisms of a primal subgraph of H_1 onto a primal subgraph of H_2. Each of these represents a way in which G_1 and G_2 might overlap in G, and so determines a possible structure of G. This must be checked against the other G_j. Whatever the optimist may say in favor of this algorithm, it offers no clue to any proof of the Reconstruction Conjecture.

The upshot of these speculations is that, if we are trying to find a proof of the Reconstruction Conjecture, there is no point at the present stage in trying to make that proof also a good algorithm. So in the following sections we shall discuss properties of graphs that could not conceivably be determined by any good algorithm, but which have the advantage of being reconstructible according to our definition.

V.3. KELLY'S LEMMA

If H and K are graphs, we write $\gamma(H, K)$ for the number of subgraphs of H isomorphic with K. It seems unlikely that there is a good algorithm for determining this number. However, in a theoretical argument, we can suppose it known whenever the isomorphism classes of H and K are given.

Theorem V.9 (Kelly's Lemma). *The number $\gamma(G, K)$ is a reconstructible property of G, for any graph K such that $|V(K)| \neq |V(G)|$.*

Proof. If K has more vertices than G, then clearly $\gamma(G, K) = 0$. We may therefore assume that $|V(K)| < |V(G)|$.

A subgraph of G isomorphic with K occurs as a subgraph in exactly $|V(G)| - |V(K)|$ of the primal subgraphs G_j. Hence $\gamma(G, K)$ is determined as follows:

$$(|V(G)| - |V(K)|)\gamma(G, K) = \sum_j \gamma(G_j, K). \qquad (V.3.1)$$

We can regard Theorem V.1 as asserting the special cases of Kelly's Lemma in which K is a loop-graph or a link-graph.

We go to see if any information can be obtained in the case $|V(K)| = |V(G)|$.

Consider a sequence

$$S = (H_1, H_2, \ldots, H_h) \qquad (V.3.2)$$

of $h > 1$ graphs H_i. A *realization* of S in a graph K is a sequence $(K_1, K_2, \ldots, K_h)$ of h subgraphs of K such that $H_j \cong K_j$ for each j. The *carrier* of this realization is the subgraph of K defined as the union of the h subgraphs K_j. If J is any subgraph of K, we write $\delta(J, S)$ for the number of realizations of S in K having J as carrier. We now have

$$\prod_j \gamma(G, H_j) = \sum_{(K)} \gamma(G, K) \delta(K, S), \qquad (V.3.3)$$

where the symbol (K) indicates that we are summing over a complete set of nonisomorphic subgraphs K of G. This is because each side of the equation gives the total number of realizations of S in G. □

Theorem V.10. *Let S be a sequence of two or more graphs, each of which has fewer vertices than G. Then the sum*

$$\sum \gamma(G, K) \delta(K, S),$$

taken over a complete set of nonisomorphic subgraphs K of G such that $|V(K)| = |V(G)|$, is reconstructible.

This theorem follows from Eq. (V.3.3), for by Kelly's Lemma the product on the left of Eq. (V.3.3) is reconstructible, and so is the contribution to the sum, on the right, of those graphs K for which $|V(K)| < |V(G)|$.

We go on to consider some special cases of Theorem V.10. Suppose first that the sequence S of Eq. (V.3.2) is such that

$$\sum_j |V(H_j)| = |V(G)|.$$

If K is a carrier of S in G with $|V(G)|$ vertices, it can only be a disconnected subgraph of G with h components $K_1, K_2, \ldots, K_h$ isomorphic with $H_1, H_2, \ldots, H_h$, respectively. Thus the isomorphism class of K is uniquely determined. If the H_j are defined as the components of a disconnected graph H, then K is isomorphic with H. Now there is only one non-zero term $\gamma(G, K)\delta(K, S)$ in the sum of Theorem V.10. Moreover, the number $\delta(K, S)$ can be calculated. Let $C_1, C_2, \ldots, C_m$ be the isomorphism classes represented in S, and let c_i denote the number of terms of S (and components of K) belonging to C_i. Then

$$\delta(K, S) = \prod_{i=1}^{m} (c_i!). \qquad (V.3.4)$$

Thus $\gamma(G, K)$ can be determined. We state this result as a theorem.

Theorem V.11. *Let H be a disconnected graph such that $|V(H)| = |V(G)|$. Then $\gamma(G, H)$ is a reconstructible property of G.*

Theorem V.12. *Disconnected graphs are reconstructible.*

Proof. If G is disconnected, we can recognize it as such and determine the number of its components, by Theorem V.6. Given any graph H with the same numbers of edges, vertices, and components as G, we can (in principle) reconstruct $\gamma(G, H)$. If G is isomorphic with H, this number will be 1; otherwise it will be zero. So G can be reconstructed by testing all possible structures; the argument tell us that only one can work. □

Theorem V.13. *Theorem V.10 remains valid when we impose on the graphs K the extra condition that they be connected.*

Proof. The contribution to the sum in Theorem V.10 of disconnected graphs K is reconstructible, by Theorem V.11. □

For our next application of Theorem V.10, we shall need the following theorem. It is a supplement to the theory of Sec. III.3.

Theorem V.14. *Let a connected graph H be expressed as a union of h distinct 2-connected subgraphs $H_1, H_2, \ldots, H_h$, each having at least one edge. Then the H_j are the blocks of H if and only if*

$$\sum_j |V(H_j)| = |V(H)| + h - 1. \qquad (V.3.5)$$

Proof. If the H_j are the blocks of H, then Eq. (V.3.5) follows from the fact that $\mathrm{Blk}(H)$ is a tree. (See Theorems I.37 and III.23.)

Conversely, suppose Eq. (V.3.5) to hold. Let R_j denote the union of the subgraphs H_i such that $i \leq j$. Since H is connected, we can adjust the enumeration so that H_{j+1} has at least one vertex in common with R_j, for $1 \leq j < h$. Then each R_j is connected. Moreover, H_{j+1} and R_j have only one vertex in common, by Eq. (V.3.5). Repeated application of Theorem III.22 shows that the blocks of R_j are $H_1, H_2, \ldots, H_j$, for each j. □

Let us now suppose that, in the sequences S of Eq. (V.3.2), each H_j is a 2-connected graph having fewer vertices than G. Let us suppose, further, that the H_j satisfy Eq. (V.3.5). Then if K is a connected carrier of a realization of S in G, it must be a separable graph whose blocks are, to within isomorphism, the terms of S, by Theorem V.14. Formula (V.3.4) for $\delta(K, S)$ is still valid, and it gives the same number for each such K. We deduce from Theorem V.13 that the sum $\Sigma\gamma(G, K)$, taken over all connected K whose blocks are to within isomorphism the terms of S, is reconstructible. Another way of stating this result is as follows.

Theorem V.15. *The number of spanning connected subgraphs K of G such that K has a given number of blocks in each isomorphism class, but no spanning block, is a reconstructible property of G.*

If a spanning connected subgraph K of G does have a spanning block, then any other block must be a loop-graph. The requirement of no spanning block excludes, in particular, the case in which K is 2-connected.

Theorem V.16. *Graphs having no spanning block are recognizable. For each such graph, the number of blocks in any given isomorphism class is reconstructible.*

Proof. The first part follows from Theorem V.1 and V.8. In proving the second part, we can assume G to be connected, by Theorem V.12, and with no spanning block.

Consider any suggested list U of blocks of G, so many from each isomorphism class. This list U is required to give G the right number of edges and vertices. According to Theorem V.15, the number $n(U)$ of connected spanning subgraphs of G with U as block-set is reconstructible. But clearly $n(U) = 1$ if U is indeed a list of the blocks of G, and $n(U) = 0$ otherwise. So G has a determinable list of blocks. □

Theorem V.17. *Theorem V.10 remains valid, with the extra condition on the graphs K that each one has a spanning block.*

Proof. We note that this condition implies the condition of Theorem V.13. To prove Theorem V.17, we have only to observe that the contribution to the sum of Theorem V.10, as restricted by Theorem V.13, of connected graphs K with no spanning block is reconstructible, by Theorem V.15 and the preceding observation about $\delta(K, S)$. □

The results of this section can be used to show that some well-known properties of graphs, for example the characteristic and chromatic polynomials, are reconstructible. At first these results were startling, but now we can recognize them as mere exercises on Kelly's Lemma. (See [4].)

V.4. EDGE-RECONSTRUCTION

Let G be a graph with edges enumerated as $A_1, A_2, \ldots, A_k$. Let G_j now denote the subgraph of G obtained by deleting the edge A_j.

C_2 (*The Edge-Reconstruction Conjecture*). *Let the isomorphism classes of the k graphs G_j, as just defined, be given. Then if $k \geq 4$, the isomorphism class of G is uniquely determined.*

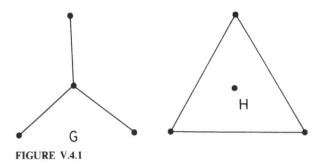

FIGURE V.4.1

The reason for the restriction to the case $k \geq 4$ can be seen from Fig. V.4.1. It shows two graphs G and H, for which the operation of deleting an edge gives the same set of three subgraphs, to within isomorphism.

A theory analogous to that of the preceding sections can be set up for the Edge-Reconstruction Conjecture, which, like C_1, is still unproved. The analogue of Kelly's Lemma asserts that the number of spanning proper subgraphs of G belonging to any given isomorphism class can be determined from the isomorphism classes of the graphs G_j. We express this by saying that the number is *edge-reconstructible*. This result seems to give a vast quantity of information about G, but so far it has not been found to be enough to determine the structure of G.

V.5. NOTES

V.5.1. Generalizations

It might be convenient to have a generalized Reconstruction Conjecture, with vertex-reconstruction and edge-reconstruction as special cases. One possibility is to define a structure on a finite set S simply by assigning a nonnegative integer $n(T)$ to each subset T of S. Given such a structure J on S, and a member x of S, we can define a structure J_x on $S - \{x\}$ by giving to each subset T of $S - \{x\}$ the same number $n(T)$ as before.

We define isomorphism of structures in the obvious way. We may now ask "If the isomorphism classes of the structures J_x, $x \in S$, are given, and if the value of $n(S)$ is stated, is the isomorphism class of J uniquely determined?" A Reconstruction Conjecture would assert that, for all structures satisfying certain stated conditions, the answer to the above question is "Yes." For example, S could be the set of vertices of a graph G, and $n(T)$ could be the number of edges with both ends in T. Our question would then pose the ordinary Reconstruction Conjecture for graphs. The information about $n(S)$ would then be redundant in nontrivial cases, by Theorem V.1. But with it we would no longer have to admit the existence of a counterexample in the case of two vertices. For the Edge-Reconstruction Conjecture S would be the set of edges of G, and $n(T)$ would be the number of vertices incident with members of T.

It would be going too far to assert a Reconstruction Conjecture for structures in general. That would imply a Reconstruction Conjecture for matroids, one for which nontrivial counterexamples have been given [2]. But perhaps we can specify a class of structures that includes those of purely graph-theoretical interest, and for which the Reconstruction Conjecture still seems plausible.

EXERCISES

1. List the primal subgraphs of the 5-arc, the 5-circuit, and the 5-wheel. In each case prove that your set of primal subgraphs can be derived from one graph only.
2. Could you reconstruct each graph of Exercise 1 if you were given only one member of each isomorphism class of primal subgraphs?
3. Can you find a good algorithm for reconstructing a disconnected graph G, based on a discussion of the components of the primal subgraphs?

REFERENCES

[1] Bondy, J. A., On Kelly's congruence theorem for trees, *Proc. Cambridge Phil. Soc.*, **65** (1969), 387–397.
[2] Brylawski, T. On the nonreconstructibility of combinatorial geometries, *J. Comb. Theory* (B), **19** (1975), 72–76.
[3] Kelly, P. J., A congruence theorem for trees, *Pacific J. Math.*, **7** (1957), 961–968.
[4] Kocay, W. L., An extension of Kelly's Lemma to spanning subgraphs, *Congressus Numerantium* **31** (1981), 109–120.

Chapter VI

Digraphs and Paths

VI.1. DIGRAPHS

A *digraph* Γ is something like a graph. It is defined by a set $V(\Gamma)$ of elements called *vertices*, a set $W(\Gamma)$ of elements called *directed edges* or *darts*, and two relations of *incidence*. These associate each dart D, the one with a vertex $h(D)$ called its *head* and the other with a vertex $t(D)$ called its *tail*. In this work we shall always suppose $V(\Gamma)$ and $W(\Gamma)$ to be finite.

Any dart D will be said to be directed *from* $t(D)$ and *to* $h(D)$. It is a *loop-dart* if its head and tail coincide and a *link-dart* if they do not. The *invalency* $\mathrm{inv}(\Gamma, x)$ of a vertex x of Γ is the number of darts of Γ having x as head. The *outvalency* $\mathrm{outv}(\Gamma, x)$ of x is the number of darts of Γ having x as tail. We say that the digraph Γ is *Eulerian* if, for every vertex, the invalency is equal to the outvalency.

From the above definitions we can deduce the following formula.

$$\sum_{x \in V(\Gamma)} \mathrm{inv}(\Gamma, x) = \sum_{x \in V(\Gamma)} \mathrm{outv}(\Gamma, x) = |W(\Gamma)|. \quad \text{(VI.1.1)}$$

Each digraph has an *underlying graph* $U(\Gamma)$ defined as follows: Its vertices are the vertices of Γ and its edges are the darts of Γ. The ends of an edge of $U(\Gamma)$ are its head and tail in Γ.

Let Γ and Δ be digraphs. We say that Δ is a *subdigraph* of Γ if

$$V(\Delta) \subseteq V(\Gamma), \quad W(\Delta) \subseteq W(\Gamma),$$

and each dart of Δ has the same head and tail in Δ as in Γ. Clearly there is a 1-1 correspondence between the subdigraphs of Γ and the subgraphs of $U(\Gamma)$ such that each subdigraph Δ of Γ corresponds to its own underlying graph $U(\Delta)$.

Suppose we are given a tree T in which one vertex r is distinguished. We can construct a digraph Γ as follows: We put

$$V(\Gamma) = V(T) \quad \text{and} \quad W(\Gamma) = E(T).$$

An edge A of T has two end-graphs H and K in T, the components of T'_A. (See Sec. I.6.) We may suppose r to be in K. We define the head and tail of A in Γ as the ends of the edge A of T in H and K, respectively. This completes the definition of Γ. A digraph Γ so defined is called an *arborescence diverging from r*.

Theorem VI.1. *Let Γ be an arborescence diverging from a vertex r. Then r is the head of no dart of Γ, and each other vertex is the head of exactly one dart of Γ.*

Proof. By the definition of an arborescence, $U(\Gamma)$ is a tree T. Let v be any vertex of Γ. We can enumerate the edges of T incident with v as $A_1, A_2, \ldots, A_k$, and the arms of T at v as $Y_1, Y_2, \ldots, Y_k$, so that A_i is an edge of Y_i for each suffix i. (See Sec. III.4.) Here we are supposing T to have at least one edge, because otherwise the theorem is trivial. One end-graph of A_i in T is the union of the arms at v other than Y_i if such arms exist, and is the vertex-graph defined by v in the remaining case. The other end-graph is obtained from Y_i by deleting v and A_i.

Suppose $v = r$. By the above observations and the definition of an arborescence, r is the tail in Γ of each of the k edges A_i, that is, of each dart of Γ incident with it. In the remaining case, r belongs to just one of the Y_i, let us say, to Y_k. We then find that v is the head of A_k and the tail of each other A_i. The theorem follows. □

Three examples of arborescences are given in Fig. VI.1.1. Here, and in our other drawings of digraphs, we show the underlying graph, with an arrow on each edge directed from tail to head.

Each arborescence in the figure diverges from a vertex labelled r.

Theorem VI.2. *Let Γ be a digraph such that $U(\Gamma)$ is a tree and no vertex is the head of more than one dart. Then Γ is an arborescence diverging from some vertex r.*

Proof. Since $U(\Gamma)$ is a tree the number of vertices of Γ exceeds the number of darts by 1, by Theorem I.37. So Γ has one vertex r that is the head of no dart, and each other vertex must be the head of exactly one dart.

Consider any dart A of Γ. Let its end-graphs in $U(\Gamma) = T$ be H and K, with r in K. Assume that the end of A in K is the head of A in Γ. Then

Digraphs

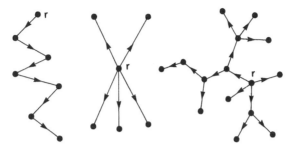

FIGURE VI.1.1 Arborescences.

the number of vertices of K must exceed the number of edges of K by 2. Hence, K is disconnected, by Theorem I.31, contrary to its definition as a component of T'_A. Accordingly, the head of A must in fact be in H. The theorem follows. □

Theorem VI.3. *Let Γ be an arborescence diverging from a vertex r. Let A be a dart of Γ with end-graphs H and K in $U(\Gamma)$, r being in K. Then the subdigraph of Γ corresponding to H is an arborescence diverging from the end of A in H, and the one corresponding to K is an arborescence diverging from r.*

Since H and K are trees, by Theorem I.38, Theorem VI.3 is an immediate consequence of Theorem VI.2.

To *reverse* a digraph Γ is to redefine the incidence relations so as to interchange the head and tail of each dart. If Γ is an arborescence diverging from r, then its reversed digraph is called an *arborescence converging to r*. The theory of converging aborescences is the same as that of diverging ones, with the terms "head" and "tail" interchanged.

A *pairing* of a digraph Γ is a mapping θ of $W(\Gamma)$ onto itself that satisfies the following three conditions.
i) For each $A \in W(\Gamma)$ the darts A and θA are distinct.
ii) $\theta(\theta A) = A$, for each $A \in W(\Gamma)$.
iii) The head and tail of A are the tail and head, respectively, of θA, for each $A \in W(\Gamma)$.

Not every digraph has a pairing. A digraph Γ and a pairing θ of Γ are said to define a *paired digraph* (Γ, θ).

A paired digraph (Γ, θ) has an *equivalent graph G* defined as follows: The vertices of G are the vertices of Γ. The edges of G are in 1-1 correspondence with the unordered pairs of darts of Γ that are the orbits of the permutation θ. The ends of an edge of G are the head and tail of each member of the corresponding pair.

There is a converse construction whereby we form from a given graph G an *equivalent paired digraph* (Γ, θ). We associate with each edge E of G two distinct elements E^+ and E^- called the *darts on E*. Darts associated with distinct edges of G are to be distinct. To each dart on E we assign one end of E as head and the other as tail. The head and tail of E^+ are to be the tail and head, respectively, of E^-. We now have our paired digraph (Γ, θ). The vertices of Γ are the vertices of G and its darts are the darts on the edges of G, with their specified heads and tails. Now θ is the permutation of $W(\Gamma)$ that interchanges the two darts on each edge of G. Hence (Γ, θ) is, by definition, an equivalent paired digraph of G. If we start with (Γ, θ) then G is an equivalent graph of this paired digraph.

We shall use the correspondence between graphs and equivalent paired digraphs to deduce theorems about graphs from theorems about digraphs. It will not be necessary to go through the definition of an equivalent paired digraph anew for each application. Instead, we shall take it for granted that, on each edge of any given graph G, we are given the two darts E^+ and E^- with heads and tails as specified above. These define Γ and θ. We call Γ the *equivalent digraph* of G, and θ the *pairing* of G. Together they constitute *the* equivalent paired digraph (Γ, θ) of G. The darts of Γ can be called the darts of G, and their heads and tails in Γ can be called their heads and tails in G, respectively.

In terms of a diagram of G, the darts E^+ and E^- represent the two directions in which an arrow can be drawn on the corresponding edge-curve. Even for a loop there are two such distinguishable directions, and even on a loop we suppose two distinct darts to exist. The *edge* of a dart D of G is the edge of G on which D is defined. The two darts E^+ and E^- on an edge E of G are said to be *opposites*. The opposite of a dart D is often denoted by D^{-1}.

If H is a subgraph of G, we normally take the darts on an edge of H to be the same as in G. We can then say that the equivalent digraph Γ_H of H is a subdigraph of Γ, and that the pairing θ_H of H is the restriction of θ to the set of darts of H.

An *orientation* or *oriented form* of G is a spanning subdigraph of Γ having exactly one dart on each edge of G. As with subgraphs, the adjective "spanning" indicates that the subdigraph includes all the vertices of Γ. Clearly, if Δ is an orientation of G, the graphs $U(\Delta)$ and G can be identified. More precisely, there is an isomorphism of $U(\Delta)$ onto G that maps each vertex onto itself and each edge of $U(\Delta)$, that is, each dart of Δ, onto its edge in G.

We conclude this section with some simple examples of graph–digraph correspondence.

Theorem VI.4. *Let T be a tree and r a vertex of T. Then T has exactly one orientation that is an arborescence diverging from r, and exactly one that is an arborescence converging to r.*

Proof. The definition of an arborescence tells us what must be the head and tail of the dart to be chosen on any edge A of T. Since A cannot be a loop this uniquely determines one of the two darts on A. □

Theorem VI.5. *Let G be a graph and r a vertex of G. Then there is a 1-1 correspondence between the spanning trees of G and the spanning arborescences of its equivalent digraph Γ diverging from (converging to) r, each such arborescence being an orientation of the corresponding tree in G.*

Proof. Let S be any spanning arborescence of Γ diverging from (converging to) r. Let T be the unique spanning subgraph of G of which S is an orientation, that is, whose edges are those of the darts of S. Then $U(S)$ is a tree, by the definition of an arborescence, and therefore T is a spanning tree of G.

Conversely, if T is a spanning tree of G, there is a unique orientation of T that is an arborescence diverging from (converging to) r, by Theorem VI.4. This is a spanning arborescence of Γ. □

VI.2. PATHS

A *nondegenerate path* in a digraph Γ is a sequence

$$P = (D_1, D_2, \ldots, D_n) \tag{VI.2.1}$$

of $n \geq 1$ darts D_j of Γ, not necessarily all distinct. It is required that the head of D_j shall be the tail of D_{j+1} whenever $1 \leq j < n$. The tail of D_j is called the jth vertex of P. The head of D_n is called the last or $(n+1)$th vertex of P. The first and last vertices of P, that is, the tail of D_1 and the head of D_n, are called the *origin* and *terminus* of P, respectively. The number n is the *length* $s(P)$ of P.

It is convenient to recognize also certain *degenerate* paths in Γ. There is exactly one of these, denoted by P_v, associated with each vertex v of Γ; P_v has no member darts. We say that its length $s(P_v)$ is zero, and v is both its origin and its terminus.

In what follows, a "path" may be either degenerate or nondegenerate. In either case it is said to be directed *from* its origin and *to* its terminus.

Consider two paths P and Q in Γ. If the terminus of P is the origin of Q, there is a path PQ in Γ formed by writing down the terms of P, in their order in P, and continuing with the terms of Q, in their order in Q. If P or Q is degenerate, we take this to mean that PQ is Q or P, respectively. We say that PQ is the *product* of P and Q, in that order. If the terminus of P is not the origin of Q, then no product PQ is defined. If PQ is defined, we have

$$s(PQ) = s(P) + s(Q). \tag{VI.2.2}$$

Moreover, the origin of PQ is the origin of P and the terminus of PQ is the terminus of Q.

It is easy to verify that multiplication of paths is associative, but not in general commutative. When we write an equation

$$P(QR) = (PQ)R$$

expressive of associativity, we imply that all the products occurring are defined. Here this means that the terminus of P is the origin of Q and that the terminus of Q is the origin of R. Associative products of three or more terms, in a definite order, may be written without brackets, without ambiguity. We may speak of a product

$$P_1 P_2 \cdots P_m$$

of paths P_j ($1 < j < m$). We shall then imply that the terminus of P_j is the origin of P_{j+1} whenever $1 < j < m$. The product is the path formed by writing first the terms of P_1 in proper order, then those of P_2, and so on. Its origin is the origin of P_1 and its terminus is the terminus of P_m.

Let v_k denote the kth vertex of the path P of Eq. (VI.2.1). If $1 \le i < j \le n+1$, we write $P[i, j]$ for the path

$$(D_i, D_{i+1}, \ldots, D_{j-1})$$

whose origin is v_i and whose terminus is v_j. We call it the *part* of P extending from the ith vertex to the jth. We also write $P[i, i]$ for the degenerate path on the ith vertex.

Let $i_0, i_1, i_2, \ldots, i_q$, where $q > 2$, be integers such that

$$1 = i_0 \le i_1 \le i_2 \le \cdots \le i_q = n+1.$$

Then we can write

$$P = \prod_{j=0}^{q} P[i_j, i_{j+1}],$$

where the factors are multiplied in the order of j increasing. We refer to this formula as the *factorization* of P imposed by the sequence $(i_1, i_2, \ldots, i_{q-1})$. If $q = 2$ we have the factorization $P = P[1, i_1]P[i_1, n+1]$, which we can say is imposed by the integer i_1.

We now define some special kinds of path. The path P of Eq. (VI.2.1) is *reentrant* if its origin and terminus coincide. Thus every degenerate path is reentrant. We say that P is *dart-simple* if no dart is repeated in it. It is *head-simple* (*tail-simple*) if no vertex occurs twice as head (tail) of a term of P. It is *simple* if it is both head-simple and tail-simple.

A head-simple or tail-simple path is necessarily dart-simple. If a tail-simple path is not simple, its terminus must be the head of two of its darts. We count degenerate paths as simple. A nondegenerate simple path is called a *circular* path if it is reentrant, and a *linear* path if its origin and terminus are distinct.

The *digraph* $\Gamma(P)$ of a path P in Γ is defined as the subdigraph of Γ whose darts are those occurring in P and whose vertices are the heads and tails of these darts, if P is nondegenerate. For a degenerate path P_v we take $\Gamma(P_v)$ to have no darts, and to have the single vertex v. The digraph of a circular path is called a *directed circuit*. The digraph of a linear path with origin x and terminus y is called a *directed arc* from x to y.

The *underlying graph* $U(P)$ of a path P of Γ is the underlying graph of its digraph $\Gamma(P)$. If P in Eq. (VI.2.1) is a linear path with origin x and terminus y, then $U(P)$ is an arc in $U(\Gamma)$ with ends x and y. For it satisfies the definition of an arc with $(v_1, v_2, \ldots, v_{n+1})$ as defining sequence of vertices and P as defining sequence of edges. If instead P is a circular path, then $U(P)$ is a circuit. We can take $(v_1, v_2, \ldots, v_n)$ as defining sequence of vertices and P as defining sequence of edges. If P is a linear path, it follows from Theorem VI.2 that $\Gamma(P)$ is an arborescence diverging from the origin of P and converging to its terminus.

Theorem VI.6. *Let P be a path in Γ. Then $U(P)$ is connected.*

Proof. We may suppose P to be a nondegenerate path, as in Eq. (VI.2.1), since otherwise the theorem is trivial.

Assume $U(P)$ disconnected. Let C be a component of $U(P)$ not including the terminus of P. Let k be the greatest integer such that the kth vertex of P belongs to C. Then D_k has its tail in C and its head not in C. So C is not detached in $U(P)$, contrary to its definition as a component. □

As with graphs we write $\Gamma_1 \subseteq \Gamma_2$ to denote that the digraph Γ_1 is a subdigraph of the digraph Γ_2. If, in addition, Γ_1 is distinct from Γ_2, we call Γ_1 a *proper* subdigraph of Γ_2 and write $\Gamma_1 \subset \Gamma_2$.

Theorem VI.7. *Let x and y be distinct vertices of a digraph Γ. Then there exists a linear path P from x to y in Γ if there is any path Q from x to y in Γ. Moreover, we can choose P so that $\Gamma(P) \subseteq \Gamma(Q)$ and $s(P) \leq s(Q)$, with equality only if Q is linear.*

Proof. If Q is not linear, there is a repeated head or repeated tail. Hence $Q = P_1 P_2 P_3$, where P_2 is reentrant and nondegenerate. But now $P_1 P_3$ is a path from x to y in Γ such that $\Gamma(P_1 P_3) \subseteq \Gamma(Q)$ and $s(P_1 P_3) < s(Q)$. If $P_1 P_3$ is not linear, we repeat the process, and so on, to termination. We then have the required path P, and $s(P) < s(Q)$. If Q is linear, we put $P = Q$. □

Theorem VI.8. *Let Γ be an arborescence diverging from a vertex r. Let v be any vertex of Γ. Then there is just one path P_{rv} in Γ from r to v, and P_{rv} is either degenerate or linear. Moreover, any path in Γ is a factor of some path P_{rv}.*

Proof. By the definition of a diverging arborescence, the $(j+1)$th vertex of any path P in Γ determines the jth dart ($j = 1, 2, \ldots$). Hence P is uniquely determined when its origin and terminus are given. (See Theorem VI.1.) So by Theorem VI.7, P is either degenerate or linear.

Let X be the set of all vertices of Γ to which there is a path in Γ from r. Let H be the subgraph of $U(\Gamma)$ induced by X. Then H is connected, by Theorem VI.6 and I.25. Suppose $U(\Gamma)$ to have an edge D with one end a in X and one end b outside X. One end-graph of D in $U(\Gamma)$ contains H and so includes both a and r, by Theorem I.24. So in Γ we have $a = t(D)$ and $b = h(D)$, by the definition of a diverging arborescence. This contradicts the definition of H, for the product $P_{ra}(D)$ is a path from r to b. Hence $H = U(\Gamma)$, by the connection of $U(\Gamma)$. (See Theorem I.21). Hence v is in X and the path P_{rv} exists.

To complete the proof, let P be any path in Γ, say from u to v. By the foregoing results, the product $P_{ru}P$ must be identical with P_{rv}. □

The structure $\text{Blk}_3(G, A)$ of Sec. IV.3, though described there as a graph, is best regarded as an arborescence diverging from the leading 3-block of G, for is not Fig. IV.3.8 a drawing of an arborescence? A virtual edge A_i is a dart, an outlet of its tail in $\text{Blk}_3(G, A)$ and the intake of its head.

More than one kind of connection is recognized for digraphs. A digraph Γ is said to be *strongly connected* if, for any ordered pair (x, y) of vertices of Γ, there is a path in Γ from x to y. If we can only say of such a pair that there is a path either from x to y or from y to x, then we say only that Γ is *connected*. Let us say also that Γ is *graph-connected* if $U(\Gamma)$ is connected. Thus an arborescence diverging from r is graph-connected. But in general it is not connected and *a fortiori* not strongly connected, for there can be no path in it joining two monovalent vertices of $U(\Gamma)$ neither of which is r. It is clear, however, that a directed arc is connected though not strongly connected, and that a directed circuit is strongly connected.

Let us now consider a graph G with its equivalent digraph Γ and its pairing θ. Paths in Γ are called also *paths in G*. A path in G is *edge-simple* if no two of its terms are darts on the same edge of G. Thus an edge-simple path is necessarily dart-simple. The *graph* $G(P)$ of a nondegenerate path P in G is the reduction of G defined by the edges to which the darts of P belong. The *graph* $G(P_v)$ of a degenerate path P_v is the vertex-graph defined by v. In any case, $G(P)$ is obtained from $\Gamma(P)$ by replacing its darts by the edges on which they are defined. If $\Gamma(P)$ has at most one dart on each edge of G, then $\Gamma(P)$ is an orientation of $G(P)$.

From the definitions of arc and circuit we have the following rules.

Theorem VI.9. *A subgraph H of G is an arc with ends x and y if and only if $H = G(P)$, where P is a linear path in G from x to y.*

Theorem VI.10. *A subgraph H of G is a circuit if and only if $H = G(P)$, where P is an edge-simple circular path in G.*

The requirement in Theorem VI.10 that P be edge-simple is redundant unless $s(P) = 2$. But the two darts D and D^{-1} on any link of G define a circular path (D, D^{-1}) that is not edge-simple. The graph of this path is not a circuit but a link-graph.

Theorem VI.11. *If P is a path in G then $G(P)$ is connected.*

To prove this we use the argument of Theorem VI.6 with $G(P)$ replacing $U(P)$.

Theorem VI.12. *If G is connected, then Γ is strongly connected, by Theorems VI.9 and I.43.*

Given a path P in G from x to y we can construct a path P^{-1} in G from y to x by writing the terms of P in reverse order, and then replacing each by its opposite. We call P^{-1} the *inverse* of P. Clearly, inverse paths satisfy the following rules:

$$(P^{-1})^{-1} = P, \quad (PQ)^{-1} = Q^{-1}P^{-1}. \qquad (VI.2.3)$$

We shall often make use of the following combination of Theorems VI.9 and I.43.

Theorem VI.13. *Two distinct vertices x and y of a graph G belong to the same component of G if and only if there is a linear path in G from x to y.*

VI.3. THE BEST THEOREM

We have defined an Eulerian digraph in Section 1. A graph is called *Eulerian* if each vertex has even valency. An *Eulerian path* in a digraph Γ is a reentrant dart-simple path P such that $\Gamma(P) = \Gamma$. An *Eulerian path* in a graph G is a reentrant edge-simple path P such that $G(P) = G$. Thus the Eulerian paths of G are the Eulerian paths of the orientations of G.

Theorem VI.14. *If a digraph Γ has an Eulerian path P, then Γ is strongly connected and Eulerian.*

Proof. Let x and y be distinct vertices of Γ. We can adjust the notation so that $P = P_1 P_2 P_3$, where P_2 has origin x and terminus y. But then P_2 is a path from x to y and $P_3 P_1$ is a path from y to x. We deduce that Γ is strongly connected.

It is clear that the number of darts in P having a given vertex v as head is equal to the number having v as tail. Thus $\text{inv}(\Gamma, v) = \text{outv}(\Gamma, v)$. Accordingly, Γ is Eulerian. □

Theorem VI.15. *If a graph G has an Eulerian path P, then G is connected and Eulerian.*

Proof. G is connected, by Theorem VI.11. P is an Eulerian path in an orientation Δ of G, and Δ is Eulerian, by Theorem VI.14. We thus have val(G, v) = inv(Δ, v) + outv(Δ, v) = 2inv(Δ, v) for each vertex v of G. Thus G is Eulerian. □

Theorems VI.14 and VI.15 have converses, and the converse of Theorem VI.14 is the main theorem of this section. The converse of Theorem VI.15 was studied by Euler, in a paper which is held to mark the beginning of graph theory [6].

Theorem VI.16. *Let G be an Eulerian graph with at least one edge. Then $G \cdot E(G)$ is a union of edge-disjoint circuits.*

Proof. If a component C of G has an edge, it is not a tree, by Theorem I.40. Hence G contains a circuit J_1, by Theorem I.45. The graph G_1 obtained from G by deleting the edges of J_1 is Eulerian. It therefore contains a circuit J_2, unless it is edgeless. Proceeding in this way, we obtain a sequence $J_1, J_2, \ldots, J_k$ of edge-disjoint circuits of G whose union is $G \cdot E(G)$. □

Theorem VI.17. *Every Eulerian graph has an Eulerian orientation.*

Proof. We may suppose our graph G to have an edge, since otherwise the theorem is trivial. Consider the sequence of circuits J_i derived in Theorem VI.16. Each is the graph of an edge-simple circular path P_j in G, by Theorem VI.10. The darts of these circular paths define an Eulerian orientation of G. □

Theorem VI.18. *Let Γ be a digraph, and let r be a vertex of Γ. Let there be a path in Γ from r to each other vertex. Then Γ has a spanning arborescence diverging from r.*

Proof. Let us say that a sequence $S = (v_0, v_1, \ldots, v_k)$ of distinct vertices of Γ has *property P* if $v_0 = r$ and, for each other vertex v_j of S, there is a dart D_j of Γ such that $h(D_j) = v_j$ and $t(D_j)$ is a vertex v_i of S such that $i < j$. Suppose that we have such a sequence S and that u is a vertex of Γ that does not belong to it. By hypothesis, there is a path in Γ from r to u. This path must include a dart D_{k+1} such that $t(D_{k+1})$ is in S but $h(D_{k+1})$ is not in S. By adjoining $h(D_{k+1})$ to S as a new last vertex v_{k+1} we obtain a sequence S' of $(k+2)$ distinct vertices of Γ, still with property P.

By starting with $S = (r)$ and repeatedly applying the above procedure, we arrive at a sequence $Z = (v_0, v_1, \ldots, v_n)$ with property P that enumerates all the vertices of Γ. The corresponding darts $D_1, D_2, \ldots, D_n$ define a spanning subdigraph Δ of Γ. Moreover, $U(\Delta)$ is connected, by repeated application of Theorem I.25. Hence $U(\Delta)$ is a tree, by Eq. (I.6.1). Accordingly, Δ is an arborescence diverging from r, by Theorem VI.2. □

The BEST Theorem

Theorem VI.19. *Let r be a vertex of a digraph Γ. Let P be an Eulerian path from r to r in Γ. For each vertex x of Γ, other than r, let D_x be the first dart of P having x as head. Then the darts D_x are those of a spanning arborescence Δ of Γ diverging from r.*

Proof. Consider the sequence Z of vertices of Γ starting with r and continuing with the heads of the darts D_x, in the order of those darts in P. Then Z is an enumeration of the vertices of Γ "with property P" as defined in the proof of Theorem VI.18. It follows, as in that proof, that the darts D_x define the required arborescence. □

We call Δ the *residual arborescence* of the Eulerian path P.

Let us now consider an Eulerian digraph Γ, a vertex r of Γ, and a spanning arborescence Δ of Γ diverging from r. Let the vertices of Γ be enumerated as $v_0, v_1, \ldots, v_q$, where $v_0 = r$, and let the invalency and outvalency of v_j be n_j, for each j. For each vertex v_j let there be given an ordering of the n_j darts with head v_j, the one in Δ being put last when $j \neq 0$. The number N of ways of so ordering the darts of Γ is

$$N = n_0 \prod_{j=0}^{q} (n_j - 1)!. \qquad (VI.3.1)$$

We now construct a sequence

$$R = (D_1, D_2, D_3, \ldots)$$

of darts of Γ by the following rules. D_1 is the first dart with head at r in the dart-order at r. Having constructed the sequence as far as D_j we define D_{j+1} as the first dart not already chosen, if such exists, in the dart-order at $t(D_j)$. We note that if $t(D_j)$ is not r, then the dart D_{j+1} exists. For up to and including D_j, the sequence R includes one more dart with tail at $t(D)$ than with head at $t(D)$, and Γ is Eulerian. We deduce that R terminates with a dart D_m having r as its tail, that its terms are all distinct, and that it includes all the darts incident with r. It is not a path in Γ, for its darts have their heads and tails in the wrong order. It is, however, a path in the reverse of Γ.

Let us say that a vertex x of Γ is *filled* if all its incident darts are in R, and is *unfilled* otherwise. Assume that at least one unfilled vertex exists. Then there is a shortest path S in Δ whose origin is r and whose terminus t is unfilled. (See Theorem VI.8.) Considering the factors of S with origin r, we see that all the vertices of S except the last are filled. But R does not, by its construction, include the last dart of S. Hence the last vertex but one of S is unfilled. From this contradiction we deduce that in fact all the vertices of Γ are filled. So by taking the terms of R in reversed order, we obtain an Eulerian path from r to r in Γ, with Δ as its residual arborescence.

For each set of dart-orders the above construction determines such an Eulerian path uniquely, and different sets of dart-orders give different Eulerian paths. On the other hand we can find a set of dart-orders that determines any given Eulerian path P having Δ as its residual arborescence. We have only to take the darts with heads at a vertex x in the reverse order of their appearance in P. We can now assert the following theorem.

Theorem VI.20. *Let r be a vertex of an Eulerian digraph Γ, and let Δ be a spanning arborescence of Γ diverging from r. Then the number of Eulerian paths from r to r in Γ having Δ as residual arborescence is the number N of Eq.* (VI.3.1).

Theorem VI.21. *Let r be a vertex of a strongly connected Eulerian digraph Γ. Then Γ has at least one Eulerian path from r to r.*

This converse of Theorem VI.14 is a consequence of Theorems VI.18 and VI.20.

For any digraph Γ, whether Eulerian or not, let us denote by $T'(\Gamma)$ the number of spanning arborescences of Γ diverging from the vertex r, and by $T_r(\Gamma)$ the number of spanning arborescences of Γ converging to r.

We can now give the number of Eulerian paths from r to r in an Eulerian digraph Γ as

$$N \cdot T'(\Gamma),$$

by Theorem VI.20. We can get a neater expression by introducing the notion of an Eulerian tour.

We define a *tour* in a digraph Γ as a cyclic sequence C of distinct darts of Γ such that the head of each member of C is the tail of its successor. The tour is *Eulerian* if it includes all the darts of Γ.

Now let Γ be an Eulerian digraph in which no outvalency is zero (and in which $V(\Gamma)$ is nonnull). Then from each Eulerian path in Γ we can derive an Eulerian tour in which the first dart of the path is the successor of the last. Conversely, given an Eulerian tour C of Γ and a dart D of Γ, we find that there is a uniquely determined Eulerian path P of Γ that starts with D and continues with the terms of C in order until it ends with the predecessor of D in C. We then say that P *belongs* to C. Thus, the number of Eulerian paths belonging to C and having a given vertex r as origin is the outvalency of r. We note also that each Eulerian path in Γ belongs to some Eulerian tour. Applying these results to Theorem VI.20, we arrive at the following theorem.

Theorem VI.22. *Let r be a vertex of an Eulerian digraph Γ in which no outvalency is zero. Let $C(\Gamma)$ denote the number of Eulerian tours of Γ. Then, in the notation of Eq.* (VI.3.1),

$$C(\Gamma) = \prod_{j=0}^{q} (n_j - 1)! T'(\Gamma). \tag{VI.3.2}$$

Theorem VI.23. *Let Γ be a nonnull Eulerian digraph. Then there exists an integer $T(\Gamma)$ that is equal to $T'(\Gamma)$ and $T_r(\Gamma)$ for each vertex r of Γ.*

Proof. Suppose first that some vertex r of Γ has zero outvalency and therefore zero invalency. If Γ has any other vertex, it can have no spanning arborescence of any kind. The theorem then holds, with $T(\Gamma) = 0$. If Γ has no other vertex the theorem is still true, with $T(\Gamma) = 1$. In the remaining case we apply Eq. (VI.3.2), noting that the expression on the left is independent of r. We find that $T^r(\Gamma)$ has the same value for each r. Now the Eulerian tours of the reverse of Γ are those of Γ with the cyclic orders of their terms reversed. So, by applying Eq. (VI.3.2) to both Γ and its reverse, we find that $T^r(\Gamma) = T_r(\Gamma)$ for each vertex r. □

We call $T(\Gamma)$ the *tree-number* of Γ. We can restate Theorem VI.22 in terms of $T(\Gamma)$ as follows:

Theorem VI.24. *Let Γ be a nonnull Eulerian digraph in which no outvalency is zero. Let the vertices of Γ be enumerated as $v_0, v_1, \ldots, v_q$, and let the outvalency of v_j be n_j. Then*

$$C(\Gamma) = \prod_{j=0}^{q} (n_j - 1)! \, T(\Gamma). \qquad \text{(VI.3.3)}$$

This result is known as the BEST Theorem. In this acronym, B and E stand for N. G. de Bruijn and T. van Aardenne-Ehrenfest, who published the theorem in 1951. [1] The S and T stand for C. A. B. Smith and W. T. Tutte, who solved the special case in which each n_j is 2 in 1941 [11]. They are also associated with the Matrix-Tree Theorem, which gives a formula for $T(\Gamma)$.

We have still to prove a converse of Theorem VI.14.

Theorem VI.25. *Let G be a connected Eulerian graph. Then it has at least one Eulerian path.*

Proof. In a nontrivial case G has an edge, and therefore a circuit and a circular path, by Theorem VI.10 and VI.16, this path being edge-simple. Hence there exists in G an edge-simple reentrant path P of maximum length $s(P)$. Let H be formed from G by deleting the edges with darts in P. If possible choose a component C of H having at least one edge.

By connection some vertex v of C is incident with a dart of P. Clearly, C is Eulerian. Hence by Theorem VI.16 there is an edge-simple circular path J in C from v to v. But we can write $P = QR$, where Q has terminus v. Then QJR is an edge-simple reentrant path in G whose length exceeds $s(P)$. This contradiction shows that H is edgeless. Hence P is an Eulerian path in G. □

The combination of Theorems VI.14 and VI.25 is called *Euler's Theorem*.

VI.4. THE MATRIX-TREE THEOREM

The theory of this section is based on the following characterization of diverging arborescences.

Theorem VI.26. *Let r be a vertex of a digraph Γ and let Δ be a spanning subdigraph of Γ. Then in order that Δ shall be an arborescence diverging from r, it is necessary and sufficient that the following conditions shall be fulfilled: There must be no circular path in Δ, the invalency of r in Δ must be 0, and the invalency of every other vertex in Δ must be 1.*

Proof. Let Δ be a spanning arborescence of Γ diverging from r. Then Δ satisfies the conditions, by Theorems VI.1 and VI.8.

Conversely, let Δ be a spanning subdigraph of Γ satisfying the stated conditions. Let v be any vertex other than r. We construct a sequence $(D_1, D_2, \ldots)$ of darts of Δ such that $v = h(D_1)$ and $t(D_{j-1}) = h(D_j)$ whenever $j > 1$. If this sequence can be continued to a dart D_k with tail r, then by reversing its terms we obtain a path in Δ from r to v. If this has not happened within $|V(\Gamma)|$ steps, there must be a first dart D_k having the same tail as some preceding dart, say D_j. But then Δ has the circular path $(D_k, D_{k-1}, \ldots, D_{j+1})$, which is impossible.

By Theorems VI.6 and I.25, $U(\Delta)$ is connected. The conditions imply that $|V(\Delta)| = |W(\Delta)| + 1$. Hence $U(\Delta)$ is a tree. Hence Δ is a spanning arborescence of Γ diverging from r, by Theorem VI.2. □

Now let Γ be any nonnull digraph. With each dart D of Γ we associate a *conductance* $c(D)$. Usually we can take $c(D)$ to be a real or complex variable. More algebraically, we can define it as an indeterminate over the ring of integers. If Δ is any subdigraph of Γ, we write $\Pi(\Delta)$ for the product of the conductances of the darts of Δ. If Δ has no darts, we write $\Pi(\Delta) = 1$.

We enumerate the vertices of Γ as $v_1, v_2, \ldots, v_n$. We define an $n \times n$ matrix $K(\Gamma)$, called the *Kirchhoff matrix* of Γ, as follows. The ith diagonal entry k_{ii} is the sum of the conductances of the link-darts with v_i as head. The entry k_{ij} in the ith row and jth column, where $i \neq j$, is minus the sum of the conductances of the darts with head v_i and tail v_j. We note that loop-darts are ignored in the construction of $K(\Gamma)$. From these definitions we have

$$\sum_j k_{ij} = 0, \qquad (VI.4.1)$$

$$\det K(\Gamma) = 0. \qquad (VI.4.2)$$

If r is any vertex of Γ, we write $K_r(\Gamma)$ for the submatrix of $K(\Gamma)$ obtained by striking out the row and column corresponding to r. Let us investigate the expansion of $\det K_r(\Gamma)$ on the assumption that $r = v_n$.

The Matrix-Tree Theorem

By the definition of a determinant, we can write, if $n > 1$,

$$\det K_r(\Gamma) = \sum_{ab \cdots s} N(ab \cdots s) k_{1a} k_{2b} \cdots k_{(n-1)s}, \qquad \text{(VI.4.3)}$$

where $(a, b, \ldots, s)$ is any permutation of $(1, 2, \ldots, n-1)$ and $N(ab \cdots s)$ is 1 or -1 according as this permutation is even or odd. Let us call each product $k_{1a} k_{2b} \cdots k_{(n-1)s}$ an *initial product* of our expansion. It can be any product of $(n-1)$ entries in $K_r(\Gamma)$ with one factor from each row and one from each column.

A given initial product can be written as PQ, where P is the product of its factors k_{ii} from the diagonal of $K_r(\Gamma)$, and Q is the product of its nondiagonal factors. If Q is not an empty product, it can be resolved uniquely into factors called *cycles*. A *cycle* is a product of the form $k_{\alpha\beta} k_{\beta\gamma} k_{\gamma\delta} \cdots k_{\mu\nu} k_{\nu\alpha}$, where the integers $\alpha, \beta, \ldots, \mu, \nu$ are all distinct. For any initial product X let the number of cycles be $q(X)$ and the number of cycles of even length $q_0(X)$. It is found that the coefficient $N(ab \cdots s)$ of X is 1 or -1 according as $q_0(X)$ is even or odd.

Let us now suppose that each k_{ij} is written in terms of the conductances and that $\det K_r(\Gamma)$ is then further expanded as a linear combination of products of conductances. We then have

$$\det K_r(\Gamma) = \sum_{\Delta} N(\Delta) \Pi(\Delta), \qquad \text{(VI.4.4)}$$

where Δ can be any spanning subdigraph of Γ, and $N(\Delta)$ is an integer whose value we must now investigate.

Now each product $\Pi(\Delta)$ corresponds to one or more of the initial products X if $N(\Delta)$ is nonzero. The correspondence means that we can form the product $\Pi(\Delta)$ by taking exactly one conductance from each factor k_{ij} of X. For this to be possible the invalency of r in Δ must be zero, and the invalency of any other vertex in Δ must be 1. So from now on we may suppose the sum in Eq. (VI.4.4) to be restricted to spanning subdigraphs Δ satisfying these conditions. We describe the above correspondence by saying that $\Pi(\Delta)$ is *picked out* of X.

Let X_0, an initial product, be the product of the diagonal elements of $K_r(\Gamma)$. Every $\Pi(\Delta)$ occurring in Eq. (VI.4.4) can be picked out of X_0. If $\Pi(\Delta)$ is picked out of any other X, then the conductances taken from any one of its cycles must correspond to the darts of a circular path in Δ. We deduce from Theorem VI.26 that if Δ is an arborescence diverging from r, then $\Pi(\Delta)$ can be picked out only from X_0. This implies that $N(\Delta) = 1$.

In the remaining case, Δ has at least one circular path. Then $\Pi(\Delta)$ can be picked out of an initial product X if and only if each cycle of X corresponds to a circular path in Δ. Now the circular paths in Δ determine $t > 0$ distinct tours such that the corresponding vertex-sets are disjoint. These define a set T of t possible cycles such that $\Pi(\Delta)$ can be picked out of

X if and only if the cycles of X constitute a subset of T. In particular, X may be X_0. Suppose such an X to have j cycles. Since conductances taken from nondiagonal entries in $K_r(\Gamma)$ come with a negative sign, the contribution of X to the coefficient $N(\Delta)$ in Eq. (VI.4.4) is $(-1)^j$. But there is just one X whose cycles are any given j elements of T. Accordingly,

$$N(\Delta) = \sum_{j=0}^{t} \binom{t}{j}(-1)^j = (1-1)^t = 0.$$

We conclude that Eq. (VI.4.4) remains valid when Δ is restricted to be a spanning arborescence of Γ diverging from r, and each $N(\Delta)$ is set equal to 1.

Let us now consider the effect of changing from one enumeration of the vertices of Γ to another. Then in each of the matrices $K(\Gamma)$ and $K_r(\Gamma)$, the rows are permuted and the same permutation is applied to the columns. So the determinants of these matrices are unchanged. Hence the validity of Eq. (VI.4.4) in its new restricted form is not really dependent on r being v_n. We may summarize our results as follows.

Theorem VI.27 (The Matrix-Tree Theorem). *If r is a vertex of a digraph Γ, and if $K(\Gamma)$ is defined in terms of an arbitrary enumeration of the vertices, then*

$$\det K_r(\Gamma) = \sum_{\Delta} \Pi(\Delta), \qquad (\text{VI.4.5})$$

where the sum is over all spanning arborescences Δ of Γ diverging from r.

Strictly we have proved this only for the case $n > 1$. We can make it true when $n = 1$ by postulating that the determinant of a 0×0 matrix is 1.

Consider the effect of substituting the value 1 for each conductance $c(D)$. The Matrix-Tree Theorem tells us that

$$T'(\Gamma) = \det K_r(\Gamma). \qquad (\text{VI.4.6})$$

Moreover, by applying Eq. (VI.4.6) to the reverse digraph Γ' of Γ, we have

$$T_r(\Gamma) = \det K_r(\Gamma'). \qquad (\text{VI.4.7})$$

Thus the numbers $T'(\Gamma)$ and $T_r(\Gamma)$ can be found by evaluating determinants.

Even with general conductances, there is a close resemblance between the matrices $K(\Gamma)$ and $K(\Gamma')$. The first can differ from the transposed matrix of the second only in its diagonal elements. In $K(\Gamma)$ the diagonal terms are chosen to make the row-sums vanish, as asserted by Eq. (VI.4.1); in $K(\Gamma')$, transposed, they are chosen to make the column-sums vanish. The matrix $K(\Gamma)$ is identical with the transpose of $K(\Gamma')$ if and only if the conductances are restricted to satisfy the following condition. For each vertex v_i the sum of the conductances of the link-darts with heads at v_i is equal to the sum of the conductances of the link-darts with tails at v_i.

For unit conductances this condition is satisfied when Γ is Eulerian. Then, from the fact that $K(\Gamma)$ is the transpose of $K(\Gamma')$, we can deduce from Eqs. (VI.4.6) and (VI.4.7) that $T_r(\Gamma) = T'(\Gamma)$. Indeed from these equations and the elementary theory of determinants, we can prove the whole of Theorem VI.23. This is done, for example, in [12]. The Matrix-Tree Theorem, combined with Theorem VI.23, gives us the following rule.

Theorem VI.28. *Let Γ be a nonnull Eulerian digraph. Then we can find its tree-number $T(\Gamma)$ by constructing the Kirchhoff matrix $K(\Gamma)$ with unit conductances, striking out from it any row and the corresponding column, and evaluating the determinant of the resulting matrix.*

Let us derive a corresponding theory for graphs. Now we associate a conductance $c(A)$ with each edge A of a nonnull graph G. We enumerate the vertices of G as $v_1, v_2, \ldots, v_n$ and define a *Kirchhoff matrix $K(G)$* of G as follows: The ith diagonal entry k_{ii} is the sum of the conductances of the links incident with v_i. The entry k_{ij} in the ith row and jth column, where $i \neq j$, is minus the sum of the conductances of the edges joining v_i and v_j. We observe that $K(G)$ is a symmetric matrix. We have

$$\sum_j k_{ij} = 0, \quad \sum_i k_{ij} = 0, \qquad (VI.4.8)$$

$$\det K(G) = 0. \qquad (VI.4.9)$$

For any vertex r of G we write $K_r(G)$ for the submatrix of $K(G)$ obtained from it by striking out the row and column corresponding to r.

Consider G in relation to its equivalent digraph Γ. We assign to each dart on an edge A the conductance $c(A)$ of A. This makes the matrices $K(G)$ and $K(\Gamma)$ identical.

For any subgraph H of G we write $\Pi(H)$ for the product of the conductances of the edges of H. We can now combine Theorems VI.5 and VI.27 to obtain the following theorem.

Theorem VI.29 (The Matrix-Tree Theorem for graphs). *If r is a vertex of a graph G, and if $K(G)$ is defined in terms of an arbitrary enumeration of the vertices, then*

$$\det K_r(G) = \sum_T \Pi(T), \qquad (VI.4.10)$$

where the sum is over all spanning trees T of G. Thus for unit conductances, $\det K_r(G) = T(G)$. (See Sec. II.2.)

The next result is a famous theorem of Cayley [5].

Theorem VI.30. *Let G be an n-clique, with $n \geq 2$. Then*

$$T(G) = n^{n-2}. \qquad (VI.4.11)$$

Proof. In $K_r(G)$, each diagonal entry is $(n-1)$, and each nondiagonal entry is -1. Subtracting the first row from each of the others and then dividing each of the $(n-2)$ rows other than the first by n, we obtain a matrix M with the following structure: Its first row is that of $K_r(G)$. In each other row, the first element is -1, the diagonal element is 1, and each other entry is zero. Adding each other row of M to the first, we obtain a matrix whose determinant is obviously 1. The theorem follows. □

VI.5. THE LAWS OF KIRCHHOFF

We continue with the theory of a nonnull digraph Γ with a conductance $c(D)$ assigned to each dart D. We will refer to the subdigraphs of a digraph Δ corresponding to the components of $U(\Delta)$ as the *components* of Δ. We call Δ a *multiple diverging arborescence* if each of its components is an arborescence diverging from a vertex. If the number of components is 2 or 3, we can speak instead of a *double* or *triple* arborescence.

We shall use symbols such as the following.

$$S = \langle abc, defg, h, ij, \ldots, yz \rangle. \qquad (VI.5.1)$$

Between the angular brackets we have k groups of letters separated by commas. S denotes a sum of conductance-products $\Pi(\Delta)$ over spanning k-tuple diverging arborescences Δ of Γ. The letters refer to vertices of Γ. Such a Δ contributes to the sum S if and only if its components can be ordered so that the qth one includes all the vertices of the qth letter-group of S and diverges from the first of them. Usually we shall be interested only in such simple cases as $\langle a, b \rangle$ and $\langle ax, by \rangle$. The first of these is the sum of the conductance-products of the double arborescences diverging from a and b, and spanning Γ. The second is defined similarly except that the component of the double arborescence diverging from a must include x and the other component must include y.

We now take note of some rules for operating with the symbols S of Eq. (VI.5.1).

Theorem VI.31. *If some vertex v appears in two distinct letter-groups of S, then $S = 0$. If some vertex w appears twice in the same letter-group of S, its second appearance there can be deleted without changing the value of S.*

Theorem VI.32. *Let v be any vertex of Γ. Let $S_j(v)$ denote the symbol formed from S by adjoining v to its jth letter-group as the new last* *letter.* Then

$$S = \sum_{j=1}^{k} S_j(v). \tag{VI.5.2}$$

The results follow immediately from the definitions, plus the observation for Theorem VI.32 that v must belong to some component of each spanning subdigraph of Γ.

For any vertices a, b, x, y of Γ, we write

$$[ab, xy] = \langle ax, by \rangle - \langle ay, bx \rangle. \tag{VI.5.3}$$

We call $[ab, xy]$ a *transpedance* of Γ. A transpedance of the form $[ab, ab]$ is called an *impedance*. From Theorem VI.31 and the definition (VI.5.3), we have the following identities.

$$[aa, xy] = [ab, xx] = 0, \tag{VI.5.4}$$

$$[ab, xy] = -[ba, xy] = -[ab, yx], \tag{VI.5.5}$$

$$[ab, ab] = \langle a, b \rangle. \tag{VI.5.6}$$

By Theorem VI.32 we have

$$\langle ax, b \rangle = \langle axy, b \rangle + \langle ax, by \rangle$$

and

$$\langle ay, b \rangle = \langle axy, b \rangle + \langle ay, bx \rangle.$$

Hence Eq. (VI.5.3) can be rewritten as

$$[ab, xy] = \langle ax, b \rangle - \langle ay, b \rangle. \tag{VI.5.7}$$

Theorem VI.33 (Kirchhoff's Second Law). *The transpedances of Γ satisfy the transitive identity*

$$[ab, xy] + [ab, yz] = [ab, xz], \quad \text{by (VI.5.7)}. \tag{VI.5.8}$$

Theorem VI.34 (Kirchhoff's First Law). *The transpedances of Γ satisfy the identity*

$$\sum_D [ab, t(D)y] c(D) = \langle a \rangle \delta(b, y) - \langle b \rangle \delta(a, y), \tag{VI.5.9}$$

where the sum is over all darts D of Γ such that $h(D) = y$.

Here $\langle a \rangle$, for example, is one of the symbols S of Eq. (VI.5.1). Its contributors are the spanning arborescences of Γ diverging from a. Here $\delta(b, y)$ is 1 if $b = y$, and 0 otherwise. Loop-darts make zero contributions to the sum, by Eq. (VI.5.4).

Proof. We may suppose $a \neq b$, for otherwise Eq. (VI.5.9) is trivial.

By Eq. (VI.5.3) we can write the sum Σ on the left of Eq. (VI.5.9) as $P - Q$, where

$$P = \sum_D \langle at(D), by \rangle c(D), \qquad Q = \sum_D \langle ay, bt(D) \rangle c(D). \qquad (VI.5.10)$$

Consider first the case $y = b$. Then by Theorem VI.31,

$$\Sigma = P = \sum_D \langle at(D), b \rangle c(D).$$

If, to a double arborescence Δ contributing to $\langle at(D), b \rangle$, we adjoin the extra dart D, we obtain a spanning arborescence of Γ diverging from a, by Theorems VI.2 and IV.28. Conversely, if we have such a spanning arborescence it will have just one dart with its head at b by Theorem VI.26, and the deletion of D from it will give a contributor Δ to $\langle at(D), b \rangle$. We conclude that in this case $\Sigma = P = \langle a \rangle$. (See Fig. VI.5.1.)

To deal with the case $y = a$, we have only to interchange a and b in the above argument, for

$$\langle ay, bt(D) \rangle = \langle bt(D), ay \rangle,$$

by the definition of the symbols S. We find $\Sigma = -Q = -\langle b \rangle$.

In the remaining case, y is neither a nor b. Any contributor T to $\langle at(D), by \rangle$ has two components, X diverging from a and including $t(D)$, and Y diverging from b and including y. There is just one dart E in Y with its head at y. If we delete E from y, we decompose Y into two arborescences, Y_1 diverging from b and including $t(E)$, and Y_2 diverging from y, by Theorem VI.3. If we now adjoin D we unite X and Y_2 into a single arborescence, diverging from a and including y. (See Fig. VI.5.2.) In this way we convert T into a contributor T' to $\langle ay, bt(E) \rangle$. Another way of putting this is to say that we have transformed a contributor T to P into a

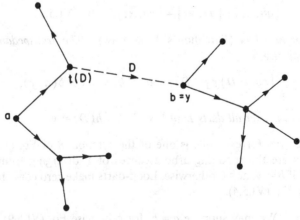

FIGURE VI.5.1

The Laws of Kirchhoff

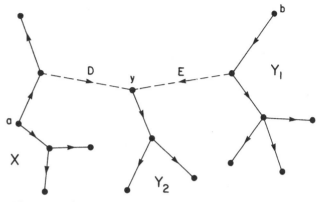

FIGURE VI.5.2

contributor T' to Q. The same procedure applied to contributors to Q changes them into contributors to P, reversing the preceding transformation. We deduce that there is a 1-1 correspondence between the contributors to P and those to Q, under which corresponding double arborescences make equal contributions to P and Q. So in this case $\Sigma = P - Q = 0$. This completes our verification of Eq. (VI.5.9). □

It is true that Theorems VI.33 and VI.54 are not quite the same as Kirchhoff's Laws for ordinary electrical networks, though we hope to exhibit them soon as a simple generalization. Nevertheless, it is now convenient to introduce some more "electrical" terminology.

Let us say that the distinct vertices a and b of Γ are chosen as *positive* and *negative poles*, respectively. It follows from Theorem VI.33 that, with respect to this choice of poles, we can associate a *potential* $V(x)$ with each vertex x so that

$$[ab, xy] = V(x) - V(y) \qquad (VI.5.11)$$

for all choices of x and y. A potential can be chosen arbitrarily at one vertex in each component of Γ. The remaining potentials are then fixed by the transpedances, by (VI.5.11). Accordingly, we may speak of a transpedance $[ab, xy]$ as the *fall of potential* from x to y, with the given poles.

The *current* in a dart D, with the given poles, is defined as

$$[ab, t(D)h(D)]c(D).$$

It is always zero if D is a loop-dart. The First Law says that the sum of the currents in the darts with a given head x distinct from a and b is zero. The corresponding sums at a and b are $-\langle b \rangle$ and $\langle a \rangle$, respectively. To balance the currents at these vertices, we say that a current $\langle b \rangle$ enters Γ at a and that a current $\langle a \rangle$ leaves it at b. We note that the current entering at a is not, in general, equal to that leaving at b, a departure from the usual electrical rule.

There is an extension of the Matrix-Tree Theorem expressing transpedances as determinants. We can substitute Eq. (VI.5.11) in Eq. (VI.5.9) to get the following linear equations for potentials:

$$\sum_D (V(t(D))-V(y))c(D) = \langle a \rangle \delta(b,y) - \langle b \rangle \delta(a,y).$$

Let us rewrite this in the notation of Sec. VI.4:

$$\sum_j k_{ij} V(v_j) = \langle b \rangle \delta(a,v_i) - \langle a \rangle \delta(b,v_i), \quad (i=1,2,\ldots,n). \quad \text{(VI.5.12)}$$

To simplify the discussion, let us put $a = v_1$ and $b = v_n$. We are free to reduce the number of unknowns by writing $V(v_n) = 0$. Correspondingly, we set aside the nth equation. We now have $(n-1)$ linear equations in $(n-1)$ unknown potentials, the matrix of our equations being $K_b(\Gamma)$. These equations have a unique solution if det $K_b(\Gamma) \neq 0$. Whether we take general or unit conductances, this happens if Γ has a spanning arborescence diverging from b, by Theorem VI.27, and therefore, if there is a path in Γ from b to each other vertex, by Theorem VI.18. Then by Cramer's Rule, $V(v_i)$ is the cofactor of k_{1i} in $K_b(\Gamma)$, if $i \neq n$.

Consider for example the digraph shown in Fig. VI.5.3. Taking a and b to be the vertices numbered 1 and 4, respectively, and using unit conductances, we have

$$K(\Gamma) = \begin{bmatrix} 1 & 0 & 0 & -1 \\ -1 & 1 & 0 & 0 \\ -1 & -1 & 2 & 0 \\ 0 & -1 & -1 & 2 \end{bmatrix}.$$

By the Matrix-Tree Theorem

$$\langle a \rangle = \begin{vmatrix} 1 & 0 & 0 \\ -1 & 2 & 0 \\ -1 & -1 & 2 \end{vmatrix} = 4, \quad \langle b \rangle = \begin{vmatrix} 1 & 0 & 0 \\ -1 & 1 & 0 \\ -1 & -1 & 2 \end{vmatrix} = 2.$$

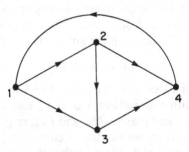

FIGURE VI.5.3

Since $\langle b \rangle$ is nonzero, we can calculate our potentials by the rule given above. Thus,

$$V(a) = \begin{vmatrix} 1 & 0 \\ -1 & 2 \end{vmatrix} = 2, \quad V(v_2) = -\begin{vmatrix} -1 & 0 \\ -1 & 2 \end{vmatrix} = 2,$$

$$V(v_3) = \begin{vmatrix} -1 & 1 \\ -1 & -1 \end{vmatrix} = 2, \quad V(b) = 0.$$

Any required transpedance, with the chosen poles, can now be found from Eq. (VI.5.11). With such a simple diagram it is easy to verify Eq. (VI.5.3) and the Matrix-Tree Theorem directly. The corresponding distribution of currents in Γ is shown in Fig. VI.5.4.

We note that there is a current of magnitude 2 entering at $a = v_1$, and one of magnitude 4 leaving at $b = v_4$.

It seems natural to divide the currents of Fig. VI.5.4 by their common factor 2. Currents as defined above, with the chosen poles, are often said to constitute the *full flow* from a to b. If they are all multiplied by a constant, the result is still called a *flow* from a to b, and the multiplied full currents are called the "currents" of this flow. When the full currents are integers, division by their highest common factor gives the *reduced flow* from a to b.

Consider the special case in which Γ is Eulerian, with unit conductances. If $T(\Gamma)$ is nonzero, Eq. (VI.5.12) has a unique solution for any choice of poles. Moreover, the current entering at the positive pole is equal to that leaving at the negative. (See Theorem VI.23.) We note that $T(\Gamma)$ is nonzero if and only if Γ is strongly connected, by Theorems VI.14, VI.18, and VI.20. There are some interesting relations between the full flows in Γ and those in its reversed digraph Γ'. We have $T(\Gamma') = T(\Gamma)$, by Theorem VI.23. Since $K(\Gamma')$ is the transpose of $K(\Gamma)$, an impedance $[ab, ab]$ has the same value in Γ' as in Γ.

We next derive an analogous theory for a graph G in which each edge A has an associated conductance $c(A)$. We enumerate the vertices as $v_1, v_2, \ldots, v_n$. The symbol S of Eq. (VI.5.1) with its k letter-groups, now

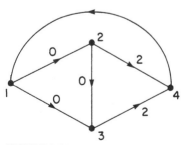

FIGURE VI.5.4

denotes a sum of conductance-products of spanning subgraphs of G, each with just k components. Such a subgraph contributes to the sum if and only if its components can be ordered so that the qth one includes all the vertices of the qth letter-group, for each q.

Consider the equivalent digraph Γ of G. As in Sec. VI.4, we give to each dart on an edge A the conductance $c(A)$ in Γ. The theories of Γ and G can be related by the following observation:

Theorem VI.35. *The symbol S of Eq.* (VI.5.1) *has the same value for Γ as for G.*

Proof. A contributor to S for Γ is an orientation of a contributor to S for G, and the two contributors have the same conductance-product. Moreover, a contributor for G has a uniquely determined orientation as a contributor for Γ, by Theorem VI.4. □

We now define the transpedances and potentials of G as the corresponding transpedances and potentials of Γ. We have already seen that G and Γ have the same Kirchhoff matrix. For both G and Γ we have the same algebra, based on the same symbols S. It is only the interpretation of these symbols that is different.

The two darts on an edge A of G can be said to define two opposite directions in A. For a given choice of poles, the current in one direction is the negative of that in the other, by Eq. (VI.5.5) and the definition of a current. Theorem VI.34 now becomes the familiar Kirchhoff's First Law. It says that the sum of the currents flowing to a vertex x of G, other than a pole, in the edges incident with x is zero. Clearly, $\langle u \rangle$ has the same value for each vertex u of G. Hence, by Theorem VI.34, the current entering G at the positive pole is equal to the current leaving it at the negative.

We now note some theorems about the transpedances of graphs that are not true for general digraphs.

Theorem VI.36. *In the case of a graph, the symbol S of Eq.* (VI.5.1) *is not altered in value when the elements of any one letter-group are permuted.*

This is true by the definition of S for a graph. The first element of a letter-group is no longer distinguished as a center of divergence.

Theorem VI.37. *The transpedances of a graph satisfy the identity*

$$[ab, xy] = [xy, ab], \qquad (VI.5.13)$$

by Eq. (VI.5.3) *and Theorem VI.36.*

Theorem VI.38. *The transpedances of a graph satisfy the identity*

$$2[ab, xy] = [ay, ay] + [bx, bx]$$
$$- [ax, ax] - [by, by]. \qquad (VI.5.14)$$

Proof. By Theorems VI.33 and VI.37, we have
$$[ab, bx] = [ab, ba] + [ab, ax],$$
$$[ab, bx] = [ax, bx] + [xb, bx].$$
Adding and using Eq. (VI.5.5), we find
$$2[ab, bx] = [ab, ax] + [ax, bx]$$
$$- [ab, ab] - [bx, bx],$$
$$2[ab, bx] = [ax, ax] - [ab, ab] - [bx, bx], \quad (VI.5.15)$$
by Theorems VI.33 and VI.37. We rewrite this with y replacing x and then subtract Eq. (VI.5.15). By Theorem VI.33, we then have Eq. (VI.5.13). □

We observe that the transpedances of a graph can be expressed in terms of its impedances, that is, in terms of its symbols of the form $\langle u, v \rangle$, by Eq. (VI.5.6).

VI.6. IDENTIFICATION OF VERTICES

Let $S = \{a_1, a_2, \ldots, a_k\}$ be a set of $k > 1$ distinct vertices of a digraph Γ. We can obtain a new digraph Γ' by identifying the members of S. By definition, Γ' has the same darts as Γ, and the same vertices except that the members of S are replaced by a single new vertex a. A dart has the same head or tail in Γ' as in Γ, except that any head or tail in S is replaced by a.

If Δ is a spanning subdigraph of Γ, the identification of the vertices of S converts it into a spanning subdigraph Δ' of Γ'.

Suppose Δ to be a k-tuple diverging arborescence such that each component includes exactly one member of S and diverges from it. By Theorem IV.28, the graph $U(\Delta')$ is a tree. Hence, Δ' is an arborescence diverging from a, by Theorem VI.2. Conversely, suppose Δ' to be such an arborescence. We can deduce from Eq. (I.6.1) that $U(\Delta)$ has at least k components. Each of these must include a member of S, since otherwise it would be a component of $U(\Delta')$. Hence, $U(\Delta)$ has exactly k components, each including exactly one vertex of S. It follows from Eq. (I.6.1) that $U(\Delta)$ is a forest, and so each of its k components is a tree. So each component of Δ is an arborescence diverging from its vertex in S, by Theorem VI.2.

Let each dart D have a conductance $c(D)$, the same in Γ' as in Γ. Then the above result is expressed by the following formula, in which the prime denotes that the symbol $\langle a \rangle$ refers to the digraph Γ':

$$\langle a_1, a_2, \ldots, a_k \rangle = \langle a \rangle'. \quad (VI.6.1)$$

Let us carry the theory of Sec. VI.5 a little further. Let $K(i, j)$ be the matrix obtained from $K(\Gamma)$ by striking out the ith row and jth column,

$K(ij, \ell m)$, the one obtained by striking out the ith and jth rows and the ℓth and mth columns, and so on. In the latter example we suppose $i \neq j$ and $\ell \neq m$, so that we shall not be called upon to strike out any row or column twice. We find that

$$\det K(i, j) = (-1)^{i+j} \langle v_i \rangle. \tag{VI.6.2}$$

If $j = i$, this is Theorem VI.27. If $j \neq i$, we add all the other columns of $K(i, j)$ to the ith, that is, the column of v_i. An entry k_{si} in that column is then replaced by $-k_{sj}$, by Eq. (VI.4.1). By $|i - j| - 1$ interchanges of consecutive columns, we can convert the resulting matrix into $K(i, i)$. Equation (VI.6.2) now follows from Theorem VI.27.

In solving Eq. (VI.5.12), we are allowed to put $V(v_n) = 0$. So, by Eq. (VI.5.11), $[v_1 v_n, v_i v_n]$ is the cofactor of k_{1i} in the matrix $K_b(\Gamma)$, that is, in $K(n, n)$. So we have

$$[v_1 v_n, v_i v_n] = (-i)^{1+i} \det K(1n, in) \tag{VI.6.3}$$

if $i \neq n$. This formula can be generalized as

$$[v_1 v_n, v_i v_j] = (-1)^{1+n+i+j} \det K(1n, ij), \tag{VI.6.4}$$

where $i < j$. If $j = n$, then Eq. (VI.6.4) merely repeats Eq. (VI.6.3). If $j \neq n$, we can prove Eq. (VI.6.4) as follows: We add all the other columns of $K(1n, ij)$ to the last, thus replacing each element k_{sn} in this column by $-k_{si} - k_{sj}$, by Eq. (VI.4.1). So

$$\det K(1n, ij) = -\det A_i - \det A_j,$$

where A_i, for example, is obtained from $K(1n, ij)$ by replacing each element k_{sn} in the last column by the corresponding k_{si}. Allowing for interchanges of columns, we find that

$$\det A_i = (-1)^{n+i} \det K(1n, jn)$$

$$\det A_j = (-1)^{n+j+1} \det K(1n, in).$$

It follows that

$$(-1)^{1+n+i+j} \det K(1n, ij)$$

$$= (-1)^j \det K(1n, jn) + (-1)^{j+1} \det K(1n, in)$$

$$= -[v_1 v_n, v_j v_n] + [v_1 v_n, v_i v_n], \quad \text{(by Eq. (VI.6.3))}$$

$$= [v_1 v_n, v_i v_j], \quad \text{(by Eq. (VI.5.8))}.$$

We have now proved Eq. (VI.6.4). We can extend to the case $i > j$ by introducing a symbol $\sigma(p, q)$ referring to distinct integers p and q. We define it to be 0 if $p < q$ and 1 if $p > q$. We make the extension by adding $\sigma(i, j)$ to the index of (-1) in Eq. (VI.6.4). As a final generalization of Eqs.

Identification of Vertices   151

(VI.6.3) and (VI.6.4), we offer

$$[v_g v_h, v_i v_j] = (-1)^\alpha \det K(gh, ij), \qquad (VI.6.5)$$

where $g \neq h$, $i \neq j$, and

$$\alpha = g + h + i + j + \sigma(g, h) + \sigma(i, j). \qquad (VI.6.6)$$

To justify this, we observe that an interchange of two consecutive vertices in the enumeration changes at most the sign on each side of Eq. (VI.6.5), and moreover makes the same change of sign on each side. Hence, Eq. (VI.6.5) follows from Eq. (VI.6.4).

One way of stating Eq. (VI.6.5) is to say that $[v_g v_h, v_i v_j]$ is the cofactor of the element k_{gi} in the matrix $K(h, j)$, multiplied by $(-1)^{h+j}$. This is put, rather loosely, in some papers as follows: The transpedance is the cofactor of k_{gi} in the cofactor of k_{hj} in $K(\Gamma)$.

Since transpedances can be expressed in terms of determinants, we can expect to get relations between them by using determinantal identities. Consider, for example, the matrix $K_S(\Gamma)$, obtained from $K(\Gamma)$ by striking out the rows and columns corresponding to the vertices in S. By Eq. (VI.6.1) and the Matrix-Tree Theorem, we have

$$\det K_S(\Gamma) = \langle a_1, a_2, \ldots, a_k \rangle = \langle a \rangle'. \qquad (VI.6.7)$$

Now $K_S(\Gamma)$ is a submatrix of $K_r(\Gamma)$, where $r = a_k$. If $\langle r \rangle$ is nonzero, we can express $\det K_S(\Gamma)$ in terms of $\langle r \rangle$ and the cofactors of $K_r(\Gamma)$ by using Jacobi's Theorem on the minors of an adjugate matrix. We find that

$$\det K_S(\Gamma) = \langle r \rangle^{-k+2} \det M, \qquad (VI.6.8)$$

where M is a $(k-1) \times (k-1)$ matrix such that the entry in the ith row and jth column is $[a_i a_k, a_j a_k]$.

Let us now discuss a graph G in which each edge A has its conductance $c(A)$. We take Γ to be the equivalent digraph of G, each dart being assigned the conductance of the corresponding edge of G. We can form a new graph G' by identifying the vertices a_i of the set S to form a new vertex a. By definition G' has the same edges as G and the same vertices, except that the members of S are replaced by a. An edge has the same ends in G' as in G, except that any end in S is replaced by a. Evidently the equivalent digraph of G' is Γ'. We can now apply Theorem VI.35. It tells us, for example, that Eq. (VI.6.1), though derived as a relation between Γ and Γ', can equally well be interpreted as a relation between G and G'. Since Γ and G have the same Kirchhoff matrix, we can also regard Eqs. (VI.6.7) and (VI.6.8) as relating G and G'.

Combining Eq. (VI.6.8) with Theorem VI.38, we arrive at the following theorem, valid for graphs but not for general digraphs.

Theorem VI.39. *Let G be a connected graph. Let G' be formed from G by identifying the members of a set $S = \{a_1, a_2, \ldots, a_k\}$ of $k \geq 3$ vertices a_i of*

S to give a new vertex a. Write $r = a_k$. Then the product

$$\langle a \rangle' \langle r \rangle^{k-2}$$

can be expressed as a polynomial in the expressions $\langle a_i, a_j \rangle$ with $1 \le i < j \le k$, in which the coefficients are integers not dependent on the structure of G.

Here the connection of G makes $\langle r \rangle$ nonzero, by Theorem I.36. Hence Eqs. (VI.6.7) and (VI.6.8) are applicable. Since the result of Theorem VI.39 concerns a polynomial identity in the conductances, it is not difficult to extend it to the case in which $\langle r \rangle = 0$.

In the case of unit conductances, $\langle r \rangle$ is $T(G)$ and $\langle a \rangle'$ is $T(G')$. By Eq. (VI.6.1), $\langle a_i, a_j \rangle$ is the tree-number of the graph obtained from G by identifying a_i and a_j.

VI.7. TRANSPORTATION THEORY

Let x and y be distinct vertices of a digraph Γ. We define a *path-bundle* X of Γ from x to y as a set of dart-simple paths from x to y in Γ, no two of them having a dart in common. The number $|X|$ of paths in X is the *order* of the path-bundle.

In this section we discuss how to find the maximum possible order for a path-bundle in Γ from x to y. It seems convenient to make the following definitions.

The *digraph* $\Gamma(X)$ of X is the subdigraph of Γ defined by the darts of the paths of X, and their incident vertices. It is the union of the digraphs of the members of P. If v is a vertex of $\Gamma(X)$ other than x or y, then the invalency of v in $\Gamma(X)$ is equal to the outvalency, and we call it the *strength* of X at v. For a vertex v of Γ that is not in $\Gamma(X)$, and is not x or y, we say that the strength of X at v is zero. If the paths of X are linear, we can describe the strength of X at v as the number of members of X passing through v.

We call X *rotational* if there is a directed circuit in $\Gamma(X)$, and *irrotational* otherwise.

Theorem VI.40. *If X is irrotational, then each member of X is linear.*

Proof. Suppose that some member P of X is a nonlinear path from x to y. Then P has a nondegenerate reentrant factor. Such a factor of minimum length is a circular path, determining a directed circuit of $\Gamma(X)$. □

Theorems VI.41. *Let $\Gamma(X)$ contain a directed 2-circuit Δ. Then there is another path-bundle X' from x to y in Γ such that $|X'| = |X|$ and $\Gamma(X')$ is obtained from $\Gamma(X)$ by deleting the two darts of Δ.*

Transportation Theory   153

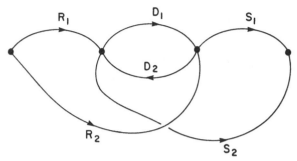

FIGURE VI.7.1

Proof. Let the darts of Δ be D_1 with tail u and head v, and D_2 with tail v and head u. (See Fig. VI.7.1.) Let P_i be the path of X including D_i ($i = 1, 2$). We can factorize P_1 and P_2 as follows:

$$P_1 = R_1(D_1)S_1, \quad P_2 = R_2(D_2)S_2.$$

We can obtain a path-bundle X' with the required properties by replacing P_1 and P_2 by $R_1 S_2$ and $R_2 S_1$ in X. □

Let us say that the path-bundle X from x to y is *reducible* if there exists another path-bundle X' from x to y in Γ such that $|X'| = |X|$, $\Gamma(X') \subseteq \Gamma(X)$, and $\Gamma(X')$ has fewer darts than $\Gamma(X)$. To *reduce* X is to replace it by such an X'.

Theorem VI.42. *If the path-bundle X is rotational, it is reducible.*

Proof. Suppose first that $\Gamma(X)$ has a loop-dart D. Then one path P of X can be factorized as $P = R(D)S$. We can reduce X by replacing P by RS.

Suppose next that $\Gamma(X)$ has a directed 2-circuit. Then X is reducible, by Theorem VI.41. From now on, we may suppose $\Gamma(X)$ to have no directed 1-circuit or 2-circuit. It must have a directed n-circuit Δ for some $n \geq 3$.

Choose a dart D of Δ. Let P be the path of X that includes D. We may suppose D to be the first dart of Δ to occur in P, perhaps by modifying the choice. Now Δ has a circular Eulerian path $Q = (D_1, D_2, \ldots, D_n)$, where $D_1 = D$. We write v_h for the head of D_h. Thus we can factorize P as RP_1, where v_n is the terminus of R and $D = D_1$ is the first dart of P_1. (See Fig. VI.7.2.)

Suppose $t(D_1)$ to be repeated in P_1 so that P can be factorized as $P = RTP_2$, where $t(D_1)$ is the terminus of R and the origin of P_2, and where D_1 is a dart of T. Then we can reduce X by replacing P by RP_2. From now on we may suppose that $t(D_1)$ is not repeated in P_1. So, in particular, D_n is not a dart of P_1.

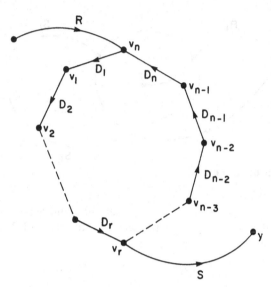

FIGURE VI.7.2

There is a greatest integer r such that $r < n$ and v_r is a vertex of P_1. Then none of the darts $D_r, D_{r+1}, \ldots, D_n$ can belong to P_1, or indeed to P. We factorize P_1 as $P_2 S$, where v_r is the origin of S.

Now to Γ we adjoin $(n - r)$ new darts $E_{r+1}, E_{r+2}, \ldots, E_n$, making the rule that the head and tail of E_j are the tail and head, respectively, of D_j ($r < j \leq n$). Let us denote the resulting digraph by Γ_1. We convert X into a path-bundle X_1 in Γ_1 by replacing P by $R(E_n, E_{n-1}, \ldots, E_{r+1})S$. Now, by $(n - r)$ applications of Theorem VI.41, we obtain from X_1 a path-bundle X' in Γ_1 such that $|X'| = |X_1| = |X|$, and such that the digraph of X' is obtained from that of X_1 by deleting all the darts D_j and E_j such that $r < j \leq n$. But then X' is a path-bundle in Γ from x to y. We have therefore succeeded in reducing X. □

By repeated application on Theorem VI.42, we have the following rule.

Theorem VI.43. *Let X be any path-bundle in Γ from x to y. Then there exists an irrotational path-bundle X' in Γ from x to y such that $|X'| = |X|$ and $\Gamma(X') \subset \Gamma(X)$.*

We now consider the problem of finding a path-bundle of highest possible order from x to y. First, however, we generalize. We introduce a function f that associates with each vertex v of Γ a nonnegative integer $f(v)$ called its *capacity*. We impose upon the required path-bundle X the condition that the outvalency in $\Gamma(X)$ of any vertex v must not exceed $f(v)$. Let us say that a path-bundle satisfying this condition is *f-limited*.

Transportation Theory

A *cut* between x and y is an ordered pair (U_x, U_y) of complementary subsets of $V(\Gamma)$ such that x is in U_x and y in U_y. If u is any vertex of U_x, we write $\beta(u)$ for the number of darts of Γ with tail at u and head in U_y. We call u an *outlet* of U_x if $\beta(u) > 0$. We define the *capacity* $C(U_x, U_y)$ of the cut (U_x, U_y) as follows:

$$C(U_x, U_y) = \sum_{u \in U_x} \text{Min}(f(u), \beta(u)). \qquad \text{(VI.7.1)}$$

Theorem VI.44. *Let (U_x, U_y) be any cut between x and y in Γ. Let X be any f-limited path-bundle from x to y. Then*

$$|X| \leq C(U_x, U_y).$$

Proof. Each path of X must have at least one dart with its tail in U_x and its head in U_y. But of the $\beta(u)$ darts with tail at $u \in U_x$ and head in U_y, the number belonging to paths of X does not exceed $f(u)$. The theorem follows. □

Let us now suppose that we have found an f-limited path-bundle X from x to y in Γ of order k. We admit the possibility that k may be zero. By Theorem VI.43 we can arrange that X is irrotational, without changing k. Then each member of X is linear, by Theorem VI.40.

It is convenient to represent Γ as an orientation of a graph G. We form G by replacing each dart D by an edge A_D with the same incident vertices. We denote the equivalent digraph of G by Λ. We take one of the two darts of Λ on any edge A_D of G to be the corresponding dart D of Γ. The other can be denoted by D^{-1}, as in Sec. VI.1. It is not a dart of Γ.

We consider those dart-simple paths Q in Λ with origin x that satisfy the following conditions:
i) *No dart of Q is in $\Gamma(X)$.*
ii) *No two darts of Q in Γ have the same tail.*
iii) *If a dart D of Q is not in Γ, then D^{-1} is in $\Gamma(X)$.*
iv) *If a dart D of Q is in Γ, then either it is the immediate successor in Q of a dart that is not in Γ or its tail u has outvalency less than $f(u)$ in $\Gamma(X)$.*

We denote the family of all such paths Q by $J(x)$. In particular, $J(x)$ includes the degenerate path at x.

We discuss the case in which $J(x)$ has a member Q from x to y. Then, by (i), we can adjoin Q to X and so obtain a path-bundle X_1 in Λ from x to y of order $k+1$. If u is any vertex of the digraph $\Lambda(X_1)$ of X_1, then its outvalency in $\Lambda(X_1)$ is at most $f(u)+1$, by (ii). Moreover, if the outvalency does attain this value, then one dart D of Q with head at u is not in Γ, by (iv). Then D^{-1} is one of the darts of $\Gamma(X)$ with tail u, by (iii).

We now apply Theorem VI.41 to all pairs (D, D^{-1}) having both members in $\Lambda(X_1)$. By (iii), there results a path-bundle X' in Γ from x to y

of order $k+1$. Moreover, X' is f-limited, by the results of the preceding paragraph.

Let us now study the remaining case, in which there is no path of $J(x)$ from x to y. We define U_x as the set of all vertices v of Γ such that there is a path of $J(x)$ from x to v. We define U_y as the complementary subset to U_x in $V(\Gamma)$. Then $x \in U_x$ and $y \in U_y$, and (U_x, U_y) is a cut between x and y in Γ.

If a member Q of $J(x)$ factorizes as $Q_1 Q_2$, it is clear that Q_1 satisfies the above Conditions (i)–(iv) and so belongs to $J(x)$. Hence, each vertex of each member of $J(x)$ must belong to U_x.

We say that a vertex v of $\Gamma(X)$ is *filled* if its outvalency in $\Gamma(X)$ is $f(v)$. An outlet u of U_x is *choked* if each dart of Γ with tail u and with head in U_y belongs to some path of X. The next three theorems are concerned with the relations between the cut (U_x, U_y) and the paths of X.

Theorem VI.45. *Let P be any path of X. Then P has exactly one dart D_P with tail in U_x and head in U_y. Moreover, when P is factorized as $P_1(D_P)P_2$, all the vertices of P_1 are in U_x and all those of P_2 are in U_y.*

Proof. Let us define D_P initially as the first dart in P with its head in U_y. Its tail must be in U_x. Then P factorizes as $P_1(D_P)P_2$, where all the vertices of P_1 are in U_x.

Suppose P_2 to have a vertex in U_x. Then some dart E of P_2 has its tail in U_y and its head in U_x. There is a path Q of $J(x)$ from x to $h(E)$. But then the path $Q(E^{-1})$ from x to $t(E)$ is in $J(x)$, contrary to the definition of U_x.

We deduce that all the vertices of P_2 are in U_y. The theorem follows. □

Theorem VI.46. *Let u be any unchoked outlet of U_x. Let $\gamma(u)$ denote the number of darts of $\Gamma(X)$ with tail at u and head in U_y. Then $\gamma(u) = f(u)$.*

Proof. Since u is not choked, there is a dart D of Γ, not in $\Gamma(X)$, such that $t(D) = u$ and $h(D) \in U_y$.

Suppose u is not filled. There is a path Q of $J(x)$ from x to u. Replacing Q by a suitable factor if necessary we can suppose no dart of Q to have tail u. Then Q can be continued along D as a path $Q(D)$ of $J(x)$ from x to $h(D)$, contrary to the definition of U_x. We conclude that, in fact, u is filled.

Suppose next that $\gamma(u) < f(u)$. Since u is filled, there is a member P of X having a dart D_1 such that $t(D_1) = u$ and $h(D_1) \in U_x$. Choose a path Q of $J(x)$ from x to $h(D_1)$.

Assume that Q has a dart E in Γ with tail u. Then Q factorizes as $Q_1 Q_2$, where E is the first dart of Q_2. By (iv), Q_1 is nondegenerate and its last dart is not in Γ. But then $Q_1(D)$ is a path in $J(x)$ from x to $h(D)$, and we have a contradiction. So from now on we may assume that Q has no such dart E.

We now find that $Q(D_1^{-1}, D)$ is a path of $J(x)$ to $h(D)$, contrary to the definition of U_x. From this contradiction we deduce that in fact $\gamma(u) > f(u)$. Since X is f-limited, it follows that $\gamma(u) = f(u)$. □

Theorem VI.47. $|X| = C(U_x, U_y)$.

Proof. By Eq. (VI.7.1) and Theorem VI.46, $C(U_x, U_y)$ is the number of darts of $\Gamma(X)$ with tail in U_x and head in U_y. By Theorem VI.45, this number is $|X|$. □

Theorem VI.48 (The Max Flow Min Cut Theorem). *The greatest integer that is the order of an f-limited path-bundle from x to y in Γ is equal to the least integer that is the capacity of a cut between x and y in Γ (for any given capacity-function f).*

Proof. Given an f-limited path-bundle X from x to y in Γ, we can, as explained above, reduce to the case in which X is irrotational, without reducing the order. If the corresponding path-family $J(x)$ has a path from x to y, we can then improve upon X by constructing from it an f-limited path-bundle X' from x to y in Γ of higher order.

Starting with the trivial path-bundle from x to y of zero order, we can repeat the above procedure until we arrive at an irrotational X, of order k, say, such that there is no path in the corresponding $J(x)$ from x to y. We then define U_x and U_y as above, and have a cut (U_x, U_y) between x and y in Γ whose capacity is k, by Theorem VI.47. By Theorem VI.44, k is both the least possible capacity for a cut between x and y and the greatest possible order for an f-limited path-bundle from x to y. □

We can claim that the above proof is constructive. It gives us a method for determining an actual example of an f-limited path-bundle of maximum order from x to y. The most difficult requirement is that of finding the path-family $J(x)$. Actually, it is not necessary to find all paths in $J(x)$. It is enough to find, for as many vertices v as possible, a single member of $J(x)$ from x to v. When we have done this for some vertices, we can either extend to another or discover the set U_x by arguments based on those of Theorems VI.45 and VI.46.

The usual form of the Max Flow Min Cut Theorem discusses not path-bundles but flows of current. We will take up this form later, in connection with the theory of 0-chains and 1-chains, showing that it is equivalent to Theorem VI.48 [7,8].

In the next two theorems we take note of some special cases.

Theorem VI.49. *Let x and y be distinct vertices in a digraph Γ. Then the maximum order for a path-bundle from x to y in Γ is the least integer k such that for some cut (U_x, U_y) between x and y in Γ the number of darts of Γ with tail in U_x and head in U_y is k.*

To prove this we choose the capacity of each vertex to be at least equal to its outvalency in Γ, and then apply Theorem VI.48.

Given a set of linear paths from x to y in Γ, we say that they are *internally disjoint* if no two of them have any common dart or any common vertex other than x and y.

Theorem VI.50 (Menger's Theorem for Digraphs). *Let x and y be distinct vertices of a digraph Γ so that there is no dart of Γ from x to y. Then the maximum number of internally disjoint linear paths from x to y in Γ is the least integer k such that for some cut (U_x, U_y) between x and y in Γ the number of outlets of U_x is k, and x is not one of these outlets.*

Proof. We put $f(v) = 1$ for each vertex v other than x and y. But for x and y we make the capacity at least equal to the outvalency in Γ. By Theorem VI.48, if h is the greatest order for an f-limited path-bundle from x to y in Γ, there is a cut (U_x, U_y) between x and y in Γ of capacity h, and no such cut of smaller capacity. By Theorem VI.43, there is an irrotational f-limited path-bundle X from x to y of order h. By Theorem VI.40, the paths of X are linear.

Let T be the subset of U_y defined by those vertices t of U_y such that there is a dart from x to t in Γ. Then y is not in T, by hypothesis. If we transfer the members of T from U_y to U_x, we do not increase the capacity of the cut, since each vertex of T has only unit capacity. Neither, of course, do we decrease this capacity. We conclude that (U_x, U_y) can be chosen so that x is not an outlet of U_x. Accordingly, $h = k$.

No two members of X have a common dart, by the definition of a path-bundle. No two of them have a common vertex other than x and y, by our choice of f. The theorem now follows from the definition of h. □

It should be noted that the form of Menger's Theorem for graphs given in Theorem I.16 can be proved by applying Theorem VI.50 to the equivalent digraph of the relevant graph G. (See Theorem VI.9.)

VI.8. NOTES

VI.8.1. Textbooks

Most books on graph theory give some attention to digraphs. For a textbook whose main concern is digraph theory, see [9].

VI.8.2. Alternating Maps

There is an interesting theory of planar Eulerian digraphs. When one of these is drawn so that incoming and outgoing edges alternate around each vertex, we have an "alternating map." Alternating maps come in sets of three, much as undirected planar maps come in dual pairs. Their theory is discussed in [12] and [4]. The three alternating maps in one of the above sets all have the same tree-number.

Reference [12] is concerned with triangulated triangles. It shows that the structure of one of these dissections can be represented by an alternating map and a set of Kirchhoff equations for the corresponding digraph.

VI.8.3. Kirchhoff's Laws for Digraphs

Computations like those carried out for Fig. VI.5.3, but for other digraphs, can be found in [14]. The theory of Kirchhoff's Laws for digraphs is carried further in [13].

VI.8.4. Rotors

Theorem VI.39 appears in Sec. 7.2 of [2], though the proof is left to the reader and only a special case is considered. The argument shows that the tree-number of a graph does not change when a "rotor" in the graph is "reversed." A rotor is a subgraph with rotational symmetry whose vertices of attachment are all equivalent with respect to this symmetry. To "reverse" a rotor is to replace it by its mirror image. The theory of rotors is discussed in [2], [3], and [13].

EXERCISES

1. Construct an Eulerian orientation for the graph of the octahedron. Find some Eulerian tours of the resulting digraph, and discuss the corresponding arborescences.
2. Use the Matrix-Tree Theorem to find the tree-number of your digraph of Example 1. Verify by an actual count of arborescences.
3. Use the Matrix-Tree Theorem to find the number of spanning trees of the 4-clique and the 5-wheel.
4. Find a general formula for the tree-number of the wheel of n spokes. (See [10].)
5. In your digraph for Exercise 1, find some examples of f-limited path-bundles from one vertex to another, $f(x)$ being 1 at each vertex.

REFERENCES

[1] van Aardenne-Ehrenfest, T., and N. G. de Bruijn, Circuits and trees in oriented linear graphs, *Simon Stevin* **28** (1951), 203–217.

[2] Brooks, R. L., C. A. B. Smith, A. H. Stone, and W. T. Tutte, The dissection of rectangles into squares, *Duke Math. J.*, **7** (1940), 312-340.
[3] _____, A simple perfect square, *Proc. Nederl. Akad. Wetensch.*, **50** (1947), 1300-1301.
[4] _____, Leaky electricity and triangulated triangles, *Phillips Res. Reports*, **30** (1975), 205-219.
[5] Cayley, A., A theorem on trees, *Quart. J. Pure Appl. Math.*, **23** (1889), 376-378.
[6] Euler, L., Solutio problematis ad geometriam situs pertinentis, *Comm. Acad. Sci. Imp. Petropol.*, **8** (1736). 128-140 = Opera omnia (1), Vol. 7, 1-10.
[7] Ford, L. R., and D. R. Fulkerson, Maximal flow through a network, *Can. J. Math.*, **8**, (1956), 399-404.
[8] _____, *Flows in networks*, Princeton Univ. Press, 1962.
[9] Harary, F., R. Z. Norman, and D. Cartwright, *Structural models*, Wiley, New York, 1965.
[10] Sedláček, J., *Lucas numbers in graph theory*. (Czech. English summary). Mathematics (Geometry and Graph and Theory) (Czech), pp. 111-115. Univ. Karlova, Prague, 1970.
[11] Smith, C. A. B., and W. T. Tutte, On unicursal paths in a network of degree 4, *Amer. Math. Monthly*, **48** (1941), 233-237.
[12] Tutte, W. T., The dissection of equilateral triangles into equilateral triangles, *Proc. Cambridge Phil. Soc.*, **44** (1948), 463-482.
[13] _____, The rotor effect with generalized electrical flows, *Ars Combinatoria*, **1** (1976), 3-31.
[14] _____, "Dissections into equilateral triangles." In *The Mathematical Gardner*, D. A. Klarner, ed. Wadsworth International, Belmont, California, 1981, pp. 127-139.

Chapter VII

Alternating Paths

VII.1. CURSALITY

In the first part of this chapter we consider a graph G with a given partition of $E(G)$ into complementary subsets E_1 and E_2, whose members are called *blue* and *red* respectively. To each dart of the equivalent digraph Γ, we ascribe the color of the corresponding edge of G.

A path in G is *alternating* if its darts are alternately red and blue. We consider all degenerate paths and all paths of length 1 to be alternating.

Given a vertex r of G, we discuss the family $J(r)$ of all edge-simple alternating paths P in G whose origin is r and whose first dart, if any, is red. We seek a necessary and sufficient condition, in terms of the structure and edge-coloring of G, for the existence of a path of $J(r)$ from r to another given vertex s. In the second part of the chapter, we use such a condition in developing a theory of factorization of graphs.

We note that if a path-product $P_1 P_2$ belongs to $J(r)$, then its first factor P_1 belongs to $J(r)$.

Our first step is a classification of the darts and edges of G with relation to $J(r)$. A dart is *cursal* if it belongs to some path of $J(r)$, and *noncursal* otherwise. An edge is *noncursal*, *unicursal*, or *bicursal* according as the number of cursal darts on it is 0, 1, or 2.

Theorem VII.1. *Let A be any loop of G. Then A is either noncursal or bicursal.*

Proof. The two darts on A have the same head and tail. If one occurs in a path of $J(r)$, we can replace it by the other, and so obtain another path of $J(r)$. □

We note that a unicursal edge must be a link, with distinct ends x and y. Hence, each dart on such an edge A can be distinguished by giving its head or tail. If its cursal dart has tail x and head y, we say that A is unicursal *from* x and *to* y.

Theorem VII.2. *Any two edges A and B unicursal to a vertex v must be of the same color. If $v = r$, then any edge unicursal to v must be blue.*

Proof. Suppose A and B to be of different colors. Interchanging their symbols if necessary, we can say that there is a path P of $J(r)$ whose terminus is v, whose last dart D is on A, and that includes no dart on B. But then $P(D_1)$, where D_1 is the noncursal dart on B, is a member of $J(r)$, and we have a contradiction.

To prove the second part of the theorem, suppose that some edge C unicursal to r is red. Then the noncursal dart on C defines a path of $J(r)$ of length 1, which is a contradiction. □

A vertex v of G is *bicursal* if it is incident with some bicursal edge. It is *noncursal* if it is distinct from r and each of its incident edges is noncursal. In the remaining case, v is *unicursal*. Thus r is unicursal if it is incident with no bicursal edge. Any other vertex is unicursal if it is incident with at least one unicursal edge and no bicursal one.

Now if v is not r and some member P of $J(r)$ includes a dart with tail v, then P must have an immediately preceding dart with head v. So we have the following rule.

Theorem VII.3. *If v is a unicursal vertex, then either $v = r$ or some edge is unicursal to v.*

A unicursal vertex v is *blue-entrant* if $v = r$ or if some blue edge is unicursal to v. It is *red-entrant* if some red edge is unicursal to v. By Theorem VII.3, each unicursal vertex is either red-entrant or blue-entrant, and by Theorem VII.2 no unicursal vertex is both. We write U for the set of unicursal vertices of G, U_1 for the set of blue-entrant ones, and U_2 for the set of red-entrant ones. In what follows "color i" means "blue" or "red" according as i is 1 or 2.

Theorem VII.4. *Let v be in U_i. Let A be an edge incident with v and not of color i. Then A is unicursal from v.*

Proof. If $v = r$, then $i = 1$ and A is red. A dart on A with tail r defines a member of $J(r)$ of length 1. Hence, A must be unicursal from v.

In the remaining case, v is not r. By Theorem VII.3 there is a path P of $J(r)$ having a dart d with head v. Taking P of minimum length, we arrange that D is its last dart, and its only dart with head v. Now D has color i since $v \in U_i$. Hence, if D' is the dart on A with tail v, then the path $P(D')$ is in $J(r)$. We deduce that A is unicursal from v. □

Theorem VII.5. *Let v be in U_i. Let A be an edge incident with v and of color i. Then either A is noncursal or it is unicursal to v.*

Proof. Suppose the theorem fails for A. Then A must be unicursal from v. Let D be its cursal dart. Some path P of $J(r)$ includes D. But D is not the first dart of P, since A has color i. Let D' be the immediately preceding dart, on an edge B. Then B is not of color i. Moreover, B, being not bicursal, is unicursal to v. But this is contrary to Theorem VII.4. □

VII.2. THE BICURSAL SUBGRAPH

The bicursal edges and vertices of G define a subgraph W. We call it the *bicursal subgraph* of G (with respect to $J(r)$). By the definition of a bicursal vertex, W is a reduction of G.

Let K be any component of W. Since each dart of K is cursal, there is a path of $J(r)$ having its terminus in K. We choose such a path P of minimum length and call it the *entry-path* of K. We refer to the terminus of P as the *entrance* of K, and denote it by e.

If r is in K, then P must be the degenerate path at r, and so $e = r$.

If r is not in K, then the last dart of P is the *entry-dart* D_e of K, and the corresponding edge A_e of G is the *entry-edge* of K. In this case, P factorizes as $R(D_e)$, where the path R has no edge or vertex in K. We note that A_e is not in K: It has one end e in K and a second end t, the terminus of R, not in K. Hence A_e is not bicursal. It must be unicursal to e.

We write $i(K) = 1$ if P is degenerate. Otherwise $i(K)$ is 1 or 2 according as the entry-edge A_e is blue or red.

We discuss the family $J(K, e)$ of all edge-simple alternating paths Q in K whose origin is e and whose first dart, if any, has color other than $i(K)$. We note that for each such Q, the path PQ is defined and is in $J(r)$. We note also that if the path-product $Q_1 Q_2$ is in $J(K, e)$, then Q_1 is in $J(K, e)$.

Theorem VII.6. *Let Q be a nondegenerate member of $J(K, e)$. Let v be its terminus, and let its last dart D be on the edge A of K. Let some edge B of G incident with v be either noncursal or unicursal to v. Then either B is the entry-edge of K or it has the same color as A.*

Proof. Assume that B is not A_e and that it differs in color from A. Let D' be a dart on B with tail v. Then the path $PQ(D')$ is in $J(r)$, contrary to the definition of B. □

Theorem VII.7. *K has at least one edge whose color is not $i(K)$ and which is incident with e.*

Proof. We know that K has at least one edge incident with e. Assume that all such edges have color $i(K)$. Let A be one of them. Let D be a dart on A with tail e. Now D belongs to a path S of $J(r)$. It is not the first dart of S, since it has color $i(K)$. Hence, it has an immediately preceding dart D' in S. The dart D' has head e. It is on an edge B of G whose color is not $i(K)$. Evidently B must be either bicursal or unicursal to e. The first alternative is contrary to our assumption. The second contradicts Theorem VII.2. □

Theorem VII.8. *Let Q be a path of $J(K,e)$ from e to another vertex v. Then v is incident with both a red and a blue edge of K.*

Proof. Assume that all the edges of K incident with v have the same color i. One of them, A say, carries the last dart of Q. Call this dart D.

The dart D^{-1} belongs to some path S of $J(r)$, in which there is an immediately preceding dart D'. (If $v = r$, then $v = e$.) The edge B of G carrying D' is incident with v and not of color i. Now B is either bicursal or unicursal to v. The first alternative puts B in K and contradicts our assumption. The second is ruled out by Theorem VII.6. □

Theorem VII.9. *Let D be a dart of K belonging to some path Q of $J(K,e)$. Then D^{-1} belongs to some path of $J(K,e)$.*

Proof. Let D be on the edge A of K. We may suppose D to be the last dart of Q. If A is a loop, we can replace D by D^{-1} in Q and so obtain a path of $J(K,e)$ that includes D^{-1}. We may assume therefore that A is a link with ends x and y, and that D is a dart from x to y. (See Figs. VII.2.1 and VII.2.)

There is a member S of $J(r)$ having D^{-1} as its last dart. If S is a path in K, we must have $e = r$, $i(K) = 1$. But then S belongs to $J(K,e)$. We can therefore assume that S has a last dart D_1 that is not in K. The vertex $h(D_1)$ is then in K. Moreover, we can factorize S as $S_1(D_1)S_2$, where S_2 is a path in K with D^{-1} as its last dart. We note that the edge A_1 of D_1 is unicursal to $h(D_1)$.

Suppose $h(D_1) = e$. Thus D_1 is blue if $e = r$ and otherwise D_1 has the color of the entry-dart of K, by Theorem VII.2. In either case, S_2 is a path of $J(K,e)$, and the theorem holds. From now on we may assume that $h(D_1)$ is not e.

There is a first dart D_2 of Q such that either D_2 or D_2^{-1} belongs to S_2. (For D is in Q and D^{-1} is in S_2.) We factorize Q as $Q_1(D_2)Q_2$ and S_2 as either $S_3(D_2)S_4$ or $S_3(D_2^{-1})S_4$. With the first factorization of S_2, we find that $Q_1(D_2)S_4$ is a path of $J(K,e)$ including D^{-1}. (See Fig. VII.2.1.) With the second possible factorization of S_2 we find that $Q_1(D_2)S_3^{-1}(D_1)^{-1}$ is a

The Bicursal Subgraph

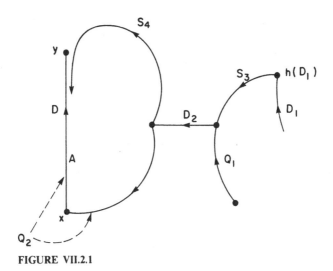

FIGURE VII.2.1

path of $J(r)$. But this is impossible since A_1 is unicursal to $h(D_1)$. (See Fig. VII.2.2.) □

Theorem VII.10. *Each dart of K belongs to some path of $J(K, e)$.*

Proof. By Theorem VII.9, we can partition $E(K)$ into complementary subsets X and Y such that each dart on each member of X belongs to some path of $J(K, e)$, and no dart on any member of Y belongs to any path of $J(K, e)$.

From Theorem VII.7, we infer that X is not null.

Assume that Y is nonnull. By connection there is a vertex v of K incident with at least one edge of X and at least one edge of Y. We can find

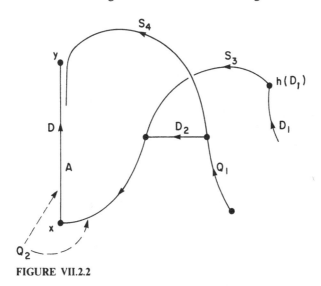

FIGURE VII.2.2

a member of $J(K,e)$ from e to v having its last dart on an arbitrary member of X incident with v. Since no such path can be continued along an edge of Y at v, we infer that all the edges of K incident with v have the same color. Hence, $v = e$, by Theorem VII.8. It follows that no edge of K incident with e has color $i(K)$ by Theorem VII.7. Hence, all the edges of K incident with $v = e$ are in X. But this contradicts the definition of v.

We infer that in fact Y is null, i.e., that the theorem is true. □

If H is a subgraph of G, let us define an *edge of attachment* of H as an edge of G that is not in H but has at least one end in H. Thus the ends in H of the edges of attachment of H are the vertices of attachment of H. If H is an induced subgraph of G, then each of its edges of attachment is a link, having one end in $V(H)$ and one in $V(G) - V(H)$.

Theorem VII.11. *Let K be any component of W. Then K is an induced subgraph of G. Moreover, any edge of attachment of K is either the entry-edge of K or unicursal from its end in K.*

Proof. Let A be any edge of $E(G) - E(K)$ that has an end x in K and is not the entry-edge of K. It carries a dart D with head x. Clearly, A is not bicursal: that would put it in K.

By Theorem VII.10 we can find a nondegenerate member of $J(K,e)$ from e to x, with its last dart D' on an arbitrary edge of K incident with x. If A is not unicursal from x, it follows from Theorems VII.1, VII.2, and VII.3 that $x = e$ and that A is not of color $i(K)$. But now we have a path $P(D^{-1})$ of $J(r)$, where P is the entry-path of K. Hence A is still unicursal from x, a contradiction.

We now know that A is not a loop, by Theorem VII.1. It remains to show that the link A does not have both ends x and y in K. But if it does, then the above argument proves it unicursal from y as well as x, which is impossible.

We remark in conclusion that the entry-edge of K, if any, is by its definition a link having only the one end e in K. The proof of Theorem VII.11 is now complete. □

It follows from Theorem VII.11 that we have no freedom of choice for the entry-edge and entry-dart of K. If r is in K, there is no entry-dart, and every edge of attachment of K is unicursal from its end in K. If r is not in K, then all but one of the edges of attachment of K are unicursal from their ends in K. The exception has to be taken as the entry-edge. It is unicursal to its end in K.

Theorem VII.12. *Each vertex of a component K of W is at least divalent in K.*

Proof. For a vertex other than e this follows from Theorem VII.8.

The entrance e is incident with some edge A of K on which there is a dart D with tail e. The dart D^{-1} belongs to some path Q of $J(K,e)$, by Theorem VII.10. If the tail of D^{-1} is e, then e is at least divalent in K. In the remaining case, D^{-1} is not the first dart of Q. But the first dart of Q is on an edge B of K incident with e, and B is not A because Q is edge-simple.
□

VII.3. BICURSAL UNITS

We continue with the theory of Sec. VII.2. Let us define Z as the set of all edges A of G such that A is not in W but has both ends in W.

By Theorem VII.11 the ends of A are in two distinct components K_1 and K_2 of W, say x in K_1 and y in K_2. By Theorem VII.11, A is unicursal, let us say from x to y. So, again by Theorem VII.11, A is the entry-edge of K_2.

Adjoining the members of Z to W, we obtain an induced subgraph W' of G. The components of W' are the *bicursal units* of G, with respect to $J(r)$.

Let B be any bicursal unit of G. We define the *constituents* of B as the components of W contained in it. Then B is formed from the union of its constituents by adjoining the edges in some subset Z_B of Z. Let $m(B)$ be the number of constituents of B. It is nonzero.

Theorem VII.13. *There is at least one constituent of B having no entry-edge in Z_B.*

Proof. Choose a constituent K of B with an entry-path P of minimum length. If $s(P) = 0$, the theorem holds; K includes r and has no entry-edge. In the remaining case, the last dart of P has a tail not in B. Hence the entry-edge of K is not in Z_B. □

Since each edge of Z_B is the entry-edge of exactly one constituent of B, we infer that

$$|Z_B| < m(B). \qquad (\text{VII}.3.1)$$

Theorem VII.14. *The bicursal unit has exactly one constituent with no entry-edge in Z_B. Moreover, each edge of Z_B is an isthmus of B.*

Proof. Let us start with the union of the constituents of B, a graph with just $m(B)$ components. Let us adjoin to it the edges of Z_B, one by one. This reduces the number of components to 1. But each adjunction of an edge reduces the number of components by at most 1, by Theorem I.29.

Hence, by Eq. (VII.3.1), we have

$$|Z_B| = m(B) - 1, \qquad (VII.3.2)$$

and the first part of the theorem is proved. The last edge to be adjoined must be an isthmus of B, by Theorem I.29. Since the edges Z_B can be adjoined in any order, the theorem follows. □

Theorem VII.15. *Let B be a bicursal unit of G. With at most one exception, each edge of attachment of B is unicursal from its end in B to some unicursal vertex of G. The exception occurs if and only if r is not in B. It is then unicursal from some unicursal vertex of G to its end in B.*

Proof. Let A be any edge of attachment of B. Since W' is an induced subgraph of G, this edge has one end x in B and one end y not in B. But A is unicursal, by Theorem VII.11. Hence y is a unicursal vertex of G; otherwise, A would be in Z and therefore in B.

If A is unicursal to x, then by Theorem VII.11, A is the entry-edge of the single constituent of B with no entry-edge in Z_B. Hence, there is just one such edge A if r is not in B, and none at all if r is in B. □

If r is not in B, it is natural to refer to the single edge unicursal to a vertex of B as the *entry-edge* of B, and to its cursal dart as the *entry-dart* of B.

Consider again the set U of unicursal vertices of G. Let us call $G[U]$ the *unicursal graph* of G. The components of $G[V(G)-U]$ are the nuclei of the bridges of $G[U]$ in G. Among them we can recognize the bicursal units of G, by Theorem VII.15. The other components of $G[V(G)-U]$ are made up of noncursal edges and vertices. Let us call them the *noncursal units* of G. Their edges of attachment are all noncursal and have each one end in U.

VII.4. ALTERNATING BARRIERS

We proceed to prove the connection theorem foreshadowed in Sec. VII.1. It is convenient to extend the definition of U_1 and U_2. We now take them to be any two disjoint subsets of $V(G)$, and we write χ for the ordered pair (U_1, U_2).

The union of U_1 and U_2 is the set U of *inner vertices* of χ. Its complement in $V(G)$ is the set of *outer vertices* of χ. The set $V(G)-U$ induces a subgraph of G called the *outer graph* of χ. Its components are called the *components* of χ.

Consider an edge A of attachment of some component M of χ. It has one end $x(A)$ in M and the other end $y(A)$ in U. We call A an *entry-edge* of M, with respect to χ, if A is red and $y(A)$ in U_1, or if A is blue and $y(A)$ in U_2.

We call χ an *alternating barrier* of G, with *center* r, if it satisfies the following conditions.
i) *If $r \in U$, then $r \in U_1$.*
ii) *If each end of an edge A of G is in U, then one end is in U_1 and one in U_2.*
iii) *A component of χ has at most one entry-edge. A component of χ including r has no entry-edge.*

Let us describe a component of an alternating barrier χ as *accessible* if it includes r, the center, or has an entry-edge, and is *inaccessible* otherwise. Thus for an inaccessible component M, each red edge of attachment has an end in U_2, and each blue one an end in U_1.

Theorem VII.16. *Let $\chi = (U_1, U_2)$, where the U_i are defined as in Sec. VII.1. Then χ is an alternating barrier of G with center r. Its accessible components are the bicursal units of G, with respect to r.*

Proof. Condition (i) holds by the definitions of Sec. VII.1. To prove (ii), we first observe that A must be unicursal, and so not a loop by Theorem VII.1. The condition now follows from Theorems VII.4 and VII.5.

The bicursal units are components of χ satisfying (iii), by Theorem VII.15 and the definition of W'. They are clearly accessible, by Theorem VII.15. For any other component M of χ, the vertices are all noncursal, and therefore the edges of attachment are all noncursal. Each such edge is incident with an inner vertex of its own color, by Theorem VII.4. So M satisfies Condition (iii) and is inaccessible. □

Let us refer to the alternating barrier recognized in Theorem VII.16 as the *cursal barrier* with center r.

Theorem VII.17. *Let $\chi = (U_1, U_2)$ be any alternating barrier of G with center r. Let s be a vertex of an inaccessible component M of χ. Then there is no path of $J(r)$ from r to s.*

Proof. Assume that there is a path P of $J(r)$ from r to s. Let us call a dart D of P *irregular* if its tail is an inner vertex of χ of the same color as D. This means that $t(D)$ is in U_1 or U_2 according as D is blue or red. For example, P must have a first dart whose head is in M, and this dart must be irregular, by the inaccessibility of M.

Let us now take D to be the first irregular dart in P. It is not the first dart of P, since P is in $J(r)$. Let D' be the immediately preceding dart of P. Suppose that $t(D')$ is an inner vertex. Then it differs in color from $t(D)$, by Condition (ii). But D and D' differ in color since P is an alternating path. Hence D' is irregular, contrary to the definition of D.

In the remaining case $t(D')$ belongs to some component N of χ. Moreover, D' is on an entry-edge A of N. Hence, N is accessible, and it does not include r, by (iii). Now P has a first dart D'' whose head is in N, and D''

must come before D and D' in P. Since D'' is not on the entry-edge A of N, it must be irregular in P, contrary to the definition of D. □

Theorem VII.18 (Theorem of Alternating Connection). *Let r and s be distinct vertices of G. In order that there shall be no path in $J(r)$ from r to s, it is necessary and sufficient that G shall have an alternating barrier whose center is r and one of whose inaccessible components contains s.*

Proof. Suppose there is no path in $J(r)$ from r to s. Let χ be the cursal barrier with center r. With respect to $J(r)$, s is a noncursal vertex and so belongs to a noncursal unit, that is to an inaccessible component of χ.

Conversely, suppose G to have an alternating barrier χ with center r and with s in an inaccessible component. Then there is no path of $J(r)$ from r to s, by Theorem VII.17. □

Let us define the *accessible* vertices of an alternating barrier of G as those not belonging to its inaccessible components. Consider the cursal barrier with center r. Its accessible vertices are those we can reach from r by paths of $J(r)$. By Theorem VII.17 all of them are accessible with respect to any other alternating barrier of G with center r. Hence, the cursal barrier is minimal with respect to the number of accessible vertices.

VII.5. f-FACTORS AND f-BARRIERS

We now come to an application of the theory of alternating paths. We consider a graph G and a function f that associates with each vertex x an integer $f(x)$ such that

$$0 \le f(x) \le \mathrm{val}(G, x). \qquad \text{(VII.5.1)}$$

A spanning subgraph F of G will be called f-*limited* if

$$\mathrm{val}(F, x) \le f(x) \qquad \text{(VII.5.2)}$$

for each $x \in V(G)$. For such an F, we write also

$$\delta(F, x) = f(x) - \mathrm{val}(F, x), \qquad \text{(VII.5.3)}$$

$$\delta(F) = \sum_{x \in V(G)} \delta(F, x). \qquad \text{(VII.5.4)}$$

We note that $\delta(F, x)$ is always nonnegative. We say that a vertex x is *filled* or *unfilled* by F according as $\delta(F, x)$ is zero or positive.

We call $\delta(F)$ the *deficiency* of F. If it is zero, that is, if every vertex of G is filled by F, we call F an f-*factor* of G. In another terminology, f-limited spanning subgraphs are "f-matchings" and f-factors are "perfect f-matchings."

f-Factors and *f*-Barriers   171

In what follows we seek necessary and sufficient conditions for the existence in G of an *f*-factor. One rather trivial necessary condition should be noted now.

Theorem VII.19. *If G has an f-factor, then*

$$\sum_{x \in V(G)} f(x) \equiv 0 \quad (\text{mod } 2). \tag{VII.5.5}$$

We prove this by applying Eq. (I.1.1) to an *f*-factor F.
It is convenient to define another function f' on $V(G)$ by

$$f'(x) = \text{val}(G, x) - f(x). \tag{VII.5.6}$$

Then Eq. (VII.5.1) also holds with f' replacing f. With each *f*-factor F of G, there is associated a unique f'-factor F', whose edges are those edges of G not belonging to F. The correspondence $F \to F'$ imposes a kind of duality on the theory of graph factors.

Here is a very simple factor theorem.

Theorem VII.20. *Let G be a connected Eulerian graph, and let $f(x) = \frac{1}{2}\text{val}(G, x)$ for each vertex x. Then G has an f-factor if and only if $|E(G)|$ is even.*

Proof. If $|E(G)|$ is odd, there is no *f*-factor, by Theorem VII.19 and Eq. (I.1.1). There is an Eulerian path P in G by Theorem VI.25. If $|E(G)|$ is even, then by taking every other edge in P, we obtain the edge-set of an *f*-factor F. □

It is sometimes possible to find in G a structure, called an *f-barrier*, that imposes a lower bound on the deficiencies of the *f*-limited spanning subgraphs. In describing an *f*-barrier, we shall use the following definitions.

Let Y be a subset of $V(G)$, and let x be a vertex of G. If x is not in Y, we define $\lambda(Y, x)$ as the number of links of G joining x to vertices of Y. If x is in Y, then $\lambda(Y, x)$ is the number of links of G joining x to other vertices of Y, plus twice the number of loops of G on x; for disjoint subsets X and Y of $V(G)$, we define $\lambda(X, Y)$ as the number of links of G with one end in X and one in Y.

Our definition of an *f*-barrier is rather like that of an alternating barrier in Sec. VII.4. We start with an ordered pair $B = (S, T)$ of disjoint subsets of $V(G)$. We write $U = V(G) - (S \cup T)$. The set U induces a subgraph $G[U]$ of G that we call the *outer graph* of B. We refer to the components of this outer graph as the *components* of B. We proceed to classify these components as *even* or *odd*.

Let K be any component of B. We write

$$J(K) = \sum_{b \in V(K)} f(b) + \lambda(V(K), T), \tag{VII.5.7}$$

and we say that K is even or odd according as the integer $J(K)$ is even or odd.

We denote the number of odd components of B by $h(B)$. We write also

$$\delta(B) = h(B) - \sum_{s \in S} f(s) - \sum_{t \in T} f'(t) + \lambda(S,T). \qquad \text{(VII.5.8)}$$

We call $\delta(B)$ the *deficiency* of B. We say that B is an *f-barrier* if $\delta(B)$ is positive.

Theorem VII.21. *If G has an f-barrier, it has no f-factor.*

Proof. Assume that G has both an f-factor F and an f-barrier $B = (S, T)$.

Let K be an odd component of B. Clearly, the number of links of F joining vertices of K to vertices not in K has the parity of

$$\sum_{x \in V(K)} f(x).$$

It is therefore not $\lambda(V(K), T)$, by Eq. (VII.5.7). Hence either there is an edge of F joining a vertex of $V(K)$ to one of S, or there is an edge not in F that joins a vertex of $V(K)$ to one of T. Let there be a odd components of B for which the first alternative holds and b for which the second is true. Then

$$a + b \geq h(B). \qquad \text{(VII.5.9)}$$

Let m be the number of edges of F with one end in S and one in T. Then we have

$$\sum_{s \in S} f(s) \geq a + m. \qquad \text{(VII.5.10)}$$

The number of edges of F with one end in U and one in T is at most $\lambda(T, U) - b$. The number with both ends in T, loops counted twice, is the sum of the numbers $\lambda(T, t)$ over the vertices t of T, at most. We deduce that

$$m \geq \sum_{t \in T} f(t) - \lambda(T, U) - \sum_{t \in T} \lambda(T, t) + b$$

$$= b + \sum_{t \in T} f(t) - \sum_{t \in T} \text{val}(G, t) + \lambda(S, T),$$

$$m \geq b - \sum_{t \in T} f'(t) + \lambda(S, T). \qquad \text{(VII.5.11)}$$

Combining Eqs. (VII.5.9), (VII.5.10), and (VII.5.11), we find that

$$0 \geq h(B) - \sum_{s \in S} f(s) - \sum_{t \in T} f'(t) + \lambda(S, T),$$

$$0 \geq \delta(B).$$

This contradicts the definition of B as an f-barrier. □

We conclude with some minor theorems about f-barriers.

Theorem VII.22. *Let $B = (S, T)$ be any ordered pair of disjoint subsets of $V(G)$. Then*

$$\delta(B) \equiv \sum_{x \in V(G)} f(x) \pmod{2}. \tag{VII.5.12}$$

Proof. From Eqs. (VII.5.7) and (VII.5.8), we have

$$\delta(B) \equiv \sum_{u \in U} f(u) + \lambda(T, U) + \sum_{s \in S} f(s) + \sum_{t \in T} f(t)$$

$$+ \sum_{t \in T} \mathrm{val}(G, t) + \lambda(S, T)$$

$$\equiv \sum_{x \in V(G)} f(x) + \sum_{t \in T} \mathrm{val}(G, t) + \lambda(S \cup U, T) \pmod{2},$$

and Eq. (VII.5.12) follows. □

Because of Theorem VII.19 we are usually concerned with the case in which f is *even-summing*, that is, in which the sum on the right of Eq. (VII.5.12) is even. The fact that $\delta(B)$ is then always even is often helpful.

We now have some comments on the duality of f and f'. Let $h'(B)$ and $\delta'(B)$ denote the functions $h(B)$ and $\delta(B)$ calculated in terms of f' instead of f. Then B being the ordered pair (S, T), let us write also $B' = (T, S)$.

Theorem VII.23. *If K is an odd component of $B = (S, T)$ with respect to f, then K is an odd component of $B' = (T, S)$ with respect to f'.*

Proof. B and B' have the same set U. Hence K is a component of B' as well as B. Let $J'(K)$ be defined like $J(K)$ in Eq. (VII.5.7) but with f' replacing f and with S replacing T. Then, by Eq. (VII.5.7),

$$J(K) \equiv \sum_{b \in V(K)} \mathrm{val}(G, b) + \sum_{b \in V(K)} f'(b) + \lambda(V(K), T)$$

$$\equiv \lambda(V(K), S \cup T) + \sum_{b \in V(K)} f'(b) + \lambda(V(K), T)$$

$$\equiv \lambda(V(K), S) + \sum_{b \in V(K)} f'(b)$$

$$\equiv J'(K) \pmod{2},$$

and the required result follows. □

Since $(f')' = f$, we can infer the identities

$$h'(B') = h(B), \qquad \delta'(B') = \delta(B), \tag{VII.5.13}$$

for each ordered pair $B = (S, T)$ of disjoint subsets of $V(G)$.

VII.6. THE f-FACTOR THEOREM

An algorithmic search for an f-factor in G might proceed as follows: Having found one f-limited spanning subgraph F of G, we would try to transform it into another with fewer unfilled vertices. By repetition of this process we would hope to reduce the number of such vertices to zero.

The initial choice for F could be the edgeless spanning subgraph. Suppose that, at some stage in the algorithm, we have an f-limited spanning subgraph F and are trying to improve it. We describe the edges of F as blue and the other edges of G as red, and we use the terminology of the theory of alternating paths.

We choose an unfilled vertex r of F (if F is not an f-factor). We consider the family $J(r)$ of alternating paths from r. Suppose one such path P to terminate at another unfilled vertex s. We delete from F all the blue edges of P and adjoin to it all the red edges of P, thus converting F into a new spanning subgraph F_1 of G. It is clear that F_1 is f-limited, having the following properties:

$$\delta(F_1, r) = \delta(F, r) - 1, \qquad (\text{VII.6.1})$$

$$\delta(F_1, s) = \delta(F, s) \pm 1, \qquad (\text{VII.6.2})$$

$$\delta(F_1, t) = \delta(F, t), \qquad (\text{VII.6.3})$$

if t is not r or s.

In transforming from F to F_1, we have not increased the number of unfilled vertices, and we have diminished $\delta(F, r)$. We count this as an improvement.

If there is no such path P, then every unicursal and bicursal vertex of G, other than r, is filled by F. Even in this case, there is one possibility of improvement. It occurs when r is a bicursal vertex and $\delta(F, r) \geq 2$. Then r is the entrance of some bicursal component K. By Theorems VII.7 and VII.10, there is a nondegenerate reentrant path Q of $J(K, r)$ whose first and last darts are red. We now form a spanning subgraph F_2 of G from F by deleting the blue edges of Q and adjoining the red ones. We find that F_2 is f-limited, having the following properties:

$$\delta(F_2, r) = \delta(F, r) - 2, \qquad (\text{VII.6.4})$$

$$\delta(F_2, t) = \delta(F, t) = 0 \quad \text{if } t \neq r. \qquad (\text{VII.6.5})$$

Again we have improved F by diminishing $\delta(F, r)$ without increasing the number of unfilled vertices. Indeed, if either kind of improvement reduces $\delta(F, r)$ to zero, the number of unfilled vertices is diminished.

Let us consider the case in which no improvement of either kind is possible. Then r is unfilled by F but every other unicursal or bicursal vertex, with respect to $J(r)$, is filled. Moreover, if r is bicursal we can only have $\delta(F, r) = 1$. It is convenient to write $\sigma(r) = 0$ or 1 according as r is unicursal or bicursal. We write S and T for the sets of red and blue unicursal vertices, respectively, and put $B = (S, T)$. Let a be the number of bicursal units with a red entry-edge, and b the number with a blue one.

The bicursal units are components of B. We proceed to show that they are odd. Let K be a bicursal unit. Suppose first that it has an entry-edge. Then the number of blue edges of attachment of K is either one greater or one less than $\lambda(V(K), T)$, according as the entry-edge is blue or red. This follows from Theorem VII.15. Since F fills every vertex of K, it follows that $J(K)$ is odd. Suppose next that K has no entry-edge, so that it includes r. Now the number of blue edges of attachment of K is exactly $\lambda(V(K), T)$. But now $\delta(F, r) = 1$ and r is the only unfilled vertex of K. So again $J(K)$ is odd. We deduce that

$$h(B) \geq a + b + \sigma(r). \tag{VII.6.6}$$

Let m be the number of blue edges with one end in S and one in T. Any blue edge incident with a member of S is unicursal, either to a vertex of T or to a vertex of some bicursal unit K by Theorems VII.4 and VII.5. In the latter case, the edge is the entry-edge of K. Moreover, each member of S is filled by F, for if r is a unicursal vertex, it is a blue one, by definition. We thus have

$$\sum_{s \in S} f(s) = m + b. \tag{VII.6.7}$$

Similarly, the red edges incident with members of T are all unicursal to members of S, except for red entry-edges of bicursal units. Hence the number of blue edges with one end in T and one in $U = V(G) - (S \cup T)$ is $\lambda(T, U) - a$. We deduce that

$$m = \sum_{t \in T} \mathrm{val}(F, t) - \sum_{t \in T} \lambda(T, t) - (\lambda(T, U) - a)$$
$$= \sum_{t \in T} f(t) - 1 + \sigma(r) + a - \sum_{t \in T} \mathrm{val}(G, t) + \lambda(S, T),$$
$$m = a - 1 + \sigma(r) - \sum_{t \in T} f'(t) + \lambda(S, T). \tag{VII.6.8}$$

Hence, by Eq. (VII.6.7),

$$1 = a + b + \sigma(r) - \sum_{s \in S} f(s) - \sum_{t \in T} f'(t) + \lambda(S, T).$$

It follows from Eqs. (VII.5.8) and (VII.6.6) that

$$\delta(B) \geq 1.$$

Thus B is an f-barrier.

We conclude that we can either find an f-barrier or use our algorithm to improve F. So, if G has no f-barrier, we can eventually reduce the deficiency of r to zero without introducing any new unfilled vertices. Then we can do the same for another unfilled vertex, and so on until none are left. We than have an f-factor. We can now assert that G has either an f-factor or an f-barrier. Combining this result with Theorem VII.21, we have:

Theorem VII.24 (The f-Factor Theorem). *For a given f satisfying Eq. (VII.5.1), G has either an f-factor or an f-barrier, but not both* [12].

In the solution of problems it is often useful to supplement Theorem VII.24 with further information about the family of f-barriers of G, assuming such f-barriers to exist. For example, our algorithmic study shows that one of these f-barriers is a cursal barrier.

As another example, we can choose to work with the *maximal* f-barriers, those with the greatest possible deficiency. These have some special properties derivable from the theorems that follow.

Now let $B = (S, T)$ be any ordered pair of disjoint subsets of $V(G)$. Write $U = V(G) - (S \cup T)$. If x is any vertex of S or T, we write $\mu(x)$ for the number of odd components K of B such that some edge of G incident with x has an end in K. Let us consider the effect of transferring x to U, thus replacing B by a new ordered pair B_1, either $(S - \{x\}, T)$ or $(S, T - \{x\})$. We then absorb $\mu(x)$ odd components of B into a single component L of B_1 including x. We write $\eta(x) = 0$ or 1 according as this component of B_1 is even or odd. The other components of B, apart from even ones absorbed in L, persist as components of B_1, without change of parity. From these observations and the definition of $\delta(B)$, we derive the following rules.

Theorem VII.25. *If $x \in S$, then*
$$\delta(B) - \delta(B_1) = \mu(x) - f(x) + \lambda(T, x) - \eta(x).$$

Theorem VII.26. *If $x \in T$, then*
$$\delta(B) - \delta(B_1) = \mu(x) - f'(x) + \lambda(S, x) - \eta(x).$$

These two results correspond under the duality between f and f'. In view of Theorem VII.22, we can characterize $\eta(x)$ in each case as that number 0 or 1 that makes the right of the equation even. In each case, the right of the equation is nonnegative if B is a maximal f-barrier, and nonpositive if B_1 is a maximal f-barrier. If the right side is zero and one of B and B_1 is a maximal f-barrier, then so is the other. In the next theorem we see one application of Theorems VII.25 and VII.26.

Theorem VII.27. *If G has an f-barrier, then it has a maximal f-barrier $B = (S, T)$ such that $f(x) \geq 2$ if $x \in T$, and $f(x) \geq 1$ if x is not in S.*

Proof. Let $B = (S, T)$ be a maximal f-barrier of G with the least possible number of vertices x violating the above conditions.

Suppose that some $x \in T$ satisfies $f(x) \leq 1$. Then $f'(x) \geq \text{val}(G, x) - 1$. Since $\text{val}(G, x) \geq \mu(x) + \lambda(S, x)$, the right of Theorem VII.26 is nonpositive. Hence B_1 is also a maximal f-barrier, contrary to the definition of B.

We deduce that each $t \in T$ satisfies $f(t) \geq 2$. Now suppose that some x in U satisfies $f(x) = 0$. We consider the effect of transferring x to S, so changing B into a new ordered pair B_2. We apply Theorem VII.25 with the expressions on the right defined in terms of B_2. Then the left side of Theorem VII.25 can be written as $\delta(B_2) - \delta(B)$, and the right side is nonnegative. Hence B_2 is also a maximal f-barrier of G, contrary to the definition of B. We deduce that, in fact, B satisfies the required conditions. □

Special interest attaches to the case in which $f(x) = 1$ for each vertex x. We then describe an f-factor as a 1-*factor*. Given a subset S of $V(G)$, it is then convenient to define $h(S)$ as the number of components of $G[V(G) - S]$ having an odd number of vertices.

Theorem VII.28 (The 1-Factor Theorem). *G has either a 1-factor or a subset S of $V(G)$ such that $|S| < h(S)$, but not both* [11].

Proof. By the f-factor Theorem, G has either a 1-factor or an f-barrier $B = (S, T)$, but not both. (Here $f(x) = 1$ for all x.) If there is such a B, we can arrange that T is null, by Theorem VII.27. Then the odd components of B are those with an odd number of vertices. We then have $|S| < h(S)$, by the definition of $\delta(B)$. Conversely, if $|S| < h(S)$ for some S, it is clear that the pair $(S, \emptyset)$ is an f-barrier of G. □

It should be noted that for a general f that is not even-summing, the pair $(\emptyset, \emptyset)$ is an f-barrier. We have already seen that there is no f-factor in this case.

We conclude the section with a proof of one old and famous factor theorem [10].

Theorem VII.29 (Petersen's Theorem). *Let G be a connected cubic graph which either has no isthmus or is such that all the isthmuses lie in a single arc. Then G has a 1-factor.*

Proof. The number of vertices of G is even, by Eq. (I.1.1). We therefore have a zero-summing function f.

Assume that G has no 1-factor. By Theorem VII.28, there is a subset S of $V(G)$ such that $|S| < h(S)$. Since $|V(G)|$ is even, the difference $h(S) - |S|$ is even. We therefore have

$$h(S) \geq |S| + 2. \tag{VII.6.9}$$

Consider an odd component K of $G[V(G) - S]$. It is clear that $\lambda(V(K), S)$ is odd. Let there be a cases in which this number is 1, and b in which it is 3 or more. Then $h(S) = a + b$.

Let m be the number of edges of G having one end in S and one in an odd component of $G[V(G) - S]$. By the above definitions and observa-

tions, and by the trivalency of G, we have
$$3|S| \geq m \geq a + 3b = 3h(S) - 2a \geq 3|S| + 6 - 2a.$$
It follows that $a \geq 3$.

We deduce that G has three distinct isthmuses A_1, A_2, and A_3 having end-graphs K_1, K_2, and K_3, respectively, that are distinct odd components of $G[V(G) - S]$. If all the isthmuses of G lie in an arc L of G, then each of K_1, K_2, and K_3 must contain an end of L, which is clearly impossible. (See Theorems I.41 and I.24.) □

VII.7. SUBGRAPHS OF MINIMUM DEFICIENCY

Let G and f be as in Sec. VII.5. Even if G has no f-factor, we may wish to determine the least possible deficiency for an f-limited spanning subgraph of G. We write

$$\sigma(f) = \sum_{x \in V(G)} f(x), \qquad (\text{VII}.7.1)$$

and we note the following rule.

Theorem VII.30. *Let F be any f-limited spanning subgraph of G. Then $\delta(F)$ has the parity of $\sigma(f)$.*

This follows from Eq. (VII.7.1) and an application of Eq. (I.1.1) to F.

We now choose a positive integer k having the parity of $\sigma(f)$ and try to find a condition for the existence in G of a spanning subgraph f such that $\delta(F) \leq k$. To this end, we construct from G a graph H as follows:

We choose a positive integer $m \geq \frac{1}{2}k$, and another positive integer q that is not less than any $f(x)$, $x \in V(G)$. To form H, we adjoin to G a new vertex w, and join w to each vertex of G by q new links. We then attach m loops to w. Finally we extend f by putting $f(w) = k$. Now f is even-summing in the new graph H.

Theorem VII.31. *G has an f-limited spanning subgraph F such that $\delta(F) \leq k$ if and only if H has an f-factor.*

Proof. Suppose, first, that H has an f-factor F_1. The number of links of F_1 joining w to vertices of G is at most k. Deleting w and its incident edges from F_1, we obtain an f-limited spanning subgraph F of G such that $\delta(F) \leq k$.

Suppose next that G has an f-limited spanning subgraph F such that $\delta(F) \leq k$. We form an f-limited spanning subgraph F_2 of H as follows: First we adjoin w to F. Then for each $x \in V(G)$ we adjoin exactly $\delta(F, x)$ of the edges of H joining x to w. The resulting spanning subgraph F_2 of H fills every vertex of G. Moreover, $\delta(F_2, w)$ is even, by Eq. (VII.7.1) and the choice of k. We can therefore convert F_2 into an f-factor F_1 of H by adjoining some or none of the loops of H on w. □

Theorem VII.32. *Let k be any positive integer with the parity of $\sigma(f)$. Then G has either an f-limited spanning subgraph F such that $\delta(F) \leq k$ or an f-barrier B such that $\delta(B) > k$, but not both.*

Proof. Suppose first that G has no such F. Then H, as defined above, has no f-factor, by Theorem VII.31. Hence H has an f-barrier $B = (S, T)$, by Theorem VII.24. We choose B from among all the f-barriers of H according to the following rules.

i) If possible, w is to be in S.
ii) If we cannot have w in S, then, if possible, w is to be in $U = V(G) - (S \cup T)$. Then, subject to this rule, B is to be chosen so that $|T|$ is as small as possible.

Assume, first, that neither of these choices is possible, so that we have $w \in T$. Consider the effect of transferring w to U, thus changing B into another ordered pair B_1. We apply Theorem VII.26 to H, recalling that now

$$f'(x) = \text{val}(H, x) - f(x).$$

Because of the m loops on w, the number $\text{val}(H, x)$ is at least equal to $f(w) + \lambda(S, w) + \mu(w)$. Hence the right of Theorem VII.26 is nonpositive, and B_1 is another f-barrier of H. This is contrary to our assumption.

Assume, next, that w is in U. Consider the effect of transferring to U a vertex t of T. We first note that B has now only one component, and therefore at most one odd component. Hence $h(B) \leq 1$ and $\mu(t) \leq 1$. We have also

$$\text{val}(G, t) \geq q + \lambda(S, t) \geq f(t) + \lambda(S, t).$$

Thus the right side of Theorem VII.26 is nonpositive, by our characterization of $\eta(x)$, and therefore the transfer of t changes B into another f-barrier of H. This is impossible by the choice of B, as described in (ii). There still remains the possibility that T is null. But then, since $h(B) \leq 1$, it follows from Eq. (VII.5.8) that $\delta(B) = 1$. This is impossible by Theorem VII.22, since f is even-summing in H.

We can now assert that w is in S. Deleting it from S, we obtain an f-barrier $B_1 = (S - \{w\}, T)$ of G such that $\delta(B_1) > k$ for, in going from H to G, the decrease in $\lambda(S, T)$ is exactly balanced by the decrease in $\Sigma f'(t)$, $t \in T$. □

The combination of Theorems VII.27 and VII.32 can be stated in maximum–minimum form. The minimum deficiency for an f-limited spanning subgraph of G is equal to the maximum deficiency for an ordered pair (S, T) of disjoint subsets of $V(G)$. In this connection we note that $\delta(S, T) \geq 0$ when S and T are null.

We can apply Theorem VII.32 to the special case in which $f(x) = 1$ for each x. We then say "1-limited" rather than "f-limited." By Theorem VII.27, we need only consider f-barriers for which T is null. We obtain the following result, called Berge's Extension of the 1-Factor Theorem [1].

Theorem VII.33. *Let k be a positive integer having the parity of $|V(G)|$. Then G has either a 1-limited spanning subgraph F such that $\delta(F) \leq k$ or a subset S of $V(G)$ such that $h(S) > |S| + k$, but not both.*

VII.8. THE BIPARTITE CASE

In this section we suppose G to have a bipartition (U, V). We denote the sums of $f(x)$ over U and V, respectively, by $\Sigma(U)$ and $\Sigma(U)$. We say that f is *balanced* if $\Sigma(U) = \Sigma(V)$.

Theorem VII.34. *If G has an f-factor, then f is balanced.*

We say this because each of $\Sigma(U)$ and $\Sigma(V)$ is equal to the number of edges of the f-factor.

Let $B = (S, T)$ be any ordered pair of disjoint subsets of $V(G)$. In the bipartite case we replace $\delta(B)$ by the simpler function

$$\rho(B) = \lambda(S, T) - \sum_{s \in S} f(s) - \sum_{t \in T} f'(t). \quad \text{(VII.8.1)}$$

We note that

$$\rho(U, V) = \Sigma(V) - \Sigma(U). \quad \text{(VII.8.2)}$$

From Eq. (VII.8.1) we can deduce the following equations (VII.8.3) and (VII.8.4). Here s is any vertex of S and t any vertex of T.

$$\rho(S - \{s\}, T) = \rho(S, T) + f(s) - \lambda(T, s), \quad \text{(VII.8.3)}$$

$$\rho(S, T - \{t\}) = \rho(S, T) - f(t) + \lambda(V(G) - S, t). \quad \text{(VII.8.4)}$$

These two equations would be valid even in a nonbipartite graph.

Theorem VII.35. *Let f be balanced. Then G has either an f-factor or an ordered pair $B = (S, T)$ such that $S \subseteq U$, $T \subseteq V$, and $\rho(B) > 0$, but it does not have both.*

Proof. Suppose, first, that G has such a pair $B = (S, T)$. Assume that G has an f-factor F. Let m be the number of edges of F joining S and T. Then

$$\sum_{s \in S} f(s) \geq m \geq \sum_{t \in T} \{f(t) - \lambda(U - S, t)\}$$

$$= \lambda(S, T) - \sum_{t \in T} f'(t).$$

Thus $\rho(B) \leq 0$ and we have a contradiction.

Suppose next that G has no such pair B. Let the algorithm of Sec. VII.6 be applied until we have an f-limited spanning subgraph F of G that cannot be further improved. Let us assume that F is not an f-factor. Then, since f is balanced, we can find a vertex r of V that is not filled by F. As in Sec. VII.6 we consider the path-family $J(r)$. Each nondegenerate path P of $J(r)$ has a red first dart with head in U. The second dart, if any, is blue and has its head in V. A third dart would be red and have its head in U, and so on. We deduce that there are now no bicursal edges. Moreover, if S and T are the sets of red-entrant and blue-entrant unicursal vertices, respectively, we have $S \subseteq U$ and $T \subseteq V$. The argument of Sec. VII.6 leads to the equation

$$1 = a + b + \sigma(r) - \sum_{s \in S} f(s) - \sum_{t \in T} f'(t) + \lambda(S, T).$$

But $a + b + \sigma(r)$ is the number of bicursal units, and this is now zero. So we have $\rho(B) = 1$. Thus B satisfies the stated conditions, and we have a contradiction. □

Consider the case in which $f(x) = 1$ for each x. Then if there is any $B = (S, T)$ such that $\rho(B) > 0$, we can find one such that no vertex of T is joined by an edge to any vertex not in S, by Eq. (VII.8.4). So the argument of Theorem VII.35 gives us another proof of Hall's Theorem. (See Theorem II.39.)

Problems about general f-limited spanning subgraphs can often be reduced to problems about f-factors. Suppose, for example, that in the bipartite graph G we seek a condition for the existence of an f-limited spanning subgraph F whose "U-deficiency" does not exceed some positive integer k. Here the U-deficiency $\delta(F, U)$ is the sum of the numbers $\delta(F, x)$ for all the vertices x of U. We may assume that $\Sigma(U) \leq \Sigma(V) + k$, since otherwise no such F exists.

We form from G a new graph H by adjoining a new vertex u^* to U and a new vertex v^* to V, joining u^* to each vertex of V by a large number of new edges and then joining v^* to each vertex of U, and to u^*, by a large number of new edges. Finally we extend f by writing $f(v^*) = k$ and choosing $f(u^*)$ to balance f in H. If those large numbers are big enough, we can say that G has a subgraph F of the required kind if and only if H has an f-factor [13].

VII.9. A THEOREM OF ERDÖS AND GALLAI

We discuss partitions $(f_1, f_2, \ldots, f_p)$ of a positive integer $2q$ into p positive parts. We order the parts so that $f_1 \geq f_2 \geq \cdots \geq f_p$. We recall such a partition P *strictly graphic* if there is a strict graph G of p vertices such that the numbers f_i are the valencies of those vertices. P. Erdös and T. Gallai have proved the following [3].

Theorem VII.36. *P is strictly graphic if and only if*

$$\sum_{i=1}^{r} f_i \le r(r-1) + \sum_{i=r+1}^{p} \mathrm{Min}(r, f_i) \qquad (VII.9.1)$$

for each integer r satisfying $1 \le r \le p-1$.

To make applicable the methods of the present chapter we introduce a p-clique K, with vertices enumerated as $v_1, v_2, \ldots, v_p$. We then write $f(v_j) = f_j$ for each vertex v_j. Evidently P is strictly graphic if and only if K has an f-factor.

Suppose Eq. (VII.9.1) to fail for some r. Then we write T for the set of r vertices with suffixes from 1 to r, and S for the set of all other vertices x such that $f(x) < r$. As usual, we put $U = V(G) - (S \cup T)$. Then, by our supposition,

$$\sum_{t \in T} f(t) > \sum_{t \in T} \lambda(T, t) + \sum_{s \in S} f(s) + r|U|$$
$$= \sum_{t \in T} \mathrm{val}(K, t) - \lambda(S, T) + \sum_{s \in S} f(s),$$

whence $\delta(S, T) > 0$, K has no f-factor by Theorem VII.24, and P is not strictly graphic.

Conversely, suppose that P is not strictly graphic. Then K has an f-barrier $B = (S, T)$. Since K is a clique, we have $h(B) \le 1$ for each such B, and therefore $\mu(x)$ in Theorems VII.25 and VII.26 is either 0 or 1. We choose B to have maximum deficiency. We then put $|T| = r$. By Theorem VII.25 we can arrange that

$$\mu(x) - f(x) + \lambda(T, x) - \eta(x) \ge 0 \qquad (VII.9.2)$$

if x is in S and

$$\mu(x) - f(x) + \lambda(T, x) - \eta(x) \le 0 \qquad (VII.9.3)$$

if x is in U. We deduce that $f(x) \le r$ in the first case and $f(x) \ge r$ in the second. We use the fact that the expression on the left is even, so that if $f(x) = \lambda(T, x) = r$, then $\mu(x) = \eta(x) = 0$ or 1. We now have

$$\sum_{i=1}^{r} f_i \ge \sum_{t \in T} f(t)$$
$$\ge \delta(S, T) + \sum_{s \in S} f(s) + \sum_{t \in T} \mathrm{val}(K, t) - \lambda(S, T) \quad \text{(by Eq. (VII.5.8))}$$
$$> \sum_{s \in S} f(s) + r(r-1) + r|U|$$
$$= r(r-1) + \sum_{i=r+1}^{p} \mathrm{Min}(r, f_i).$$

Thus if P is not strictly graphic, Eq. (VII.9.1) must fail for some r. This completes the proof. □

VII.10. NOTES

VII.10.1. Cursality

The theory of the first four sections of this chapter can be extended to any number of colors. This is explained in a paper by the author that is to be published in *Combinatorica*.

VII.10.2. Algorithms

There is a good algorithm for finding 1-factors and f-factors, or demonstrating their nonexistence. See [2].

VII.10.3. Pfaffians

Consider a loopless graph G, with conductances associated with the edges as in Section VI.4. Let its vertices be enumerated as $v_1, v_2, \ldots, v_n$. We associate with it a square matrix $\langle u_{ij} \rangle$, where u_{ij} is plus or minus the sum of the conductances of the edges joining v_i and v_j. We arrange that $u_{ij} = -u_{ji}$, that is, that the matrix is skew-symmetric. If n is even, then the determinant of $\langle u_{ij} \rangle$ is the square of a polynomial P in the conductances called the *Pfaffian* of $\langle u_{ij} \rangle$. This Pfaffian can be defined explicitly as follows

$$P = \sum \varepsilon u_{ij} u_{kl} u_{mn} \cdots u_{rs},$$

where $i < j$, $k < l$, etc., and $i < k < m < \cdots < r$, and where $(i, j, k, l, \ldots, r, s)$ is a permutation of the sequence of integers from 1 to n. The coefficient ε is 1 or -1 according as the permutation is even or odd. It can be seen that the terms of P, each the product of $\frac{1}{2}n$ distinct conductances, are in 1-1 correspondence with the 1-factors of G, but some have a positive sign and some a negative.

The theory of 1-factors given in [11] is based on the properties of Pfaffians. F. G. Maunsell showed in [9] that the appeals to Pfaffians could be replaced by simple graph-theoretical arguments. Later, P. Kasteleyn showed that, for a planar graph with unit conductances, the signs could be adjusted so as to ensure that the value of the Pfaffian would be the number of 1-factors. He thus found the numbers of 1-factors for some planar graphs of interest to physicists [7]. His method can be extended to some nonplanar graphs ([5, 6, 8]).

EXERCISES

1. Color the edges of a cube-graph red and blue. Choose a vertex r and classify the edges as unicursal, bicursal, or non cursal with respect to r.

2. Find a 9-circuit in the Petersen graph. Color its edges blue and the remaining edges in the graph red. Find an alternating path that begins and terminates at the remaining vertex r (and is not degenerate). Use this path to change the complement of the 9-circuit into a 1-factor.
3. Are the sequences (3,2,2,1) and (5,5,5,3,3,2) strictly graphic? Test your conclusions against the Erdös–Gallai Theorem.
4. Show that any 4-connected 5-regular graph with an even number of vertices has a 1-factor.

REFERENCES

[1] Berge, C., *Graphes et hypergraphes* (p. 154), Dunod, Paris, 1970.
[2] Edmonds, J., Paths, trees and flowers, *Can. J. Math.*, **17** (1965) 449–467.
[3] Erdös, P., and T. Gallai, Graphs with prescribed valencies, *Mat. Lapok* **11** (1960), 264–274 (Hungarian).
[4] Gallai, T., Neuer Beweiss eines Tutte'schens Satzes, *Magyar Tud. Akad. Mat. Kutato Int. Kozl.* **8** (1963), 135–139.
[5] Gibson, P. M., The Pfaffian and 1-factors of graphs, *Trans. New York Acad. Sci.*, (2) **34** (1972), 52–57.
[6] _____, The Pfaffian and 1-*Factors of Graphs — II*; "Graph theory and applications," Springer, Berlin, 1972, pp. 89–98.
[7] Kasteleyn, P., Dimer statistics and phase transitions, *J. Math. Phys.* (1963), 287–293.
[8] Little, C. H. C., "An Extension of Kasteleyn's Method of Enumerating the 1-Factors of Planar Graphs," *Combinatorial Mathematics*, Springer, Berlin, 1974.
[9] Maunsell, F. G., A note on Tutte's paper "The factorization of linear graphs," *J. London Math. Soc.*, **27** (1952), 127–128.
[10] Petersen, J., Die Theorie der regulären Graphs, *Acta. Math.* **15** (1891), 193–220.
[11] Tutte, W. T., The factorization of linear graphs, *J. London Math. Soc.* **22** (1947), 107–111.
[12] _____, The factors of graphs, *Can. J. Math.*, **4** (1952), 314–328.
[13] _____, Graph factors, *Combinatorica* **1** (1981), 79–97.

Chapter VIII

Algebraic Duality

VIII.1. CHAIN-GROUPS

In this chapter we apply linear algebra to graph theory. We start with a finite set S whose members are called *cells*, and with a commutative ring R having a unit element. We denote the zero and unity of R by 0 and 1, respectively. In applications, S is likely to be the vertex-set or edge-set of a graph.

A *chain* on S to R is a mapping f of S into R. If $x \in S$, then the element $f(x)$ of R is called the *coefficient* of x in f. If S is nonnull, there is a *zero chain* on S in which each coefficient is 0, and a *unit chain* in which each coefficient is 1. When S is null, we say that there is just one chain on S to R, and we call it a zero chain.

If f is a chain on S to R and if $\lambda \in R$, then we define the product λf as the chain on S to R given by the following rule: The coefficient $(\lambda f)(x)$ of x in λf is $\lambda . f(x)$ for each $x \in S$. If f_1 and f_2 are chains on S to R, then their *sum* $f_1 + f_2$ is the chain on S to R defined as follows: The coefficient $(f_1 + f_2)(x)$ of x in $f_1 + f_2$ is $f_1(x) + f_2(x)$ for each $x \in S$. Addition thus defined is commutative and associative. We can extend it in the usual way to sums of three or more chains. Moreover, multiplication by an element of R is distributive over addition. A *difference* $f_1 - f_2$ of chains is defined as $f_1 + (-1)f_2$.

Consider a nonnull collection $F = \{f_1, f_2, \ldots, f_n\}$ of chains on S to R, not necessarily all distinct. Suppose that some chain f on S to R satisfies

$$f = \sum_{j=1}^{n} \lambda_j f_j = \lambda_1 f_1 + \lambda_2 f_2 + \cdots + \lambda_n f_n, \qquad \text{(VIII.1.1)}$$

where the λ_j are elements of R. Then we say that f is expressed as a *linear combination* of the members of F, with the λ_j as *coefficients*. If the zero chain can be so expressed, with coefficients not all zero, then we say that the members of F are *linearly dependent*. This happens, for example, when one member of F is zero, or when two of them are equal. If the zero chain cannot be so expressed, then the members of F are *linearly independent*. In that case they are nonzero and distinct.

The *support* $\mathrm{Sp}(f)$ of a chain f on S to R is the set of all $x \in S$ such that $f(x)$ is nonzero. If f and g are two chains such that $\mathrm{Sp}(g) \subseteq \mathrm{Sp}(f)$, we say that f *encloses* g. If, moreover, $\mathrm{Sp}(g)$ is properly contained in $\mathrm{Sp}(f)$, we say that f *properly* encloses g.

A *chain-group* on S to R is a nonnull set N of chains on S to R that is closed with respect to addition and with respect to multiplication by an element of R. Considering the zero element of R, we see that N must contain the zero chain. With any chain f, the chain-group must include also its negative $-f$, that is $(-1)f$, which satisfies $f + (-f) = 0$. Here and elsewhere we use the symbol 0 to denote the zero chain. More generally, we observe that any linear combination of members of N must be a member of N.

An *elementary* chain of N is a nonzero chain of N that properly encloses no other nonzero chain of N.

Theorem VIII.1. *If $f \in N$ has the same support as some elementary chain g of N, then f is itself elementary* (by definition).

Theorem VIII.2. *Each nonzero chain f of N encloses an elementary chain of N.*

Proof. Of the nonzero chains of N enclosed by f, f being one, choose one with minimal support. This is elementary. □

A *cell-base* of N is a subset D of S that meets the support of each nonzero chain of N, and that is minimal with respect to this property.

Theorem VIII.3. *Each chain-group N on S to R has at least one cell-base.*

Proof. Let F be the class of all subsets T of S that meet the support of each nonzero chain of N. Then F is nonnull since it has S as a member. Choose a member D of F with as few cells as possible. Then D is a cell-base of N. □

Theorem VIII.4. *Let D be a cell-base of a chain-group N on S to R, and let x be any cell of D. Then there is an elementary chain g of N such that* Sp(g) *meets D only in x.*

Proof. By the minimality of D there is a nonzero chain f of N such that Sp(f) does not meet $D - \{x\}$. But f encloses an elementary chain g of N, by Theorem VIII.2. The support of g must meet D, and it can do so only in x. □

Consider any subset T of S. If f is any chain on S to R, we define its restriction $f \cdot T$ to T as that chain on T to R for which each cell of T has the same coefficient as in f. Evidently, if N is any chain-group on S to R, then the restrictions to T of the chains of N constitute a chain-group $N \cdot T$ on T to R. We call this the *reduction* of N to T. It is also clear that the restrictions to T of those chains f of N for which Sp(f) $\subseteq T$ constitute a chain-group $N \times T$ on T to R. We call this the *contraction* of N to T. From these definitions, we can deduce the following general rules, for $U \subseteq T \subseteq S$:

$$(N \cdot T) \cdot U = N \cdot U, \qquad \text{(VIII.1.2)}$$

$$(N \times T) \times U = N \times U, \qquad \text{(VIII.1.3)}$$

$$(N \times T) \cdot U = (N \cdot (S - (T - U))) \times U. \qquad \text{(VIII.1.4)}$$

The last of these is perhaps less obvious than the other two. To prove it, we observe that each side is the set of restrictions to U of those chains of N whose supports do not meet $S - T$.

We refer to a chain-group of the form $(N \times T) \cdot U$ as a *minor* of N. Evidently $N \times T$ and $N \cdot T$ are minors $(N \times T) \cdot T$ and $(N \times S) \cdot T$ of N, respectively. N itself is the minor $(N \times S) \cdot S$ of N. Moreover, each minor of N can be written in the form $(N \cdot T) \times U$, by Eq. (VIII.1.4).

Theorem VIII.5. *A minor of a minor of N is a minor of N.*

This follows from Eqs. (VIII.1.2), (VIII.1.3), and (VIII.1.4). Compare the proof of Theorem II.8. □

If f and g are chains on S to R, we define their *scalar product* $(f \cdot g)$ as follows

$$(f \cdot g) = \sum_{x \in S} f(x)g(x). \qquad \text{(VIII.1.5)}$$

We note the following identities:

$$(f \cdot g) = (g \cdot f), \qquad \text{(VIII.1.6)}$$

$$\lambda(f \cdot g) = (\lambda f \cdot g), \qquad \text{(VIII.1.7)}$$

$$(f + g) \cdot h = (f \cdot h) + (g \cdot h) \qquad \text{(VIII.1.8)}$$

If $(f \cdot g) = 0$, we say that the chains f and g are *orthogonal*.

Now let N be any chain-group on S to R, and let N^* be the set of those chains on S to R that are orthogonal to all the chains of N. The *zero chain* is one such. By Eqs. (VIII.1.6), (VIII.1.7), and (VIII.1.8), N^* is a chain-group on S to R.

Theorem VIII.6. *Let N be any chain-group on S to R. Then $N \subseteq N^{**}$ and $N^* = N^{***}$.*

Proof. Each chain of N is orthogonal to every chain of N^*, and so belongs to N^{**}. Thus $N \subseteq N^{**}$.

Similarly, $N^* \subseteq N^{***}$. On the other hand, each chain of N^{***} is orthogonal to each chain of N^{**}, and so to each chain of N, by the result already proved. So we have also $N^{***} \subseteq N^*$. □

If $N^{**} = N$, we say that N^* is the *dual* chain-group of N. Then N is the dual chain-group of N^*. For each of N^* and N^{**} is the dual of the other, by Theorem VIII.6. Here we encounter the "algebraic duality" of the chapter title.

Theorem VIII.7. *If $T \subseteq S$, then $(N \cdot T)^* = N^* \times T$.*

Proof. Consider any chain f on T to R that is in $(N \cdot T)^*$. Let f_1 be the chain on S to R such that $f_1 \cdot T = f$ and the coefficients outside T are zero. Then f_1 is orthogonal to each member of N and therefore $f \in N^* \times T$.

Conversely, if f is in $N^* \times T$, it is the restriction to T of a chain f_1 of N^* specified as above. Hence f is orthogonal to the restriction to T of every chain of N. That is, $f \in (N \cdot T)^*$. □

VIII.2. PRIMITIVE CHAINS

We write R' for the set of regular elements of R, that is, the elements of R having reciprocals in R.

Theorem VIII.8. *The product of a regular element λ and a nonzero element μ of R is nonzero.*

For suppose $\lambda\mu = 0$. Then $\mu = (\lambda^{-1}\lambda)\mu = \lambda^{-1}(\lambda\mu) = 0$.

Theorem VIII.9. *The product of two regular elements λ and μ of R is regular.*

For $\lambda\mu$ has the reciprocal $\lambda^{-1}\mu^{-1}$.

Now let N be any chain-group on S to R. We define a *primitive chain* of N as an elementary chain of N whose coefficients are all in R'.

Theorem VIII.10. *Let g be any primitive chain of N. Then the chains f of N such that $\mathrm{Sp}(f) = \mathrm{Sp}(g)$ are the multiples of g by nonzero elements of R.*

Primitive Chains

Proof. If μ is a nonzero element of R, then μg is a nonzero chain of N with support $\operatorname{Sp}(g)$, by Theorem VIII.8.

Conversely, suppose $\operatorname{Sp}(f) = \operatorname{Sp}(g)$. Choose $x \in \operatorname{Sp}(g)$ and write $\nu = f(x)(g(x))^{-1}$. Then the support of $f - \nu g$ is contained in $\operatorname{Sp}(g) - \{x\}$. Hence $f = \nu g$, since g is elementary. □

Theorem VIII.11. *Let g be a primitive chain of N, and let x be a cell of $\operatorname{Sp}(g)$. Then the multiples of g by elements of R' are primitive chains of N. In one of them, the coefficient of x is 1.*

Proof. The multiples are elementary chains of N, by Theorems VIII.1 and VIII.8. They are primitive, by Theorem VIII.9. Since $(g(x))^{-1}$ is regular, the theorem follows. □

We call a cell-base D of N *primitive* if, for each $x \in D$, there is a primitive chain g of N such that $D \cap \operatorname{Sp}(g) = \{x\}$. By Theorem VIII.11 we can choose g so that $g(x) = 1$. Let us denote the chosen chain by $g(D, x)$. Then $g(D, x)$ is uniquely determined; if two chains satisfied its definition, the support of their difference would not meet D.

The set of chains $g(D, x)$, one for each cell x of D, is the *chain-base* $B(D)$ corresponding to the primitive cell-base D.

Theorem VIII.12. *Let D be a primitive cell-base of N. Then the members of $B(D)$ are linearly independent. Moreover, each chain of N has a unique expression as a linear combination of members of $B(D)$.*

Proof. Suppose

$$\sum_{x \in D} \lambda_x g(D, x) = 0,$$

where the λ_x are in R. Then the coefficient of x in the sum on the left is λ_x. Hence $\lambda_x = 0$ for each x. This proves the first part of the theorem.

Now let f be any chain of N. The chain

$$f - \sum_{x \in D} f(x) g(D, x)$$

has a support not meeting D and is therefore zero. Hence

$$f = \sum_{x \in D} f(x) g(D, x); \qquad \text{(VIII.2.1)}$$

f is expressed as a linear combination of members of $B(D)$.

Assume that f has a second such expression

$$f = \sum_{x \in D} \mu_x g(D, x).$$

Then

$$\sum_{x \in D} (\mu_x - f(x)) g(D, x) = 0.$$

Hence $\mu_x = f(x)$ for each x, by the result already proved. The expression in Eq. (VIII.2.1) for f is thus unique. □

Corollary VIII.13. *If a chain f of N has a support meeting D only in the cell x, then f is a nonzero multiple of $g(D, x)$.*

Theorem VIII.14 (The Interchange Theorem). *Let D be a primitive cell-base of N, and let x be a cell of D. Let y be a cell of $g(D, x)$ distinct from x. Write $D_1 = (D - \{x\}) \cup \{y\}$. Then D_1 is a cell-base of N.*

Proof. By the Corollary VIII.13 the only nonzero chains of N with supports not meeting $D - \{x\}$ are the nonzero multiples of $g(D, x)$. Hence the support of each nonzero chain of N meets D_1. So D_1 satisfies the first condition for a cell-base. It remains to prove its minimality.

We note that the support of $g(D, x)$ does not meet $D_1 - \{y\}$, that is, $D - \{x\}$. Now let z be any cell of $D - \{x\}$. Write $g(D, x) = h$ and $g(D, z) = k$. Since these two chains are primitive, we can construct the chain

$$f = k - k(y)(h(y))^{-1}h.$$

This is nonzero; we have $f(z) = 1$. But the support of f does not include y, and it does not meet $D_1 - \{z\}$. We can now assert that no proper subset of D_1 meets the support of every nonzero chain of N. □

The operation of constructing D_1 from D is called an *interchange*. If D_1 happens to be primitive, it may be possible to apply a second interchange, going from D_1 to a third cell-base D_2.

Theorem VIII.15. *Let all the cell-bases of N be primitive. Let D and E by any two of them. Then it is possible to convert D into E by a finite sequence of interchanges.*

Proof. From the class of cell-bases of N consisting of D and all cell-bases derivable from it by finite sequences of interchanges, we select one, D', having the greatest possible number of elements in common with E. If $D' \subseteq E$ then $D' = E$, by minimality. In the remaining case, D' has a cell x not in E. But then the support of $g(D', x)$ meets E in a cell y not belonging to D'. It follows from the Interchange Theorem that $(D' - \{x\}) \cup \{y\}$ is a cell-base of N. But this contradicts the definition of D'. □

Theorem VIII.16. *Let all the cell-bases of N be primitive. Let g be an elementary chain of N. Then N has a cell-base D such that g is a nonzero multiple of a member of $B(D)$.*

Proof. Choose a cell-base D of N having as few cells as possible in $\mathrm{Sp}(g)$. (It must have at least one.) Choose $x \in D \cap \mathrm{Sp}(g)$. Suppose $g(D, x)$ to have a cell y not in $\mathrm{Sp}(g)$. Then $(D - \{x\}) \cup \{y\}$ is a cell-base of N by the Interchange Theorem, and the definition of D is contradicted. We deduce

Primitive Chains

that $\operatorname{Sp}(g(D, x)) \subseteq \operatorname{Sp}(g)$, and therefore $\operatorname{Sp}(g(D, x)) = \operatorname{Sp}(g)$ since g is elementary. The theorem follows, by Corollary VIII.13. □

Let us say that the chain-group N is *primitive* if each of its elementary chains is a multiple of a primitive chain.

Theorem VIII.17. *N is primitive if and only if each of its cell-bases is primitive.*

Proof. Suppose N is primitive. Let D be any cell-base of N. For each $x \in D$ there is an elementary chain g of N such that $D \cap \operatorname{Sp}(g) = \{x\}$, by Theorem VIII.4. Since each such g is a multiple of a primitive chain of N, it follows that D is primitive.

Conversely, suppose that each cell-base of N is primitive. Then N is primitive, by Theorem VIII.16. □

Theorem VIII.18. *If N is primitive, then each nonzero chain f of N can be expressed as a sum of nonzero multiples of primitive chains, each of these primitive chains being enclosed by f.*

Proof. If possible, let f be a nonzero chain of N for which the theorem fails. Let it be chosen so that $|\operatorname{Sp}(f)|$ has the least value consistent with this condition. Now f encloses a primitive chain g of N, by Theorem VIII.2. If x is some cell of $\operatorname{Sp}(g)$, we can arrange that $g(x) = 1$, by Theorem VIII.11. Write

$$h = f - f(x)g. \qquad \text{(VIII.2.2)}$$

Then $\operatorname{Sp}(h)$ is a proper subset of $\operatorname{Sp}(g)$. (It does not include x.) So by the choice of f, the theorem is true for h. But then it is true also for f, by Eq. (VIII.2.2). This contradiction establishes the required result. □

If N is primitive, there is a number $r(N)$ such that each cell-base of N and each associated chain-base has exactly $r(N)$ elements. This follows from Theorems VIII.15 and VIII.17. We call $r(N)$ the *rank* of N. By its definition, we have

$$0 \le r(N) \le |S|. \qquad \text{(VIII.2.3)}$$

Theorem VIII.19. *If N is primitive, then every minor of N is primitive.*

Proof. Consider an elementary chain g of $N \times T$, where $T \subseteq S$. There is a chain g_1 of N such that $\operatorname{Sp}(g_1) = \operatorname{Sp}(g)$ and $g = g_1 \cdot T$. But g_1 encloses a primitive chain h_1 of N, by Theorem VIII.18. The restriction $h_1 \cdot T$ is a chain h of $N \times T$ such that $\operatorname{Sp}(h) \subseteq \operatorname{Sp}(g)$ and the nonzero coefficients in h are all in R'. Moreover, h is nonzero. Since g is elementary in $N \times T$, we must have $\operatorname{Sp}(h) = \operatorname{Sp}(g)$. Hence h is elementary in $N \times T$, and therefore primitive in $N \times T$, by Theorem VIII.1. By Theorem VIII.10, g is a nonzero multiple of h.

Consider next an elementary chain g of $N \cdot T$. There is a chain g_1 of N such that $g_1 \cdot T = g$. Using Theorem VIII.18, we find that there is a nonzero chain h of $N \cdot T$ whose nonzero coefficients are all in R' and whose support is contained in that of G. As with $N \times T$, it follows that g is a multiple of the primitive chain h of $N \cdot T$.

Suppose $U \subseteq T \subseteq S$. Then, by the preceding results, $(N \times T) \cdot U$ is primitive. □

Theorem VIII.20. *Let N be primitive. Let T and U be complementary subsets of S. Let D_1 and D_2 be cell-bases of $N \times T$ and $N \cdot U$, respectively. Then their union D is a cell-base of N.*

Proof. Let f be any nonzero chain of N. If $f \cdot U$ is nonzero, then Sp(f) meets D_2 and therefore D. If $f \cdot U$ is zero, then $f \cdot T$ is a nonzero chain of $N \times T$, and therefore Sp(f) meets D_1 and D. So D meets the support of every nonzero chain of N.

Choose $x \in D$. If x is in D_1, there is a chain of $N \times T$ in $B(D_1)$ whose support does not meet $D_1 - \{x\}$. This corresponds to a nonzero chain of N whose support does not meet $D - \{x\}$. If $x \in D_2$, there is a nonzero chain f of N whose support does not meet $D_2 - \{x\}$. By subtracting chains of N corresponding to suitable multiples of members of $B(D_1)$ in $N \times T$, we can transform f into a nonzero chain of N whose support does not meet $D - \{x\}$. We conclude that no proper subset of D meets the support of every nonzero chain of N. Accordingly, D is a cell-base of N. □

Theorem VIII.21. *Let N be primitive. Let D be a set of $r(N)$ elements of S with the following property: For each $x \in D$, there is a chain f of N such that $D \cap \text{Sp}(f) = \{x\}$. Then D is a cell-base of N.*

Proof. D is clearly a cell-base, and the only one, of $N \cdot D$. There is a cell-base D' of $N \times (S - D)$, by Theorem VIII.3. But then $D \cup D'$ is a cell-base of N, by Theorem VIII.20. Since each cell-base of N has exactly $r(N)$ elements, D' must be null. Hence D is a cell-base of N. □

Theorem VIII.22. *If N is primitive, then N^* is primitive. Moreover, the cell-bases of N^* are the complements in S of the cell-bases of N.*

Proof. Let D be any cell-base in N, and D^* its complement in S. For each $y \in D^*$, we define a chain $g(D^*, y)$ on S to R as follows: The coefficient of y is 1 and the coefficient of each other member of D^* is zero. If $x \in D$, then its coefficient in $g(D^*, y)$ is minus the coefficient of y in $g(D, x)$.

It follows that each chain $g(D^*, y)$ is orthogonal to each chain of $B(D)$, and therefore to each chain of N by Theorem VIII.12. Hence the chains $g(D^*, y)$ belong to N^*.

Now any nonzero chain of N^* has a support meeting D^*, for otherwise it could not be orthogonal to every chain of $B(D)$. We can now deduce that D^* is a cell-base of N^*.

We observe that all the nonzero coefficients in $g(D^*, y)$ belong to R'. But, for each y, $g(D^*, y)$ encloses an elementary chain g of N^*, by Theorem VIII.2. Since $\mathrm{Sp}(g)$ must meet D^*, by the preceding result, the coefficient $g(y)$ is nonzero. Consider the chain

$$h = g(y)g(D^*, y) - g$$

of N^*. It is zero because its support does not meet D^*. Since g is elementary, we deduce that $g(D^*, y)$ is an elementary and therefore a primitive chain of N^*, by Theorem VIII.8. Accordingly, the complements of the cell-bases of N are primitive cell-bases of N^*.

Now consider any cell-base E of N^*. Let D be a cell-base of N with as few cells as possible in common with E. If possible, choose a cell x in $D \cap E$. Then $\mathrm{Sp}(g(D, x))$ has a cell y not in D or E; otherwise, $g(D, x)$ could not be orthogonal to a chain f of N^* with a support meeting E only in x. (See Theorem VIII.) But then $(D - \{x\}) \cup \{y\}$ is a cell-base of N, by the Interchange Theorem, and this is contrary to the definition of D. We infer that D and E are disjoint. Hence E is a subset of the cell-base D^* of N^*. But now $E = D^*$, by the minimality of cell-bases.

To complete the proof, we have only to show that N^* is primitive. This follows from Theorem VIII.17. □

Theorem VIII.23. *If N is primitive, then $N^{**} = N$, that is, N and N^* are dual chain-groups.*

Proof. Let D be a cell-base of N. It is a cell-base of N^{**}, by Theorem VIII.22. But the chains $g(D, x)$, for a given $x \in D$, in the two primitive chain-groups N and N^{**} must be identical. Otherwise, by Theorem VIII.6, their difference would be a chain of N^{**} with a support not meeting D. Hence the chains of N are the chains of N^{**}, by Theorem VIII.12. □

Theorem VIII.24. *If N is primitive and $T \subseteq S$, then $(N \cdot T)^* = N^* \times T$ and $(N \times T)^* = N^* \cdot T$.*

Proof. We have the first equation in Theorem VIII.7. Applying it to N^*, we have $(N^* \cdot T)^* = N \times T$, equivalent to the second equation, by Theorems VIII.19, VIII.22, and VIII.23. □

To conclude this section, we note two identities concerning rank, valid for all primitive chain-groups:

$$r(N) = r(N \times T) + r(N \cdot (S - T)), \quad T \subseteq S; \quad \text{(VIII.2.4)}$$

$$|S| = r(N) + r(N^*). \quad \text{(VIII.2.5)}$$

They follow from Theorems VIII.20 and VIII.22, respectively.

We should also note, as a consequence of Theorem VIII.24, that, if N is a primitive chain-group, then the minors of N^* are the duals of the minors of N.

VIII.3. REGULAR CHAIN-GROUPS

In this section we write I for the ring of integers and I_n for the ring of residues modulo a positive integer n. We note that the only regular elements of I are 1 and -1. A primitive chain-group to I is called a regular chain-group. (See [1].)

Consider two chains f and g on S to I. We say that f *conforms* to g if $\text{Sp}(f) \subseteq \text{Sp}(g)$ and for each $x \in \text{Sp}(f)$ the numbers $f(x)$ and $g(x)$ have the same sign. The two following rules are easily verified for chains on S to I.

Theorem VIII.25. *If f conforms to g and g to h, then f conforms to h.*

Theorem VIII.26. *If $\text{Sp}(f) \subseteq \text{Sp}(g)$ and the coefficients in f are restricted to the values 0, 1, and -1, then $g - f$ conforms to g.*

Such a restriction on the coefficients applies whenever f is a primitive chain of a regular chain-group.

Theorem VIII.27. *Let N be a regular chain-group on S. Then each nonzero chain f of N has a conforming primitive chain of N.*

Proof. If possible, choose f so that the theorem fails and $|\text{Sp}(f)|$ has the least value consistent with this condition. Choose $x \in \text{Sp}(f)$. By Theorems VIII.11 and VIII.18, f encloses a primitive chain g of N such that $g(x) = 1$. Write $\sigma = 1$ or -1 according as $f(x)$ is positive or negative. Consider the chain $h_1 = f - \sigma g$ of N, which conforms to f, by Theorem VIII.26. We note that $h_1(x)$ is one less than $f(x)$ in absolute value. If $\text{Sp}(h_1) = \text{Sp}(f)$, we consider $h_2 = h_1 - \sigma g$, and so on. Each of the resulting chains $h_1, h_2, h_3, \ldots$ conforms to f, by Theorems VIII.5 and VIII.26. But after at most $f(x)$ steps, we come to an h_r whose support is a proper subset of $\text{Sp}(f)$.

If $h_r = 0$, then f is a positive integral multiple of σg, and therefore f satisfies the theorem. Otherwise, some primitive chain k of N conforms to h_r, by the choice of f, and therefore to f by Theorem VIII.25. These contradictions establish the theorem. □

Theorem VIII.28. *Let N be a regular chain-group on S. Then each nonzero chain f of N can be represented as a sum of primitive chains of N, each conforming to f.*

Proof. For each chain q of N, let $M(q)$ denote the sum of the absolute values of its coefficients.

If possible, choose f so that the theorem fails and $M(f)$ has the least value consistent with this condition. By Theorem VIII.27, f has a conforming primitive chain g. Then $f - g$ conforms to f, by Theorem VIII.26. Moreover, it is clear that $M(f - g) < M(f)$.

By the choice of f, the chain $f - g$ is a sum of primitive chains of N, each conforming to $f - g$ and therefore to f, or it is zero. But $f = (f - g) + g$, and therefore f satisfies the theorem, which is a contradiction. □

Let n be an integer ≥ 2. To *reduce* $\bmod\, n$ a chain f on S to I is to replace each coefficient by its residue class mod n. The operation transforms f into a chain f_n on S to I_n. We call f_n the *residual chain of f* mod n.

If N is a chain-group on S to I and we replace each chain of N by its residual chain mod n, we obtain the chains of a chain-group N_n on S to I_n. This is the *residual chain-group of N* mod n. A given chain of N_n may correspond to infinitely many chains of N.

Theorem VIII.29. *Let N be regular and let q be a chain of N_n. Then there is a chain f of N satisfying the two following conditions*:
i) *q is the residual chain of f mod n.*
ii) *Each coefficient in f has absolute value $< n$.*

Proof. By the definition of N_n, there exists a chain f of N satisfying (i). For each such f, let $Z(f)$ denote the number of cells of $\mathrm{Sp}(f)$ having coefficients not less than n in absolute value. If $Z(f) = 0$, there is nothing to prove. We may assume, therefore, that $\mathrm{Sp}(f)$ has a cell x such that $|f(x)| > n$.

By Theorem VIII.28 there is a primitive chain g of N, conforming to f, such that $x \in \mathrm{Sp}(g)$. Consider the chain $f_1 = f - ng$ of N, whose residual chain mod n is still q. If $f(y) = 0$, then $f_1(y) = 0$. If $0 < |f(y)| < n$, then $|f_1(y)|$ is $|f(y)|$ or $n - |f(y)|$ and, in either case, $0 < |f_1(y)| < n$. But $|f_1(x)| = |f(x)| - n$. So, if we replace f by f_1, we preserve (i), we do not introduce any new cells y with coefficients n or more in absolute value, and we diminish the absolute value of $f(x)$ by n. By sufficiently many repetitions of this procedure, we can reduce $Z(f)$, while still preserving (i).

Continuing in this way we can reduce $Z(f)$ to zero. We then have a chain f of N satisfying (i) and (ii). □

It is often convenient to use matrices in the study of a regular chain-group N on a set S. We assume that S is nonnull and that $r(N) > 0$. We fix an enumeration $(x_1, x_2, \ldots, x_s)$ of the s cells of S. For any chain f on S to I, we define its *representative vector* as the row-vector

$$V(f) = \{f(x_1), f(x_2), \ldots, f(x_s)\}.$$

In the preceding discussions using the general ring R of Sec. VIII.1, we did not assume any previous knowledge of the related theory of linear dependence. But for the ring I, especially if it is regarded as embedded in the field of rational or real numbers, the theory of linear dependence is familiar to all likely readers. In particular, we can recognize $r(N)$, for a regular N, as the maximum number of linearly independent chains of N. (See Theorem VIII.12.)

Consider a matrix M of $r(N)$ rows and s columns. Let its rows be representative vectors of chains of N. If they are linearly independent, we call M a *representative matrix* of N.

Let D be any cell-base of N. We can write the representative vectors of the $r(N)$ members of $B(D)$ as the rows of a representative matrix M of N. We can order the rows so that the columns of M corresponding to the members of D constitute a unit matrix, of order $r(N)$. We then call M the *standard representative matrix* of N with respect to D.

For any representative matrix M of N and any subset T of S, we write $M(T)$ for the submatrix of M made up of the columns corresponding to the members of T.

Theorem VIII.30. *Let M be the standard representative matrix of N with respect to a cell-base D. Let T be any set of $r(N)$ members of S. Then $\det M(T)$ is 1 or -1 if T is a cell-base of N, and $\det M(T) = 0$ otherwise.*

Proof. Let T be any cell-base of N. Let M_1 be the standard representative matrix of N with respect to T. By Theorem VIII.12, there is a square matrix Q of integers, of order $r(N)$, such that $QM_1 = M$. Considering the unit matrix $M_1(T)$, we deduce that $Q = M(T)$. Considering $M(D)$, we find that $\det(QM_1(D)) = 1$. Hence, $\det Q$, that is, $\det M(T)$, is 1 or -1.

Now suppose that T is not a cell-base. If it meets the support of every nonzero chain of N, it has a minimal proper subset T_1 with this property. But then T_1 is a cell-base of N with fewer than $r(N)$ elements, which is impossible. We deduce that in fact there is a non-zero chain f of N whose support does not meet T. But the representative vector of f is a linear combination of rows of M, by Theorem VIII.12, and therefore the rows of $M(T)$ are linearly dependent. Hence $\det M(T) = 0$. □

Theorem VIII.31. *Let M_1 be any representative matrix of N. Let T be any set of $r(N)$ members of S. Then there is a positive integer m_1 with the following property: If T is a cell-base of N, then $\det M_1(T) = m_1$ or $-m_1$, and in the remaining case, $\det M_1(T) = 0$.*

Proof. Let D be a cell-base of n and let M be the standard representative matrix of M with respect to D. By Theorem VIII.12, there is a square matrix Q of integers such that $QM = M_1$. Since the rows of M_1 are linearly independent, Q must be nonsingular. Let m_1 be the absolute value of its determinant. We have $M_1(T) = QM(T)$ for each T. Theorem VIII.31 now follows from Theorem VIII.30. □

VIII.4. CYCLES

Let G be a graph, and Ω one of its orientations. If D is a dart of Ω, we denote the corresponding edge of G by D'. We write S for the dart-set of Ω. If $T \in S$, then T' denotes the set of edges of G corresponding to the members of T. In particular, $S' = E(G)$.

For each dart D of Ω and vertex v of G, we define an *incidence-number* $\eta(D, v)$ as follows: If D' is a loop, or if D is not incident with v, then $\eta(D, v) = 0$. In the remaining case, with D' a link of G, we write $\eta(D, v) = 1$ or -1, according as v is the head or tail of D.

A chain on S to the ring R of Sec. VIII.1 is a 1-*chain* of Ω to R. A chain on $V(G)$ to R is a 0-*chain* of Ω, or of G, to R.

Let f be any 1-chain of Ω to R. Its *boundary* ∂f is the 0-chain of G defined as follows: For each vertex v,

$$(\partial f)(v) = \sum_{D \in S} \eta(D, v) f(D). \qquad \text{(VIII.4.1)}$$

Thus if $f(D)$ is called the "current" in D, we can describe $(\partial f)(v)$ as the total current in the darts with head v minus the total current in those with tail v, or simply as the *net current* to v in Ω.

From Eq. (VIII.4.1), we deduce the following identities:

$$\partial(\lambda f) = \lambda \cdot \partial f, \qquad \text{(VIII.4.2)}$$

$$\partial(f + g) = \partial f + \partial g. \qquad \text{(VIII.4.3)}$$

Here f and g are 1-chains of Ω to R, and $\lambda \in R$.

A 1-chain f of Ω to R is called a *cycle* of Ω to R if $\partial f = 0$. For example, the zero chain on S to R is a cycle of Ω. By Eqs. (VIII.4.2) and (VIII.4.3), the cycles of Ω to R are the elements of a chain-group $\Gamma = \Gamma(\Omega, R)$ on S to R. We call Γ the *cycle-group* of Ω, with respect to R.

We go on to give graph-theoretical interpretations of the elementary chains and the cell-bases of Γ. These will be in terms of G, rather than Ω, for, from the point of view of the present chapter, a change of orientation of G, say from Ω to Ω_1, is but a trivial readjustment. In a certain subset U of S, each dart D is replaced by its opposite. An incidence number $\eta(D, v)$ for such a dart is replaced by $\eta(D^{-1}, v)$, that is, by $-\eta(D, v)$. So each dart of U is replaced by its opposite as a cell of the chain-group, and the corresponding coefficient in each member-chain of Γ is replaced by its negative. We say that $\Gamma(\Omega_1, R)$ is derived from $\Gamma(\Omega, R)$ by a *reorientation* of U'.

Sometimes $1 = -1$ in R, as when $R = I_2$, and then changes of orientation become more trivial still. There is no alteration of coefficients. In this case, it is unnecessary to distinguish between a dart and its opposite, or between either and the corresponding edge of G. We can regard chains on S as chains on $E(G)$, and we can speak of the 0-chains, 1-chains, and cycles "of G," without reference to any particular orientation.

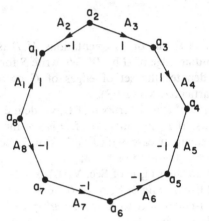

FIGURE VIII.4.1

Consider an n-circuit C of G, with the notation of Sec. I.1. For each edge A_j, let D_j be its dart in S, shown by an arrow from tail to head in Fig. VIII.4.1 for the case $n = 7$.

We define a chain p_C on S as follows: For any dart D of S not on an edge of C, we write $p_C(D) = 0$. For D_j on C we write $p_C(D_j) = 1$ or -1, according as a_j is the head or the tail of A_j. These coefficients are written inside the circuit in Fig. VIII.4.1. For $n = 1$, when A_1 is a loop, it is trivial that p_C is a cycle of Ω. For $n > 1$, we have the rules:

$$\eta(D_j, a_j)p_C(D_j) = 1, \qquad \text{(VIII.4.4)}$$

$$\eta(D_j, a_{j-1})p_C(D_j) = -1. \qquad \text{(VIII.4.5)}$$

From these and Eq. (VIII.4.1), we deduce that $\partial p_C = 0$. Summarizing, we have the following theorem.

Theorem VIII.32. *For each circuit C of G, the chain p_C is a nonzero cycle of Ω to R. Its coefficients are restricted to the values 0, 1, and -1, and its support consists of the darts of C that are in S.*

Note that reversing the defining sequences of a circuit C replaces p_C by its negative.

Theorem VIII.33. *Let f be any nonzero chain of $\Gamma(\Omega, R)$. Then there is a circuit C of G such that f encloses p_C.*

Proof. In the reduction $G \cdot \text{Sp}(f)$, each vertex v has valency at least 2, since otherwise $(\partial f)(v)$ could not be zero. Hence,

$$p_1(G \cdot \text{Sp}(f)) > 0,$$

by Theorem I.39. Hence $G \cdot \text{Sp}(f)$ has a circuit C, by Thoerems I.35 and I.45. But then C is a circuit of G, and f encloses p_C. □

Cycles

Theorem VIII.34. *A chain f of $\Gamma(\Omega, R)$ is elementary if and only if $f = \sigma p_C$, where C is a circuit of G and σ is a nonzero element of R.*

Proof. Suppose $f = \sigma p_C$. Assume it is not elementary. By Theorem VIII.33, there is a circuit U of G such that p_C properly encloses p_U. But then U is a detached proper subgraph of C, contrary to the connection of C, by Theorem I.27.

Conversely, suppose f elementary. By Theorem VIII.33 there is a circuit C of G such that f encloses p_C. Since f is elementary, we must have $\mathrm{Sp}(f) = \mathrm{Sp}(p_C)$. Hence p_C is elementary, and therefore primitive, by Theorem VIII.32. So f is of the required form, by Theorem VIII.10. □

Theorem VIII.35. $\Gamma(\Omega, R)$ *is a primitive chain-group.*

Proof. Any elementary chain f of $\Gamma(\Omega, R)$ is of the form σp_C, for some circuit C of G, by Theorem VIII.34. But p_C is primitive, by Theorem VIII.32. □

We refer to the chains p_C as the *elementary cycles* of Ω.

Let us define a *frame* of a graph G as a spanning subgraph of G whose intersection with each component H of G is a tree, necessarily a spanning tree of H.

Theorem VIII.36. *Let T be a subset of S. Then T is a cell-base of $\Gamma(\Omega, R)$ if and only if $G:(S' - T')$ is a frame F of G.*

Proof. Suppose first that $G:(S' - T')$ is a frame F of G. It contains no circuit. Hence, by Theorem VIII.33, each nonzero chain of $\Gamma(\Omega, R)$ has a support meeting T. But now let D be any dart of T, with head h and tail t, say. If D is on a loop of G, it constitutes by itself the support of an elementary chain of $\Gamma(\Omega, R)$, by Theorem VIII.34. In the remaining case, h and t are distinct vertices of a component H of G. Hence they are joined by an arc L in $F \cap H$, by Theorem I.44. Adjoining the edge of D to L, we obtain a circuit C such that $\mathrm{Sp}(p_C)$ meets T only in D. We can now assert that T is a cell-base of $\Gamma(\Omega, R)$.

Conversely, suppose we are given a cell-base T of $\Gamma(\Omega, R)$. Write $F = G \cdot (S' - T')$. Then F has no circuit, by Theorem VIII.32. Assume that it has two or more components contained in the same component of G, H say. Then there are complementary nonnull subsets U and W of $V(H)$ that are joined only by links of T'. Since H is connected, we can find one such link A of T'. By Theorems VIII.4 and VIII.34, there is a circuit C of G such that $E(C) \cap T' = \{A\}$. By its connection, C is a subgraph of H. The complementary arc of A in C, being connected, must include an edge that joins U and W, and this edge is not in T'. We have a contradiction. So the intersection of F with each component of G is connected and has no circuit: F is a frame. □

We note that if G is connected, the number of cell-bases of $\Gamma(\Omega, R)$ is equal to the tree-number of G.

Theorem VIII.37. *The rank of $\Gamma(\Omega, R)$ is $p_1(G)$.*

Proof. Let T be any cell-base of $\Gamma(\Omega, R)$. (See Theorem VIII.3.) If H is any component of G, then the number of edges of H not in T' is $|V(H)| - 1$, by Theorems VIII.36 and I.37. Hence

$$|S' - T'| = |V(G)| - p_0(G),$$

$$r(\Gamma) = |T'| = |E(G)| - |V(G)| + p_0(G)$$

$$= p_1(G),$$

by Eq. (I.6.1). □

VIII.5. COBOUNDARIES

Let g be any 0-chain of G (and Ω) to R. We define its *coboundary* δg as the 1-chain of Ω to T defined as follows:

$$(\delta g)(D) = \sum_{v \in V(G)} \eta(D, v) g(v), \qquad \text{(VIII.5.1)}$$

for each dart D of Ω. This is analogous to Eq. (VIII.4.1). It simplifies as

$$(\delta g)(D) = g(h(D)) - g(t(D)), \qquad \text{(VIII.5.2)}$$

where $h(D)$ and $t(D)$ are the head and tail of D, respectively.

From Eq. (VIII.5.1) we can deduce the following identities:

$$\delta(\gamma g) = \lambda \cdot \delta g, \qquad \text{(VIII.5.3)}$$

$$\delta(g + h) = \delta g + \delta h, \qquad \text{(VIII.5.4)}$$

where g and h are 0-chains of Ω to R, and $\lambda \in R$. We note that the coboundary of the zero 0-chain is the zero 1-chain.

By Eqs. (VIII.5.3) and (VIII.5.4), the coboundaries of the 0-chains of Ω to R are the elements of a chain-group $\Delta = \Delta(\Omega, R)$ on S to R. Reorientation is a trivial operation for this chain-group, also. If we change to a new orientation Ω_1 of G, then $\Delta(\Omega_1, R)$ is derived from $\Delta(\Omega, R)$ just as $\Gamma(\Omega_1, R)$ is derived from $\Gamma(\Omega, R)$, by a reorientation of the appropriate subset of $E(G)$. We go on to discuss the elementary chains and cell-bases of $\Delta(\Omega, R)$. We call Δ the *coboundary-group* of Ω.

Suppose the vertex-set of G to be partitioned into complementary subsets X and Y. Here we write $J(X, Y)$ for the set of all edges of G having one end in X and one in Y.

If G is connected and the two induced subgraphs $G[X]$ and $G[Y]$ are also connected, we call $J(X, Y)$ a *bond* of G. We then call $G[X]$ and $G[Y]$

the *end-graphs* of this bond. This definition implies that $J(X,Y)$ must be nonnull. If G is not connected, we define a *bond* of G as a bond of a component of G. We note that an isthmus of G is the single edge of a bond of G. The end-graphs of the isthmus are the end-graphs of the corresponding bond.

Theorem VIII.38. *Let X and Y be complementary subsets of $V(G)$, and let $J(X,Y)$ be nonnull. Then there is a bond B of G such that $B \subseteq J(X,Y)$.*

Proof. Out of all pairs (X_0, Y_0) of complementary subsets of $V(G)$ such that $\emptyset \subset J(X_0, Y_0) \subseteq J(X,Y)$, choose one so that $|J(X_0, Y_0)|$ has the least possible value.

Suppose first that G is connected. If $G[X_0]$ is not connected, let its components be $K_1, K_2, \ldots, K_m$ ($m > 1$). By the connection of G, each of the sets $J(V(K_i), V(G) - V(K_i))$ is nonnull, and each is a proper subset of $J(X_0, Y_0)$. From this contradiction, we deduce that, in fact, $G[X_0]$, and similarly $G[Y_0]$, is connected. So $J(X_0, Y_0)$ is the required bond B.

If G is not connected, let H be a component of G including some edge A of $J(X,Y)$. By the result already proved, the set $J(X \cap V(H), Y \cap V(H))$ contains a bond B of H. But then B is a bond of G. □

Given a partition (X,Y) of $V(G)$, we can define a 0-chain g of Ω by writing $g(x) = 0$ or 1, according as x is in X or Y. We then have

$$\text{Sp}(\delta g) = J(X,Y).$$

Moreover the nonzero coefficients in δg are restricted to the values 1 and -1 of R. If $J(X,Y)$ is a bond of G, we write δg also as q_B, where $B = J(X,Y)$.

Theorem VIII.39. *Let f be any nonzero chain of $\Delta(\Omega, R)$. Then there is a bond B of G such that f encloses q_B.*

Proof. There is a 0-chain g such that $\delta g = f$. Since f is nonzero, there is a dart D of S, with head h and tail t, say, such that $g(h) \neq g(t)$, by Eq. (VIII.5.2). Let X be the set of all vertices x of G such that $g(x) = g(h)$, and let Y be the set of all other vertices of G. It is clear that $J(X,Y)$ is a nonnull subset of Sp(f). But it contains a bond B of G, by Theorem VIII.38. Hence f encloses q_B. □

Theorem VIII.40. *A nonzero chain of $\Delta(\Omega, R)$ is elementary if and only if it is of the form σq_B, where B is a bond of G and σ is a nonzero element of R.*

Proof. Suppose f of the form σq_B. If it is not elementary, then, by Theorems VIII.38 and VIII.39, B must contain another bond B' as a proper subset. But this is impossible, for let H be the component of G that includes the edges of B; then H is formed from the union of two disjoint connected

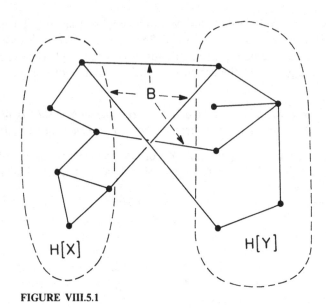

FIGURE VIII.5.1

subgraphs $H[X]$ and $H[Y]$ of H by adjoining the edges of B, each such edge having one end in X and one in Y (see Fig. VIII.5.1).

If we delete from H the edges of a proper subset of B, we leave a connected graph, by Theorem I.25. It follows that no proper subset B' of B can be a bond of G.

Conversely, suppose f elementary. By Theorem VIII.39, there is a bond B of G such that f encloses q_B. Since f is elementary, we have $\mathrm{Sp}(f) = \mathrm{Sp}(q_B)$. But q_B is elementary, and therefore primitive, by the result already proved. Hence f is of the required form σq_B, by Theorem VIII.10. This completes the proof. □

We have shown, incidentally, that each elementary chain of $\Delta(\Omega, R)$ is a multiple of a primitive chain. So we have the following corollary.

Corollary VIII.41. $\Delta(\Omega, R)$ *is a primitive chain-group.*

Theorem VIII.42. $\Gamma(\Omega, R)$ *and* $\Delta(\Omega, R)$ *are dual chain-groups.*

Proof. Let f be an arbitrary 1-chain and g an arbitrary 0-chain, both of Ω to R. Then, by Eqs. (VIII.1.5) and (VIII.5.1),

$$(f \cdot \delta g) = \sum_{D \in S} \left\{ f(D) \sum_{x \in V(G)} \eta(D, x) g(x) \right\}$$

$$= \sum_{x \in V(G)} \left\{ g(x) \sum_{D \in S} \eta(D, x) f(D) \right\}.$$

In order that f shall be orthogonal to the coboundary of every 0-chain, it is therefore necessary and sufficient that

$$\sum_{D \in S} \eta(D, x) f(D) = 0$$

for each vertex x. But this is the condition for f to be a cycle of Ω, a member of $\Gamma(\Omega, R)$.

We now have

$$(\Delta(\Omega, R))^* = \Gamma(\Omega, R). \qquad \text{(VIII.5.5)}$$

Since the chain-groups concerned are primitive, by Theorems VIII.35 and VIII.41 we have also

$$(\Gamma(\Omega, R))^* = \Delta(\Omega, R) \qquad \text{(by Theorem VIII.23)}.$$

$$\text{(VIII.5.6)} \quad \square$$

Theorem VIII.43. *Let T be a subset of S. Then T is a cell-base of $\Delta(\Omega, R)$ if and only if $G : T'$ is a frame of G.*

Proof. By Theorems VIII.33 and VIII.42, T is a cell-base of Δ if and only if $S - T$ is a cell-base of Γ. So our theorem follows, from Theorem VIII.36. $\square$

Let D be a member of a cell-base T of $\Delta(\Omega, R)$. Let f_D be the corresponding member of the corresponding chain-base. Then, by Theorem VIII.40, f_D is of the form q_B for some bond B of G. We can write $B = J(X, Y)$ and suppose the head of D to be in X. It is easy to verify that X is the set of vertices of the appropriate end-graph of the isthmus D' in the forest $G : T'$.

Given a vertex v of G, we can consider the 0-chain g_v of Ω in which the coefficient of v is 1 and the coefficient of every other vertex is 0. We refer to δg_v as the *coboundary* of v in Ω and write it also as δv. It is a *vertex-coboundary* of Ω.

Theorem VIII.44. *Each $f \in \Delta$ can be expressed as a linear combination of vertex-coboundaries.*

We have indeed

$$f = \sum_{v \in V(G)} f(v) \delta v, \qquad \text{(VIII.5.7)}$$

by Eqs. (VIII.5.3) and (VIII.5.4).

Theorem VIII.45. *A 0-chain g of Ω to R has a zero coboundary if and only if for each component K of G all the vertices of K have the same coefficient in g.*

This is an obvious consequence of Eq. (VIII.5.2).

Theorem VIII.46. *Let X be a subset of $V(G)$ obtained from $V(G)$ by deleting exactly one vertex $x(K)$ from each component K of G. Then the coboundaries of the members of X are linearly independent, and each member of Δ can be expressed as a linear combination of them.*

Proof. Suppose we have a linear relation

$$\sum_{x \in X} a(x)\, \delta x = 0,$$

where the $a(x)$ are elements of R, not all zero. We extend this definition by writing $a(y) = 0$ if $y = v(K)$ for some component K. Consider the 0-chain

$$f = \sum_{v \in V(G)} a(v) g_v.$$

Then $\delta f = 0$, by Eqs. (VIII.5.3) and (VIII.5.4). Hence each $a(v)$ is zero, by Theorem VIII.45, a contradiction.

On the other hand, each $\delta v(K)$ can be expressed as a linear combination of members of X, by Theorem VIII.45, and therefore each $f \in \Delta$ can be so expressed, by Theorem VIII.44. □

We remark in conclusion that

$$r(\Delta) = |E(G)| - p_1(G) = |V(G)| - p_0(G), \qquad \text{(VIII.5.8)}$$

by Theorems VIII.37 and VIII.42. So in what follows we may call $p_1(G)$ the *cycle-rank* and $|V(G)| - p_0(G)$ the *coboundary-rank* of the graph G.

VIII.6. REDUCTIONS AND CONTRACTIONS

Let G, Ω, and S be as in Secs. VIII.4 and VIII.5. The notions of spanning subgraph, reduction, and contraction of G extend to Ω. Thus, let T be any subset of S. We define $\Omega:T$ and $\Omega \cdot T$ as the orientations of $G:T'$ and $G \cdot T'$, respectively, in which the darts are the members of T, and each dart has the same head and tail as in Ω. Similarly, $\Omega \times T$ is an orientation of $G \times T'$. Its darts are those of T. The head and tail of a dart D in $\Omega \times T$ are those components of $G:(E(G) - T')$, that is, $G:(S' - T')$, which include the head and tail, respectively, of D in Ω.

The distinction between $\Omega:T$ and $\Omega \cdot T$ disappears when we go over to the cycle-group or coboundary-group.

Theorem VIII.47. $\Delta(\Omega:T, R) = \Delta(\Omega \cdot T, R)$ *and* $\Gamma(\Omega:T, R) = \Gamma(\Omega \cdot T, R)$.

Proof. Let W be the set of vertices of $G \cdot T'$. It follows from Eq. (VIII.5.2) that, if f is any 0-chain of Ω, then its coboundary in $\Omega:T$ is

Reductions and Contractions

identical with the coboundary in $\Omega \cdot T$ of its restriction to W, and in each case the coboundary is $\delta f \cdot T$. This establishes the first equation. We can obtain the second by dualizing and using Theorem VIII.42. □

Theorem VIII.48. $\delta(\Omega \cdot T, R) = \Delta(\Omega, R) \cdot T$ and $\Delta(\Omega \times T, R) = \Delta(\Omega, R) \times T$.

Proof. Consider any 0-chain f of Ω. By Eq. (VIII.5.2), its coboundary in $\Omega : T$ is the restriction to T of its coboundary in Ω. We deduce that $\Delta(\Omega : T, R) = \Delta(\Omega, R) \cdot T$, and the first part of the theorem follows from Theorem VIII.47.

Now consider any 0-chain h of $\Omega \times T$. Let h' be that 0-chain of Ω in which each vertex has the same coefficient as does the vertex of $G \times T'$ to which it belongs in h. Then $\mathrm{Sp}(\delta h') \subseteq T$. Moreover, δh in $\Omega \times T$ is the restriction of $\delta h'$ to T. On the other hand, if g is a 0-chain of Ω such that $\mathrm{Sp}(\delta g) \subseteq T$, then for each component of $G:(S'-T')$, all the vertices must have the same coefficient in g, by Eq. (VIII.5.2). Hence there is a 0-chain h of $\Omega \times T$ such that $g = h'$. From these observations we deduce the second part of the theorem. □

Theorem VIII.49. $\Gamma(\Omega \cdot T, R) = \Gamma(\Omega, R) \times T$ and $\Gamma(\Omega \times T, R) = \Gamma(\Omega, R) \cdot T$.

To prove this, we take dual chain-groups in Theorem VIII.48, using Theorem VIII.24, and then apply Theorem VIII.42.

The notation of G'_A and G''_A can be extended to orientations and to general chain-groups. Thus, if D is a dart of S, we write:

$$\Omega'_D = \Omega : (S - \{D\}), \qquad \text{(VIII.6.1)}$$

$$\Omega''_D = \Omega \times (S - \{D\}). \qquad \text{(VIII.6.2)}$$

These digraphs are orientations of G'_A and G''_A, respectively.

For any chain-group N on S to R, we make the following definitions:

$$N'_D = N \cdot (S - \{D\}), \qquad \text{(VIII.6.3)}$$

$$N''_D = N \times (S - \{D\}). \qquad \text{(VIII.6.4)}$$

Applying the foregoing theorems to these definitions, we obtain the following identities:

$$(\Delta(\Omega))'_D = \Delta(\Omega'_D), \qquad \text{(VIII.6.5)}$$

$$(\Delta(\Omega))''_D = \Delta(\Omega''_D), \qquad \text{(VIII.6.6)}$$

$$(\Gamma(\Omega))'_D = \Gamma(\Omega''_D), \qquad \text{(VIII.6.7)}$$

$$(\Gamma(\Omega))''_D = \Gamma(\Omega'_D). \qquad \text{(VIII.6.8)}$$

VIII.7. ALGEBRAIC DUALITY

Having constructed a theory Z of primitive chain-groups N from S to R, we can apply it to $\Gamma(\Omega, R)$, where Ω is an arbitrary digraph with dart-set S. We can regard Ω as an orientation of an associated graph G. The definitions and theorems of Z then become statements about Ω. They constitute a partial theory Z_Γ of digraphs Ω. Similarly, we can get another partial theory of digraphs Ω by using $\Delta(\Omega, R)$ instead of $\Gamma(\Omega, R)$. We denote this by Z_Δ. There is a 1–1 correspondence, called *algebraic duality*, relating the definitions and theorems of Z_Γ to those of Z_Δ. Corresponding statements under this duality derive from the same statement of Z. Sometimes corresponding statements in Z_Γ and Z_Δ are identical. We then speak of a *self-dual* concept or theorem in the theory of digraphs.

The darts of Ω are the cells both of $\Gamma(\Omega, R)$ and of $\Delta(\Omega, R)$. We say, therefore, that "dart" is a self-dual concept (for every R). Now consider the 1-chain f_D to R in which a given dart D has coefficient 1 and every other dart of S has coefficient 0. We see from Eq. (VIII.4.1) that $\partial f_D = 0$ if and only if D is a loop-dart of Ω, a dart on a loop of G. So $\{D\}$ is the support of a chain of $\Gamma(\Omega, R)$ if and only if D is on a loop of G. On the other hand, $\{D\}$ is the support of a chain of $\Delta(\Omega, R)$ if and only if D is an isthmus-dart of Ω, that is, a dart on an isthmus of G. This follows from Theorem VIII.40. We may now assert that "loop-dart" and "isthmus-dart" are dual concepts in the theory of digraphs.

Let T be any subset of S. Consider the reduction $N \cdot T$ in the theory Z. The corresponding concepts in Z_Γ and Z_Δ are $\Gamma(\Omega, R) \cdot T$ and $\Delta(\Omega, R) \cdot T$, that is, $\Gamma(\Omega \times T, R)$ and $\Delta(\Omega : T, R)$, respectively, by Theorems VIII.47, VIII.48, and VIII.49. Similarly, $N \times T$ in Z gives rise to $\Gamma(\Omega : T, R)$ in Z_Γ and to $\Delta(\Omega \times T, R)$ in Z_Δ. It therefore seems proper to assert that contracting Ω to T and taking the spanning subdigraph $\Omega : T$ are dual operations. In particular, if D is any dart of Ω, then taking Ω'_D and taking Ω''_D are dual operations. We recall that $\Omega : T$ and $\Omega \cdot T$ are indistinguishable in terms of the structures of their cycle-groups and coboundary-groups, by Theorem VIII.47.

Sometimes duality relates two statements about Ω that can be stated in terms only of G (and perhaps R). We can then speak of dual concepts or theorems in graph theory. For example, "loop" and "isthmus" are such dual concepts. So are "circuit" and "bond," or, more precisely, "edge-set of a circuit" and "bond." The two latter derive from "elementary chain" in Z, by way of Theorems VIII.34 and VIII.40. For an example of dual theorems, we may take the assertions that no set of edges of G defining a circuit, or secondly a bond, has a proper subset with that property. These theorems are related, by way of Theorems VIII.34 and VIII.40, to the definition of an elementary chain in Z.

Our remarks about contractions and spanning subdigraphs can be transferred from digraphs to graphs. We can regard the constructions of $G \times T$ and $G:T$ from G as algebraically dual operations. In particular G'_A and G''_A are related by duality, for any edge A. It is this kind of duality that was foreshadowed in Sec. II.1.

In order to give a new example, we make some new definitions. Consider any chain-group N on S to R. A subset T of S is *detached*, with respect to N, if the following condition is satisfied: Each chain f of N can be expressed as a sum $f_1 + f_2$, where f_1 and f_2 are chains of N such that $\text{Sp}(f_1) \subseteq T$ and $\text{Sp}(f_2) \subseteq S - T$. We note that f encloses both f_1 and f_2 if this condition holds, and that the enclosure is proper if both f_1 and f_2 are nonzero.

We observe that S and its null subset are detached, that any union or intersection of detached subsets of S is detached, and the complement in S of any detached subset of S is detached.

A *block-set* of S, with respect to N, is a minimal nonnull detached subset of S—minimal in the sense of containing no smaller nonnull detached subset. Evidently each nonnull detached subset of S contains a block-set. From the preceding observations, we can deduce that the block-sets of S with respect to N are disjoint and that their union is S. So if S is nonnull, it has at least one block-set with respect to N. If it has only one we call N *nonseparable*; otherwise, N is *separable*.

Theorem VIII.50. *Let N be primitive. Then a subset T of S is detached with respect to N if and only if each primitive chain f of N satisfies either* $\text{Sp}(f) \subseteq T$ *or* $\text{Sp}(f) \subseteq S - T$.

Proof. Suppose T detached. The primitive chain f satisfies $f = f_1 + f_2$, where f_1 and f_2 are chains of N such that $\text{Sp}(f_1) \subseteq T$ and $\text{Sp}(f_2) \subseteq S - T$. But either f_1 or f_2 must be zero, for otherwise the elementary chain f would properly enclose both f_1 and f_2. Hence $f = f_1$ or f_2, and the theorem is satisfied.

Conversely, suppose each primitive chain of N to have its support contained either in T or in $S - T$. Then, by Theorem VIII.18, each chain g of N is a sum $g_1 + g_2$, where g_1 and g_2 are chains of N such that $\text{Sp}(g_1) \subseteq T$ and $\text{Sp}(g_2) \subseteq S - T$. □

Theorem VIII.51. *Let N be any chain-group on S to R. Then a subset T of S is detached with respect to N if and only if*

$$N \cdot T = N \times T. \qquad \text{(VIII.7.1)}$$

Proof. We note that $N \times T \subseteq N \cdot T$, by the definitions of reductions and contractions.

Suppose T detached. Consider any $g \in N \cdot T$. Then $g = f \cdot T$ for some $f \in N$. But $f = f_1 + f_2$, where f_1 and f_2 are chains of N, $\mathrm{Sp}(f_1) \subseteq T$ and $\mathrm{Sp}(f_2) \subseteq S - T$. It follows that $g = f_1 \cdot T \in N \times T$. Hence Eq. (VIII.7.1) holds.

Conversely, assume Eq. (VIII.7.1). Let f be any chain of N. Then $f \cdot T \in N \times T$ and therefore $f \cdot T$ is the restriction to T of a chain f_1 of N such that $\mathrm{Sp}(f_1) \subseteq T$. We now have

$$f = f_1 + (f - f_1), \quad \mathrm{Sp}(f_1) \subseteq T \quad \text{and } \mathrm{Sp}(f - f_1) \subseteq S - T.$$

Accordingly, T is detached. □

Theorem VIII.52. *Let N and N^* be dual primitive chain-groups on S to R. Then they determine the same detached subsets of S, and therefore the same block-sets.*

Proof. Let a subset T of S be detached with respect to N. Then it satisfies Eq. (VIII.7.1), by Theorem VIII.51. Hence

$$(N \times T)^* = (N \cdot T)^*,$$

$$N^* \cdot T = N^* \times T \quad \text{(by Theorem VIII.24)}.$$

Hence T is detached with respect to N^*, by Theorem VIII.51.

Replacing N by N^* in the above argument and using Theorem VIII.23, we find that, if T is detached with respect to N^* then it is detached with respect to N. □

If T is a block-set of S with respect to a chain-group N, we say that $N \cdot T$ is a *block* of N. This block can be written also as $N \times T$, by Theorem VIII.51. If S is null, N has no block. Otherwise, it has exactly one block, N itself, if it is nonseparable, and more than one if it is separable.

Theorem VIII.53. *If N is primitive, then the blocks of N^* are the duals of the blocks of N.*

This follows from Theorems VIII.51 and VIII.52, with the help of Theorem VIII.24.

Theorem VIII.54. *Let Ω, with dart-set S, be an orientation of a graph G. Then a subset T of S is a block-set of S with respect to $\Gamma(\Omega, R)$ if and only if T' is the set of edges of a block of G.*

Proof. Suppose first that $G \cdot T'$ is a block of G. Let D and E be any two darts of T. Then there is a circuit C of G such that $\{D', E'\} \subseteq E(C) \subseteq T'$, by Theorems III.12 and III.18. Hence there is a primitive chain of $\Gamma(\Omega, R)$ whose support includes both D and E, by Theorem VIII.34. Hence by Theorem VIII.50, some block-set U with respect to $\Gamma(\Omega, R)$ must contain T.

Suppose T is not the whole of U. Then by Theorem VIII.50, some primitive chain of $\Gamma(\Omega, R)$ has a support meeting both T and $U - T$. The corresponding circuit of G has one edge in T' and one not. By Theorem III.12, this contradicts the definition of $G \cdot T'$ as a block of G. We conclude that T is the block-set U.

Conversely, suppose T to be a block-set of S with respect to $\Gamma(\Omega, R)$. Then T includes a dart D, and D' is an edge of some block $G \cdot U'$ of G. By the preceding argument, U is a block-set. Since it meets the block-set T, it must be identical with T. □

Theorem VIII.54 exhibits a 1-1 correspondence between the blocks of $\Gamma(\Omega, R)$ and those blocks of G that have edges, that is, the blocks of $G \cdot E(G)$. In view of Theorem VIII.53 and the duality of $\Gamma(\Omega, R)$ and $\Delta(\Omega, R)$, we can now assert that "block of $G \cdot E(G)$" and "2-connection of $G \cdot E(G)$" are self-dual concepts in graph theory.

We note that a chain-group N on S to R is uniquely determined by its blocks. The blocks determine all chains of N with support confined to a single block, and the chains of N are simply the sums of chains of that special kind. (See Theorem VIII.51.) It follows that the cell-bases of N are the union of cell-bases of the blocks, one cell-base from each block. We thus have the following theorem.

Theorem VIII.55. *The rank of a primitive chain-group N is the sum of the ranks of its blocks.*

We further observe that it is impossible to distinguish, in terms of the properties of cycle groups and coboundary groups of the orientations, between two separable graphs having the same set of blocks, even to within a vertex-isomorphism. We can use Theorem VIII.54 to identify the blocks of the graphs concerned, but we cannot tell how the blocks fit together in each case. One of the graphs might be connected and the other disconnected. Clearly, connection is not one of the concepts of graph theory to which algebraic duality applies. We ought not to ask what is the algebraic dual of the concept "component."

VIII.8. CONNECTIVITY

Having established the self-duality of 2-connection in graphs, we naturally ask if n-connection is self-dual for each integer $n \geq 2$. We exclude the case $n = 1$, for 1-connection is simply connection, and we have decided that algebraic duality does not apply to this.

We begin by defining the connectivity of a primitive chain-group N. We then relate connectivity of primitive chain-groups to connectivity of graphs.

Theorem VIII.56. *Let N be a primitive chain-group on S to R, and let T be any subset of S. Then*

$$r(N \times T) \leq r(N \cdot T), \qquad \text{(VIII.8.1)}$$

with equality if and only if T is detached with respect to N.

Proof. Let D be any cell-base of $N \times T$. It is clearly also a cell-base of $N \cdot D$, that is, of $(N \cdot T) \cdot D$, by Eq. (VIII.1.2). By Eq. (VIII.2.4),

$$r(N \times T) = r(N \cdot D) = r(N \cdot T) - r((N \cdot T) \times (T - D)).$$

If $r((N \cdot T) \times (T - D))$ is nonzero, then Eq. (VIII.8.1) holds with the strict inequality. In the remaining case, the support of each nonzero chain of $N \cdot T$ meets D. But then D, being a cell-base of $N \times T$, satisfies the definition of a cell-base of $N \cdot T$ (for $N \times T \subseteq N \cdot T$). The corresponding chain-base of $N \times T$ is thus a chain-base of $N \cdot T$, and therefore $N \times T = N \cdot T$ by Theorem VIII.12. Our theorem now follows from Theorem VIII.51. □

If T and U are complementary subsets of S, we write

$$\xi(N; T, U) = r(N) - r(N \times T) - r(N \times U) + 1. \qquad \text{(VIII.8.2)}$$

We note the rules

$$\xi(N; T, U) = \xi(N; U, T), \qquad \text{(VIII.8.3)}$$

$$\xi(N; T, U) \geq 1. \qquad \text{(VIII.8.4)}$$

In VIII.8.4 *there is equality if and only if T and U are detached with respect to N.*

To prove Eq. (VIII.8.4) and the following remark about equality we first use Eq. (VIII.2.4) to rewrite Eq. (VIII.8.2) as

$$\xi(N; T, U) = r(N \cdot T) - r(N \times T) + 1. \qquad \text{(VIII.8.5)}$$

We then apply Theorem VIII.56.

Theorem VIII.57. $\xi(N^*; T, U) = \xi(N; T, U)$.

Proof. We apply Eq. (VIII.8.5) to N^* and use Theorem VIII.24 to get

$$\xi(N^*; T, U) = r((N \times T)^*) - r((N \cdot T)^*) + 1.$$

Hence

$$\xi(N^*; T, U) = r(N \cdot U) - r(N \times U) + 1 \quad \text{(by (VIII.2.5))}$$

$$= \xi(N; T, U) \quad \text{(by Eqs. (VIII.8.3) and (VIII.8.5))}. \quad \square$$

We say that the primitive chain-group N is *k-separated*, where k is a positive integer, if there are complementary subsets T and U of S such that

$$\xi(N; T, U) = k \qquad \text{(VIII.8.6)}$$

$$\text{Min}(|T|, |U|) \geq k. \qquad \text{(VIII.8.7)}$$

If there is a least k such that N is k-separated, we call it the *connectivity* $\kappa(N)$ of N. If there is no such k, we say that the connectivity $\kappa(N)$ is infinite. As a consequence of Theorem VIII.57, we have

$$\kappa(N^*) = \kappa(N) \qquad \text{(VIII.8.8)}$$

for any primitive chain-group N.

Now let Ω be an orientation, with dart-set S, of a graph G. For simplicity we postulate that G is connected and has at least one edge, so that $G \cdot E(G) = G$. If $T \subseteq S$, then, as before, we write T' for the corresponding set of edges of G. We note that $\kappa(G) \geq 1$.

If T and U are complementary subsets of S, we write

$$\xi(G; T', U')$$

for the number of common vertices of $G \cdot T'$ and $G \cdot U'$.

Theorem VIII.58. *If there is a least positive integer k such that, for some choice of T and U,*

$$\xi(G; T', U') = k \qquad \text{(VIII.8.9)}$$

and

$$\text{Min}(|T|, |U|) \geq k, \qquad \text{(VIII.8.10)}$$

then $k \geq \kappa(G)$. In the remaining case, $\kappa(G) = \infty$.

Proof. Suppose that some positive integer k satisfies Eq. (VIII.8.9) and (VIII.8.10). Since G is without isolated vertices, Conditions (i), (ii), and (iii) of Sec. IV.1 are satisfied with $n = k$, $H = G \cdot T'$, and $K = G \cdot U'$. Hence, G has a k-separation, and so $k \geq \kappa(G)$.

On the other hand, if $\kappa(G)$ is finite, there are subgraphs H and K of G that satisfy (i)–(iii) of Sec. IV.1 with $n = \kappa(G)$. If H had isolated vertices, we could transfer them to K without changing $E(H)$ or $E(K)$, but with a decrease in the number of common vertices of H and K. But this would contradict the definition of $\kappa(G)$. We deduce that H and K are reductions $G \cdot T'$ and $G \cdot U'$, respectively, of G, and that they satisfy Eqs. (VIII.8.9) and (VIII.8.10) with $k = \kappa(G)$.

From these results, we conclude that if $\kappa(G)$ is finite it is the least positive integer k for which Eqs. (VIII.8.9) and (VIII.8.10) can be satisfied, and that if $\kappa(G) = \infty$, there is no positive k satisfying the conditions. □

We now write $\Gamma = \Gamma(\Omega, R)$, and establish a relation between $\xi(\Gamma; T, U)$ and $\xi(G; T', U')$.

Theorem VIII.59. $\xi(\Gamma; T, U) = \xi(G; T', U') - p_0(G \cdot T') - p_0(G \cdot U') + 2.$

Proof. We put $N = \Gamma$ in Eq. (VIII.8.2) and apply Theorems VIII.48 and VIII.49, finding that

$$\xi(\Gamma; T, U) = r(\Gamma(\Omega, R)) - r(\Gamma(\Omega \cdot T, R)) - r(\Gamma(\Omega \cdot U, R)) + 1$$
$$= p_1(G) - p_1(G \cdot T') - p_1(G \cdot U') + 1 \quad \text{(by Theorem VIII.37)}$$
$$= \{|E(G)| - |V(G)| + 1\} - \{|T| - |V(G \cdot T')| + p_0(G \cdot T')\}$$
$$\quad - \{|U| - |V(G \cdot U')| + p_0(G \cdot U')\} + 1$$
$$= \xi(G; T', U') - p_0(G \cdot T') - p_0(G \cdot U') + 2. \qquad \square$$

From this theorem, we deduce that

$$\xi(\Gamma; T, U) \leq \xi(G; T', U') \qquad \text{(VIII.8.11)}$$

for any complementary nonnull subsets T and U of S. It follows that

$$\kappa(\Gamma) \leq \kappa(G). \qquad \text{(VIII.8.12)}$$

We now try to prove the reversed inequality of Eq. (VIII.8.12). We approach it by way of the following theorem.

Theorem VIII.60. *Let T and U be complementary nonnull subsets of S. Let k be a positive integer such that*

$$\xi(G; T', U') \leq k + p_0(G \cdot T') + p_0(G \cdot U') - 2, \qquad \text{(VIII.8.13)}$$
$$\text{Min}(|T|, |U|) \geq k. \qquad \text{(VIII.8.14)}$$

Then $k \geq \kappa(G)$.

Proof. Assume $\kappa(G) > k$. Let T and U be chosen so that $\xi(G; T', U')$ has the least value consistent with the above assumption.

Now G has at least two edges since T and U are nonnull; hence it is loopless. If it had a loop, it would have a 1-separation, and so would satisfy $\kappa(G) = 1 \leq k$.

Each of $p_0(G \cdot T')$ and $p_0(G \cdot U')$ is at least 1. If each of them is equal to 1, we have $\xi(G; T', U') \leq k$, by Eq. (VIII.8.13). But then $\kappa(G) \leq k$, by Theorem VIII.58 and Eq. (VIII.8.14), contrary to our assumption. So we have

$$p_0(G \cdot T') + p_0(G \cdot U') \geq 3. \qquad \text{(VIII.8.15)}$$

Now each component of $G \cdot T'$ and $G \cdot U'$ has at least one edge and therefore at least two vertices, since G is loopless. Hence,

$$|V(G)| \geq 4. \qquad \text{(VIII.8.16)}$$

If H is any component of $G \cdot T'$ or $G \cdot U'$, we write $x(H)$ for the number of its vertices in the set W of common vertices of $G \cdot T'$ and $G \cdot U'$. We note that $x(H)$ is nonzero, by the connection of G. We have

$$1 \leq x(H) \leq V(H) \leq |E(H)| + 1, \qquad \text{(VIII.8.17)}$$

Connectivity

by Eq. (I.6.1) and Theorem I.35. If $|E(H)| > x(H)-1$, we say that H is of Type I. In the remaining case, that is, if $|E(H)| = x(H)-1$, we say that H is of Type II.

If H is of Type II, it is a tree, by Eq. (I.6.1) and Theorem I.35.

Lemma VIII.61. *If H is a component of $G \cdot T'$, then $|T' - E(H)| < k$.*

Proof. Assume $|T' - E(H)| \geq k$. Then

$$p_0(G \cdot (T' - E(H))) = p_0(G \cdot T') - 1 \geq 1. \quad \text{(VIII.8.18)}$$

The components of $G \cdot (U' \cup E(H))$ are those of $G \cdot U'$, except that those components of $G \cdot U'$ that meet H, at most $x(H)$ in number, are united with H into a single new component of $G \cdot (U' \cup E(H))$. Hence,

$$p_0(G \cdot (U' \cup E(H))) \geq p_0(G \cdot U') - x(H) + 1. \quad \text{(VIII.8.19)}$$

We have also

$$\xi(G; T' - E(H), U' \cup E(H)) = \xi(G; T', U') - x(H).$$

Applying Eqs. (VIII.8.13), (VIII.8.18), and (VIII.8.19) to this result, we find that

$$\xi(G; T' - E(H), U' \cup E(H))$$
$$\leq k + p_0(G \cdot (T' - E(H))) + p_0(G \cdot (U' \cup E(H))) - 2. \quad \text{(VIII.8.20)}$$

But $|T' - E(H)|$ and $|U' \cup E(H)|$ are each at least k. Hence Eq. (VIII.8.20) contradicts the choice of T and U. The lemma follows. △

Lemma VIII.62. *If $|T| > k$, then one of the components of $G \cdot T'$ is of Type I.*

Proof. Suppose all the components of $G \cdot T'$ to be of Type II. Since they are trees, we can find a vertex v of $G \cdot T'$ that is monovalent in $G \cdot T'$, by Theorem I.40. Let A be the edge of T' incident with v, and let w be the other end of A.

Here w is not monovalent in $G \cdot T'$, for otherwise $G \cdot \{A\}$ would be a component of $G \cdot T'$ and this would contradict Lemma VIII.61. Hence

$$p_0(G \cdot (T' - \{A\})) = p_0(G \cdot T'), \quad \text{(VIII.8.21)}$$

for the components of $G \cdot (T' - \{A\})$ are those of $G \cdot T'$, except that the one containing A is replaced by the single edge-containing end-graph of its isthmus A. We have also:

$$p_0(G \cdot (U' \cup \{A\})) = p_0(G \cdot U') \quad \text{or} \quad p_0(G \cdot U') - 1, \quad \text{(VIII.8.22)}$$

according as the two ends of A belong to the same component or to different components of $G \cdot U'$. It is clear that

$$\xi(G; T' - \{A\}, U' \cup \{A\}) = \xi(G; T', U') - 1.$$

Applying Eqs. (VIII.8.13), (VIII.8.21), and (VIII.8.22) to this, we find that

$$\xi(G; T' - \{A\}, U' \cup \{A\})$$
$$\leq k + p_0(G \cdot (T' - \{A\})) + p_0(G \cdot (U' \cup \{A\})) - 2. \quad \text{(VIII.8.23)}$$

But $T' - \{A\}$ and $U' \cup \{A\}$ have each at least k edges. Hence, Eq. (VIII.8.23) contradicts the choice of T and U. The lemma follows. △

Lemma VIII.63. *One of $G \cdot T'$ and $G \cdot U'$ has a component of Type I.*

Proof. Suppose the contrary. Then $|T| = k$, by Lemma VIII.62, and, similarly, $|U| = k$ (by Eq. (VIII.8.3)). Since only Type II components occur, the average valency in each of $G \cdot T'$ and $G \cdot U'$ is less than 2. Hence the average valency in G is < 4.

Let w be a vertex of G of minimal valency. Then $\text{val}(G, w) \leq 3$. Moreover, $|E(G)| \geq 2 \text{val}(G, w)$, by Eqs. (VIII.8.16) and (I.1.1). Hence, $\kappa(G) \leq \text{val}(G, w) \leq 3$, by Theorem IV.3. It now follows, from our initial assumption, that $k = 1$ or 2.

Since $|T| = |U| = k$, the assumption that $k = 1$ contradicts Eqs. (VIII.8.15). Hence $k = 2$ and, by Eq. (VIII.8.16), each of $G \cdot T'$ and $G \cdot U'$ must be the union of two disjoint link-graphs. Hence G, being connected, must be a 4-circuit. But then $\kappa(G) \leq \text{val}(G, w) = 2$, which is contrary to our initial assumption. △

Lemma VIII.64. *If $G \cdot T'$ has a component H of Type I, then H is the whole of $G \cdot T'$.*

Proof. Suppose the contrary. Then $G \cdot T'$ has a component K other than H. We have

$$|E(G)| \geq x(H) \geq 1,$$
$$|E(G) - E(H)| > |U'| \geq k,$$
$$\xi(G; E(H), E(G) - E(H)) = x(H).$$

If $x(H) \leq k$, these relations imply that $\kappa(G) \leq k$, for $\kappa(G) \leq x(H)$, by Theorem VIII.58. This is contrary to our initial assumption. But if $x(H)$, and therefore $|E(H)|$, exceeds k, then the component K violates Lemma I. △

By Eq. (VIII.8.3) and Lemmas VIII.63 and VIII.64, we can adjust the notation so that $G \cdot T'$ is connected, its only component being of Type I. Then $G \cdot U'$ has at least two components, all of Type II, by Eq. (VIII.8.15) and Lemma VIII.64. Moreover, $|U| = k$, by Lemma VIII.62. We now have

$$\xi(G;T',U') = |V(G \cdot U')| \quad \text{(by Eq. (VIII.8.17))}$$
$$= |U| + p_0(G \cdot U')$$
$$= k + p_0(G \cdot T') + p_0(G \cdot U') - 1.$$

But this is contrary to hypothesis. The proof of Theorem VIII.60 is now complete. □

Theorem VIII.65. *G being connected,* $\kappa(G) \leq \kappa(\Gamma)$.

Proof. If $\kappa(\Gamma) = \infty$, the theorem is trivial. In the remaining case, we can find complementary subsets T and U of S such that

$$\xi(\Gamma;T,U) = \kappa(\Gamma),$$
$$\mathrm{Min}(|T|,|U|) \geq \kappa(\Gamma).$$

Then, by Theorem VIII.59, we have

$$\xi(G;T',U') = \kappa(\Gamma) + p_0(G \cdot T') + p_0(G \cdot U') - 2.$$

Hence, $\kappa(G) \leq \kappa(\Gamma)$, by Theorem VIII.60. □

Theorem VIII.66. *Let G be a connected graph with at least one edge, Γ an orientation of G, and R a ring satisfying the conditions of Sect. VIII.1. Then the connectivities of G, $\Gamma(\Omega,R)$, and $\Delta(\Omega,R)$ are all equal.*

Proof. We have $\kappa(G) = \kappa(\Gamma(\Omega,R))$, by Eq. (VIII.8.12) and Theorem VIII.65. Since $\Gamma(\Omega,R)$ and $\Delta(\Omega,R)$ are duals, we complete the proof by applying Eq. (VIII.8.8). □

This theorem exhibits connectivity as an algebraically self-dual property in the realm of connected graphs.

VIII.9. ON TRANSPORTATION THEORY

Consider a digraph Ω with dart-set S and with two distinguished vertices x and y. Let g be a 1-chain of Ω to I such that $(\partial g)(v) = 0$ for each vertex v of Ω, other than x and y. By Eq. (VIII.4.1) we have

$$(\partial g)(x) = -(\partial g)(y). \qquad \text{(VIII.9.1)}$$

Here we call g a *flow* in Ω, of *magnitude* $(\partial g)(x)$, from x to y. (The "flows" of Sec. VI.5 obey different rules.) We can, of course, also call g a flow of magnitude $(\partial g)(y)$ from y to x. Usually, if g is not a cycle, we choose the description that makes the magnitude positive.

The *outflow* $d(g,v)$ of a vertex v in g is the sum

$$\eta(D,v)g(D),$$

taken over all darts D of Ω such that $\eta(D, v)g(D)$ is positive. We say that a 1-chain k of Ω is *nonnegative* if each of its coefficients is nonnegative. So if g is nonnegative, we can say that $d(g, v)$ is the sum of $g(D)$ over all darts D with head v. In all cases, $d(g, v)$ is nonnegative.

In a typical transportation problem, we assign an integral *capacity* $c(D)$ to each dart D, and an integral *capacity* $f(v)$ to each vertex v. We then ask for a flow g in Ω from x to y having the greatest possible magnitude subject to the two following conditions:
i) $0 \le g(D) \le c(D)$ for each $D \in S$.
ii) $d(g, v) \le f(v)$ for each vertex v.
We shall say that a flow g satisfying these conditions is (c, f)-*limited*. If it is understood that $c(D) = 1$ for each D, we say "f-limited" instead of "(c, f)-limited." A (c, f)-limited flow of greatest possible magnitude is called *maximal*.

A (c, f)-limited flow can be visualized as a flow of traffic in a network Ω of one-way streets. The capacities measure maximum amounts per second of a commodity of interest that can be transported along a given street or through a given junction. If the restriction to integral capacities is thought artificial, there are standard devices for extending the theory from I to the ring of rational numbers, and thence to the real numbers. If a two-way street occurs, we suppose that it can be adequately represented as a pair of oppositely directed one-way streets.

From now on, we suppose that $c(D) = 1$ for each D. We can reduce the general integral case to this simplified one: We have only to replace each dart D by $c(D)$ darts of capacity 1, with the same head and tail. We can assume each $f(v)$ nonnegative. An f-limited flow g is now nonnegative. Indeed, each of its coefficients is restricted to the values 0 and 1.

We *complete* Ω by adjoining a new dart E with head y and tail x. Evidently the restrictions to S of the cycles k of the completed digraph such that $k(E) = m$ are the flows in Ω from x to y of magnitude m. Let us denote the completed digraph by Ω_1.

For our next theorem, we draw upon the terminology of Sec. VI.7.

Theorem VIII.67. *There is an f-limited flow g in Ω from x to y of magnitude m if and only if there is an f-limited path-bundle X in Ω from x to y of order m.*

Proof. If X is such a path-bundle, let g be the 1-chain of Ω in which each dart of each path of X has coefficient 1 and each other dart of S has coefficient 0. It is easily verified that g is a flow in Ω from x to y of magnitude m, and that it is f-limited.

Conversely, suppose g to be as specified in the enunciation. We convert it into a cycle k of $\Gamma(\Omega_1, I)$ by adjoining E with coefficient m. By Theorem VIII.28, we can represent k as a sum of primitive chains of $\Gamma(\Omega_1, I)$, each conforming to k. In each of these, the coefficients can be only

0 or 1, and each dart of S has coefficient 1 in at most one of these primitive chains. Exactly m of the primitive chains include E in the support, in each case with coefficient 1. Let these m be enumerated as $h_1, h_2, \ldots, h_m$. At each vertex v we find at most $f(v)$ of them, by the f-limitation of g.

Referring to Theorem VIII.34 and the definition of p_C in Sec. VIII.4 we see that each h_i corresponds to a circular path Q_i and Ω_1 in which E can be taken as the last dart. The darts of Q_i have coefficient 1 in h_i, and the other darts of S coefficient 0. Restricting attention to Ω, in which each Q_i reduces to a linear path from x to y, we find that the h_i determine an f-limited path-bundle in Ω from x to y of order m. □

We deduce from this theorem that the maximum magnitude of an f-limited flow from x to y in Ω is equal to the maximum order of f-limited path-bundles from x to y in Ω. Given such a path-bundle of maximum order, a corresponding maximal flow can be obtained by the method of the first paragraph of the above proof. So we can attack the problem of finding an f-limited flow of maximum magnitude by the methods of Sec. VI.7. In particular, we can use the Max Flow Min Cut Theorem, VI.48.

VIII.10. INCIDENCE MATRICES

We consider an orientation Ω, with dart-set S, of a graph G. Let the vertices of Ω be enumerated as $v_1, v_2, \ldots, v_n$, and its darts as $D_1, D_2, \ldots, D_m$. Let $H(\Omega)$ be the matrix of n rows and m columns such that the entry in the ith row and jth column is in every case $\eta(D_j, v_i)$. We call $H(\Omega)$ the *incidence matrix* of Ω, with respect to the above enumerations.

If D_j is a link-dart, its column in $H(\Omega)$ has just two nonzero entries, one being $+1$ and the other -1. But if D_j is a loop-dart, its column has only zero entries. The rows of $H(\Omega)$ are the representative vectors of the vertex-coboundaries of Ω (over I). If we select one vertex in each component of G and then strike out from $H(D)$ the rows corresponding to the chosen vertices, we get a representative matrix M of $\Delta(\Omega, I)$, by Theorem VIII.46.

Theorem VIII.68. *Suppose $r(\Delta(\Omega, I)) > 0$. Let T be any subset of S such that $|T| = r(\Delta(\Omega, I)) = r(M)$, where M is defined as above. Then $\det M(T)$ is 1 or -1 if T is a cell-base of $\Delta(\Omega, I)$, and is zero otherwise.*

Proof. We use Theorem VIII.31. It is only necessary to show that the integer m_1 of that theorem is now unity. To do this, we consider any cell-base T. We have $\det M(T) = m_1$ or $-m_1$.

Write $Q_0 = M(T)$. If Q_0 has a row with exactly one nonzero element, we strike out from Q_0 one such row and the column of its nonzero element, so obtaining a smaller square matrix Q_1. If Q_1 has a row with only one nonzero element, we repeat the process with Q_1 replacing Q_0, getting a still smaller square matrix Q_2, and so on. Suppose we arrive in this way at a matrix Q_k, where $0 \leq k < r(M)$. Then we have

$$\det M(T) = \pm \det Q_k \neq 0. \qquad (\text{VIII}.10.1)$$

It is clear from Eq. (VIII.10.1) that Q_k can have no row of zeros only. Suppose, however, that each of its rows has at least two nonzero elements. But each column has at most two such elements since Q_k is a submatrix of $H(\Omega)$. Hence each row and column of Q_k must have exactly two nonzero elements; and, moreover, the entries in each column sum to zero. This implies that $\det Q_k$ is zero, contrary to Eq. (VIII.10.1).

We conclude that the sequence $Q_0, Q_1, Q_2, \ldots$ can be continued to a Q_k of order 1. The single entry of Q_k is nonzero, by Eq. (VIII.10.1), and so is either 1 or -1. Hence $\det M(T) = \pm 1$, by Eq. (VIII.10.1) and therefore $m_1 = 1$, as required. □

Let us write P^t to denote the transpose of a matrix P. We can verify that

$$H(\Omega) \cdot H^t(\Omega) = K(G), \qquad (\text{VIII}.10.2)$$

the Kirchhoff matrix of G.

Theorem VIII.69. *Let M be a representative matrix of a regular chain-group N of nonzero rank. Let the integer m_1 of Theorem VIII.31 be unity. Then $\det MM^t$ is the number of cell-bases of N.*

Proof. By the theory of determinants,

$$\det MM^t = \sum_T \left\{ \det M(T) \cdot \det(M(T))^t \right\}$$

$$= \sum_T \left\{ \det M(T) \right\}^2,$$

where the sum is over all $T \subseteq S$ satisfying $|T| = r(N)$. An application of Theorem VIII.31 completes the proof. □

Let us return to Ω, with G connected and having at least two vertices, To obtain the representative matrix M from $H(\Omega)$, we strike out the row of a single vertex, let us say v_1. We then find that $MM^t = K_1(G)$. So by Theorems VIII.68 and VIII.69, $\det K_1(G)$ is the number of cell-bases of $\Delta(\Omega, I)$. Using Theorem VIII.43, we discover another proof of the theorem that $\det K_1(G)$ is the number of spanning trees of G. (See Theorem VI.29.)

VIII.11. MATROIDS

Given a primitive chain-group N on S to R, let us consider the class Q of supports of primitive chains of N. Then Q is found to have the following

properties:
i) *If X and Y are in Q, then Y is not a proper subset of X.*
ii) *If X and Y are in Q, a is a common cell of X and Y, and b is a cell of X but not of Y, then there exists $Z \in Q$ such that $b \in Z \subseteq (X \cup Y) - \{a\}$.*

Property (i) follows from the definition of an elementary chain. To prove (ii), we take an appropriate linear combination of primitive chains on X and Y, to obtain a chain f of N whose support includes b but not a. We then apply Theorem VIII.18.

More generally, we may consider any family Q of nonnull subsets of a finite set S such that (i) and (ii) hold. We then say that Q is the set of *circuits* of a *matroid* on S. Each subset of S not containing a circuit is called an *independent* set of the matroid. In the usual definition of a matroid, (i) and (ii) are replaced by equivalent axioms about independent sets. Matroids can be regarded as generalizing both graphs and chain-groups.

There is an extensive theory of matroids, also called the abstract theory of linear dependence. We cannot develop it in this work, but we ought to take note of its existence. (See [2,5].)

VIII.12. NOTES

VIII.12.1. Unoriented Coboundaries

The coboundaries of Sec. VIII.5 have an analogue in the theory of undirected graphs. There we consider "0-chains" on $V(G)$ and "1-chains" on $E(G)$. In the coboundary of a 0-chain f, the coefficient of an edge A is the sum of the coefficients of its two ends in f. A cycle can be defined as a 1-chain orthogonal to every coboundary. A theory of such coboundaries is given in [3]. It is shown that they are closely related to the f-factors of G. (See also [4].)

EXERCISES

1. A chain-group N over the ring of integers is defined by the following matrix:
$$M = \begin{bmatrix} 2 & 2 & 3 & 3 \\ 3 & 3 & 2 & 3 \end{bmatrix}.$$

There are four cells. Each row of M represents a chain of N with the coefficient of the jth cell appearing in the jth place; N consists of the chains defined by the two rows, and of all the linear combinations of these two chains over I. Find the elementary chains and the cell-bases of N.

2. Repeat Exercise 1, but with I replaced by I_5 and all the elements of M interpreted as residues mod 5.
3. Show that the number $T(N)$ of cell-bases of a primitive chain-group N satisfies

$$T(N) = T(N'_A) + T(N''_A),$$

where A is any cell of N that belongs to the support of some chain but does not by itself constitute the support of any chain.
4. Let M be a standard representative matrix of a regular chain-group N. Show that M is totally unimodular, that is, that the determinant of each square submatrix is 1, -1, or 0.
5. The transpedances $[ab, xy]$ of a digraph Ω, with a and b fixed, can be interpreted as the coefficients of a 1-chain K (and their negatives). State Kirchhoff's Laws for them in the terminology of cycles and coboundaries.
6. Let T be a block-set of a primitive chain-group N. Let A and B be two cells of T. Show that there is an elementary chain of N having both A and B in its support. (This is a generalization of Theorem III.18.)

REFERENCES

[1] Tutte, W. T., A class of Abelian groups, *Can. J. Math.*, **8** (1956), 13–28.
[2] _____, *Introduction to the theory of matroids*, American Elsevier, New York, 1971.
[3] _____, On chain-groups and the factors of graphs, *Colloq. Math. Soc. János Bolyai*, **25**; Algebraic methods in graph theory, Szeged (Hungary), 1978, pp. 793–818.
[4] _____, Les facteurs des graphes, *Annals of Discrete Mathematics*, **8** (1980), 1–5.
[5] Welsh, D. J. A., *Matroid theory*, Academic Press, New York, 1976.

Chapter IX

Polynomials Associated with Graphs

IX.1. *V*-FUNCTIONS

We retain the ring R of Chapter VIII. Here it is usually either the ring I of integers or some ring of polynomials over I.

Let f be a correspondence that associates with each graph G a unique element $f(G)$ of R. We call f a *graph-function to* R if $f(G) = f(H)$ whenever G and H are isomorphic graphs. Thus a graph-function is really a function of isomorphism classes.

A graph-function f is called a *V-function* if it satisfies the following three conditions.
i) *If G is null, then $f(G) = 1$.*
ii) *If G is a union of two disjoint subgraphs H and K, then*

$$f(G) = f(H) \cdot f(K). \qquad (IX.1.1)$$

iii) *If A is any link of G, then*

$$f(G) = f(G'_A) + f(G''_A). \qquad (IX.1.2)$$

We note that (i) and (ii) are consistent, even though each graph is the union of itself and a null graph. As a consequence of (ii), we have the following theorem.

Theorem IX.1. *Let f be a V-function. Then its value for a graph G is the product of its values for the components of G.*

Several graph-functions of very common occurrence are V-functions. The tree-number $T(G)$ of G satisfies (iii) but not (ii), by Theorem II.18.

We can define one very general V-function in terms of the numbers denoted by $R(G; i_0, i_1, i_2, \ldots)$ in Sec. II.2. We take an arbitrary infinite sequence

$$S = (s_0, s_1, s_2, \ldots), \qquad \text{(IX.1.3)}$$

and we define a graph-function $f(G; S)$ by the following equation:

$$f(G; S) = \sum_{(i_0, i_1, \ldots)} \left\{ R(G; i_0, i_1, \ldots) \prod_{m=0}^{\infty} s_m^{i_m} \right\}. \qquad \text{(IX.1.4)}$$

The sum is over all sets of nonnegative integers $i_0, i_1, i_2, \ldots$. Formally, we have an infinite sum of infinite products on the right of Eq. IX.1.4. But, since G is finite, only a finite number of the subgraph-numbers $R(G; i_0, i_1, \ldots)$ can be nonzero, and so the right side of Eq. (IX.1.4) reduces to a finite sum for each G.

Theorem IX.2. *The graph-function $f(G; S)$ defined in Eq. (IX.1.4) is a V-function.*

Proof. If G is null, then $R(G; i_0, i_1, \ldots)$ is zero unless each i_j is zero, and in that case it is 1. Hence $f(G; S)$ satisfies Condition (i).

Condition (ii) follows from the evident fact that $R(G; i_0, i_1, i_2, \ldots)$ is the sum of all products

$$R(H; j_0, j_1, j_2, \ldots) \cdot R(K; k_0, k_1, k_2, \ldots)$$

such that $j_q + k_q = i_q$ for each suffix q. Condition (iii) is a consequence of Theorem II.20. □

Sometimes we interpret the symbols s_j of Eqs. (IX.1.3) and (IX.1.4) not as elements of R but as indeterminates over R. That is, we replace R by one of its extensions.

Let us write X_n for a graph made up of a single vertex and exactly n loops ($n = 0, 1, 2, \ldots$). Then each spanning subgraph of X_n has exactly one component, and the number of components of cyclomatic number j is $\binom{n}{j}$. So, by Eq. (IX.1.4),

$$f(X_n : S) = \sum_{j=0}^{\infty} \binom{n}{j} s_j. \qquad \text{(IX.1.5)}$$

We continue with an existence theorem and a uniqueness theorem for V-functions.

Theorem IX.3. *Let $T = (t_0, t_1, t_2, \ldots)$ be any infinite sequence of members of R. Then there exists a V-function f such that $f(X_m) = t_m$ for each nonnegative integer m.*

V-Functions

Proof. We can choose the sequence S of Eq. (IX.1.3) so as to satisfy

$$t_n = \sum_{j=0}^{\infty} \binom{n}{j} s_j \qquad (\text{IX.1.6})$$

for each suffix n. We arrange this by writing

$$s_n = \sum_{j=0}^{\infty} (-1)^{n-j} \binom{n}{j} t_j. \qquad (\text{IX.1.7})$$

By Theorem IX.2 we have the V-function $f(G; S)$, and $f(X_m; S) = t_m$ for each m, by Eqs. (IX.1.5) and (IX.1.6). □

Theorem IX.4. *If the value of a V-function f is known for each X_m, then it is uniquely determined.*

Proof. Let $\ell(G)$ denote the number of links of a graph G. If $\ell(G) = 0$, then $f(G)$ is uniquely determined by (ii). In the remaining case, let A be any link of G. Then

$$f(G) = f(G'_A) + f(G''_A)$$

by (iii). But each of $\ell(G'_A)$ and $\ell(G''_A)$ is less than $\ell(G)$. So the theorem follows, by induction over $\ell(G)$. □

Theorem IX.5. *Let A be an isthmus of G. Let its end-graphs be H and K, K being a vertex-graph. Then, for any V-function f,*

$$f(G) = \{1 + f(X_0)\} f(H). \qquad (\text{IX.1.8})$$

Proof. We use Eq. (IX.1.2), noting that G'_A is the union of disjoint subgraphs H and $X_0 = K$, and that G''_A is vertex-isomorphic to H. (See Fig. IX.1.1.) □

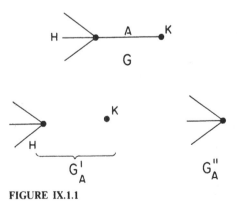

FIGURE IX.1.1

Theorem IX.6. *Let x be a divalent vertex of G, incident with two links A and B joining it to vertices y and z, respectively. Let G_1 be formed from G by deleting x, A, and B. Let G_2 be formed from G_1 by adjoining a new edge C with ends y and z. Then, for any V-function f,*

$$f(G) = \{1 + f(X_0)\}f(G_1) + f(G_2). \qquad (IX.1.9)$$

Proof. The case $y = z$, in which C is a loop of G_2, is not excluded. Portions of the graphs G, G_1, and G_2 are sketched in Fig. IX.1.2 for the case $y \neq z$.

Let H be the component of G'_A including B. Then B is an isthmus of H. One of its end-graphs in H consists solely of the vertex x. Denote the other by H_0. Evidently the complementary subgraph K of H in G'_A is the complementary subgraph of H_0 in G_1. So we have

$$f(G'_A) = f(H)f(K) = \{1 + f(X_0)\}f(H_0)f(K) = \{1 + f(X_0)\}f(G_1),$$

where K is H^c in $G'_{A'}$, by (ii) and Theorem IX.5.

We note further that $G''_A = {}_vG_2$. The theorem follows, by (iii). □

A graph-function is called *topologically invariant* if it is invariant under the operation of subdivision defined in Sec. IV.2. If graph X is obtained from graph Y by subdividing an edge, then we can take X and Y to be the graphs G and G_2, respectively, of Theorem IX.6. So from that theorem we have the following:

Theorem IX.7. *Let f be a V-function such that $f(X_0) = -1$. Then f is topologically invariant.*

There is a converse.

Theorem IX.8. *Let f be any topologically invariant V-function. Then either $f(X_0) = -1$ or $f(K) = 0$ for every nonnull graph K.*

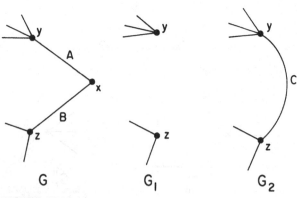

FIGURE IX.1.2

Proof. Suppose $f(K) \neq 0$ for some K. We can arrange for K to be the graph G_1 of Theorem IX.6. Then $f(G) = f(G_2)$ in Eq. (IX.1.9), by topological invariance. Hence, $f(X_0) = -1$, by Eq. (IX.1.9). □

There is indeed a trivial, topologically invariant V-function that satisfies (i) but is zero for every nonnull graph. Another rather trivial V-function f_e is defined by

$$f_e(G) = 2^{|E(G)|}. \qquad (IX.1.10)$$

It satisfies (i), (ii), and (iii). But it is not topologically invariant, by Theorem IX.1.8.

The general V-function $f(G; S)$ of Eq. (IX.1.4) can be specialized by writing

$$s_k = tz^k, \qquad (IX.1.11)$$

where t and z are fixed elements fo R, for each suffix k. We denote the resulting V-function by $Q(G; t, z)$. By Eq. (IX.1.4), with the help of Theorem II.21, we have

$$Q(G; t, z) = \sum_{S \subseteq E(G)} t^{p_0(G:S)} z^{k_1(G:S)}. \qquad (IX.1.12)$$

We usually think of t and z as indeterminates over R; this means that we should replace R by an appropriate extension in the definition of the graph function. Then $Q(G; t, z)$ is a polynomial in two indeterminates, t and z. We call it the *dichromatic polynomial* of G. Evidently, its degree in t is $|V(G)|$, corresponding to a null S; and its degree in z is $p_1(G)$, corresponding to $S = E(G)$.

In addition to the usual properties of a V-function the dichromatic polynomial satisfies the following important theorem.

Theorem IX.9. *Let G be the union of two subgraphs H and K whose intersection is a vertex-graph. Then*

$$tQ(G; t, z) = Q(H; t, z) Q(K; t, z). \qquad (IX.1.13)$$

Proof. Given $S \subseteq E(G)$, we write S_1 and S_2 for its intersections with $E(H)$ and $E(K)$, respectively. Then $G: S$ is the union of $H: S_1$ and $K: S_2$, these two graphs having no common edge and exactly one common vertex. We infer that

$$1 + p_0(G:S) = p_0(H:S_1) + p_0(K:S_2). \qquad (IX.1.14)$$

By Eq. (I.6.1), this implies

$$p_1(G:S) = p_1(H:S_1) + p_1(K:S_2). \qquad (IX.1.15)$$

Hence, by Eq. (IX.1.12),

$$tQ(G;t,z) = \sum_{S_1 \subseteq E(H)} \sum_{S_2 \subseteq E(K)} \left[t^{p_0(H:S_1)+p_0(K:S_2)} \times z^{p_1(H:S_1)+p_1(K:S_2)} \right]$$

$$= Q(H;t,z)Q(K;t,z). \qquad \square$$

Only for very simple graphs is it convenient to use Eq. (IX.1.12) in a direct determination of the dichromatic polynomial. We can check that

$$Q(X_m;t,z) = t(1+z)^m \qquad (IX.1.16)$$

for each m. (See Eq. (IX.1.5). We can also verify the following:

Theorem IX.10. *If G is a link-graph, then $Q(G;t,z) = t(1+t)$.*

For more complicated graphs, it is better to work upward from such simple cases by way of the recursion formulae (IX.1.1), (IX.1.2), and (IX.1.13). Perhaps the following variation on Eq. (IX.1.13) deserves special notice.

Theorem IX.11. *Let a connected graph G have an isthmus A with end-graphs H and K. Then*

$$tQ(G;t,z) = (1+t)Q(H;t,z)Q(K;t,z). \qquad (IX.1.17)$$

It is proved by two applications of Theorem IX.9, with the help of Theorem IX.10.

IX.2. THE CHROMATIC POLYNOMIAL

We write

$$P(G;\lambda) = (-1)^{|V(G)|}Q(G;-\lambda,-1), \qquad (IX.2.1)$$

taking R to be the ring of polynomials in an indeterminate λ, with integral coefficients. By Eq. (IX.1.12), this definition is equivalent to

$$P(G;\lambda) = \sum_{S \subseteq E(G)} (-1)^{|S|} \lambda^{p_0(G:S)}. \qquad (IX.2.2)$$

We observe that $(-1)^{|V(G)|}P(G;\lambda)$ is a V-function, by Eq. (IX.2.1). Accordingly, $P(G;\lambda)$ has the following properties, derived from Conditions (i), (ii), and (iii) of Sec. IX.1.

Theorem IX.12. *If G is null, then $P(G;\lambda) = 1$.*

Theorem IX.13. *If G is the union of disjoint subgraphs H and K, then*

$$P(G;\lambda) = P(H;\lambda)P(K;\lambda). \qquad (IX.2.3)$$

Theorem IX.14. *If A is a link of G, then*

$$P(G;\lambda) = P(G'_A;\lambda) - P(G''_A;\lambda). \qquad (IX.2.4)$$

Moreover, as a consequence of Theorem IX.9, we have:

Theorem IX.15. *If G is a union of subgraphs H and K such that $H \cap K$ is a vertex-graph, then*

$$\lambda P(G;\lambda) = P(H;\lambda)P(K;\lambda). \qquad (IX.2.5)$$

From these results, and the definition, we can derive other propositions about $P(G;\lambda)$. Some simple examples appear below as Theorems IX.16 to IX.19.

Theorem IX.16. *$P(X_m;\lambda)$ is λ if $m = 0$, and zero if $m > 0$, by Eq. (IX.1.16).*

Theorem IX.17. *If G has a loop, then $P(G;\lambda) = 0$.*

This follows from Eq. (IX.2.5) and Theorem IX.16, the latter for $m = 1$.

Theorem IX.18. *If G is loopless, then $P(G;\lambda)$ is a polynomial in λ of degree $|V(G)|$.*

This follows from Eq. (IX.2.2), since $G:S$ has exactly $|V(G)|$ components if S is null, and fewer than $|V(G)|$ components in every other case.

Theorem IX.19. *Let G have two links A and B with the same ends x and y. Then $P(G;\lambda) = P(G'_A;\lambda)$.*

This is a consequence of Eq. (IX.2.4) and Theorem IX.17, for B is a loop of G''_A.

In view of Theorems IX.17 and IX.19, we can regard the study of $P(G;\lambda)$ as basically a study of strict graphs, extensions to the nonstrict case being trivial.

When λ takes a positive integral value n, we can interpret $P(G;\lambda)$ in terms of the "n-colorings" of G.

Let S_n be the set of integers from 1 to n, here called the n "colors." We define an *n-coloring* of G as a mapping of $V(G)$ into S_n, subject to the following restriction: The two ends of any edge must map onto two distinct colors. We write $N(G;n)$ for the number of n-colorings of G. This is clearly a graph-function. The restriction just made implies that $N(G;n) = 0$ whenever G has a loop.

We adopt the convention that a null set has exactly one mapping into S_n. We therefore write $N(G;n) = 1$ when G is null.

Suppose G to be the union of two disjoint subgraphs H and K. Then the n-colorings of G are simply the combinations of one n-coloring of H

with one n-coloring of K. Thus

$$N(G;n) = N(H;n)N(K;n).$$

Suppose G to have a link A, with ends x and y. Then the n-colorings of G can be identified with those n-colorings of G'_A for which x and y have different colors. Allowing for a coalescence of x and y we can identify the n-colorings of G''_A with those n-colorings of G'_A for which x and y have the same color. Hence

$$N(G;n) = N(G'_A;n) - N(G''_A;n).$$

We can now assert that $(-1)^{|V(G)|}N(G;n)$ is a V-function. Moreover, it is clear that $N(X_0;n) = n$ and that $N(X_m;n) = 0$ if $m > 0$. Hence the V-functions $(-1)^{|V(G)|}N(G;n)$ and $(-1)^{|V(G)|}P(G;n)$ are identical, by Theorems IX.4 and IX.16. We thus have the following theorem.

Theorem IX.20. *Let n be any positive integer. Then the number of n-colorings of any graph G is $P(G;n)$.*

Theorem IX.21. *Let G be a loopless graph. Then $P(G;n) \geq 0$ for each positive integer n, with strict inequality when $n \geq |V(G)|$. Moreover, if $P(G;n) = 0$, then $P(G;m) = 0$ for each positive integer $m < n$.*

The first sentence of Theorem IX.21 follows from Theorem IX.20, together with the fact that when $n \geq |V(G)|$ we can get an n-coloring of G by giving the vertices all different colors. To complete the proof, we observe that an m-coloring is also an n-colouring.

Because of Theorem IX.20, $P(G;\lambda)$ is called the *chromatic polynomial* of G. By Theorem IX.21, there is for each nonnull loopless graph G a least positive integer n such that G has an n-coloring. This is called the *chromatic number* of G, and a graph with chromatic number k is said to be *k-chromatic*. In view of Theorem IX.12, it seems convenient to assign a zero chromatic number to the null graph.

Much work has been done by graph theorists on problems about n-colorings. Chromatic polynomials are regarded as important in graph theory mainly because of the interpretation given by Theorem IX.20. Historically, their study began as the theory of enumeration of n-colorings. From this point of view, Formula (IX.2.2), which serves us as a definition, becomes an important theorem giving an explicit formula for $N(G;n)$. As such, it is due to Hassler Whitney [17]. Many of the following theorems of this section are commonly and easily proved as theorems about $N(G;n)$. Some argument from the theory of polynomial identities may then be brought in to extend them to general λ. However, we shall use proofs in which λ is an indeterminate throughout.

We make note of a variation on the interpretation of $P(G;n)$ in Theorem IX.20. Let us say that a chain on a set S to a ring R of the type considered in Sec. VIII.1 is *nowhere zero* if its support is S. Consider an arbitrary orientation Ω of G, with dart-set S. We now take R to be finite and with exactly n elements. It may, for example, be I_n. We ask for the number of nowhere-zero coboundaries to R in Ω.

Now a 0-chain f of Ω to R has a nowhere-zero coboundary if and only if each dart of Ω has differing f-coefficients at its two ends. The number of such 0-chains f is clearly $N(G; n)$, that is, $P(G; n)$, by Theorem IX.20. But any member of $\Delta(\Omega, R)$ is the coboundary of exactly $n^{p_0(G)}$ distinct 0-chains, by Theorem VIII.45. We thus have the following:

Theorem IX.22. *If Ω is an arbitrary orientation of a graph G, and if R has just n elements, then the number of nowhere-zero coboundaries of Ω to R is*

$$n^{-p_0(G)}P(G; n).$$

In connection with this result, we observe that the polynomial $P(G; \lambda)$ always divides by $\lambda^{p_0(G)}$, by Eq. (IX.2.2).

An orientation Ω of G is called *acyclic* if it has no circular path. R. P. Stanley shows in [9] that the number of acyclic orientations of G is $P(G; -1)$.

We have $P(G; \lambda) = \lambda$ for the vertex-graph, $P(G; \lambda) = 0$ for the loop-graph, and $P(G; \lambda) = \lambda(\lambda - 1)$ for the link-graph, by Theorem IX.10. To find $P(G; \lambda)$ for other graphs, we use recursion formulae such as Eqs. (IX.2.3), (IX.2.4), and (IX.2.5). Two examples follow.

Theorem IX.23. *If T is a tree of k edges, then*

$$P(T; \lambda) = \lambda(\lambda - 1)^k. \quad \text{(IX.2.6)}$$

Proof. The theorem holds if k is 0 or 1 since T is then a vertex-graph or a link-graph. In the remaining case, T is the union of a link-graph and a tree of $(k - 1)$ edges, the intersection of these being a vertex-graph, by Theorem I.38 and I.40. So we can prove Theorem IX.23 by repeated application of Eq. (IX.2.5). □

Theorem IX.24. *Let C_k be a circuit of k edges. Then*

$$P(C_k; \lambda) = (\lambda - 1)^k + (-1)^k(\lambda - 1). \quad \text{(IX.2.7)}$$

Proof. For $k = 1$, the circuit is a loop-graph and the theorem is trivially true. For $k > 1$, we choose some edge A, necessarily a link, of C_k. We then observe that $(C_k)'_A$ is an arc of $(k - 1)$ edges and that $(C_k)''_A$ is a circuit C_{k-1} of $(k - 1)$ edges. If the theorem holds for C_{k-1}, then by Eqs. (IX.2.4) and (IX.2.5), we have

$$P(C_k; \lambda) = \lambda(\lambda - 1)^{k-1} - \{(\lambda - 1)^{k-1} + (-1)^{k-1}(\lambda - 1)\},$$

which is equivalent to Eq. (IX.2.7). So the theorem can be proved by induction over k. □

Let us define a *vertex-join* of a graph H as a graph G obtained from H by adjoining one new vertex x and then joining x to each vertex of H by one or more new links.

Theorem IX.25. *Let G be a vertex-join of H. Then*

$$P(G;\lambda) = \lambda P(H;\lambda - 1). \qquad (\text{IX.2.8})$$

Proof. In terms of n-colorings, this theorem is obvious. Keeping to a general λ, we prove it inductively as follows:

Suppose, first, that $|E(H)| = 0$. Then

$$P(G;\lambda) = \lambda(\lambda - 1)^{|V(H)|} \quad \text{(by Theorems IX.19 and IX.23)},$$

$$P(H;\lambda - 1) = (\lambda - 1)^{|V(H)|} \quad \text{(by Theorems IX.13 and IX.16)}.$$

So the theorem holds in this case. Assume, as an inductive hypothesis, that it holds when every $|E(H)|$ is less than some positive integer q. Consider the case $|E(H)| = q$.

Choose an edge A of H. If A is a loop, the theorem is trivially true, by Theorem IX.17. If A is a link, we observe that G'_A and G''_A are vertex-joins of H'_A and H''_A, respectively. We then have

$$P(G;\lambda) = P(G'_A;\lambda) - P(G''_A;\lambda) \quad \text{(by Eq. (IX.2.4))}$$

$$= \lambda P(H'_A;\lambda - 1) - \lambda P(H''_A;\lambda - 1) \quad \text{(by the inductive hypothesis)}$$

$$= \lambda P(H;\lambda - 1) \quad \text{(by Eq. (IX.2.4))}.$$

The induction succeeds. □

Theorem IX.26. *If G is an n-clique, then*

$$P(G;\lambda) = \lambda(\lambda - 1)(\lambda - 2) \cdots (\lambda - n + 1). \qquad (\text{IX.2.9})$$

Proof. If $n = 0$, we interpret the empty product on the right as unity. This is consistent with Theorem IX.12. For $n = 1$, G is a vertex-graph and so Eq. (IX.2.9) is valid. We extend to other values of n by repeated application of Theorem IX.25. □

Theorem IX.27. *Let G be the union of two subgraphs H and K whose intersection is an n-clique Q. Then*

$$\lambda(\lambda - 1) \cdots (\lambda - n + 1) P(G;\lambda) = P(H;\lambda) P(K;\lambda). \qquad (\text{IX.2.10})$$

Proof. When $n = 0$ or 1, this theorem reduces to Theorem IX.13 or Theorem IX.15, respectively. For any larger value of n, we can use the following induction:

Let w denote the number of edges of K not having two distinct ends in Q.

The Chromatic Polynomial

Suppose, first, that $w = 0$. Then K has one component K_0 containing Q. Its other components are all vertex-graphs. Let there be m of them. Then

$$P(K_0; \lambda) = P(Q; \lambda), \qquad P(K; \lambda) = \lambda^m P(Q; \lambda),$$

by Theorems IX.13 and IX.19. Similarly,

$$P(H \cup K_0; \lambda) = P(H; \lambda), \qquad P(G; \lambda) = \lambda^m P(H; \lambda).$$

These results, in combination with Eq. (IX.2.9), verify Eq. (IX.2.10).

Assume the theorem true for all w less than some positive integer q. Suppose $w = q$. Choose an edge A of K not having two distinct ends in Q. If A is a loop, then Eq. (IX.2.10) holds trivially, by Theorem IX.17. If A is a link, we observe that

$$G'_A = H \cup K'_A \qquad \text{and} \qquad G''_A = H_1 \cup K''_A,$$

where $H_1 = {}_v H$. Moreover, $H \cap K'_A$ and $H_1 \cap K''_A$ are still n-cliques. So

$$P(Q; \lambda)P(G; \lambda) = P(Q; \lambda)\{P(G'_A; \lambda) - P(G''_A; \lambda)\}$$
$$= P(H; \lambda)P(K'_A; \lambda) - P(H; \lambda)P(K''_A; \lambda)$$
$$= P(H; \lambda)P(K; \lambda),$$

by Theorem IX.14 and the inductive hypothesis. Thus the theorem holds also when $w = q$. It follows, in general, by induction. □

Let us write $P_j(G)$ for the coefficient of λ^j in $P(G; \lambda)$.

$$P(G; \lambda) = \sum_{j=0}^{|V(G)|} P_j(G) \lambda^j. \qquad (\text{IX.2.11})$$

Theorem IX.28. *Let G be a loopless graph of n vertices. Then $P_j(G)$ is nonzero if and only if $p_0(G) \le j \le n$, and it then has the sign of $(-1)^{n-j}$. Moreover, $P_n(G) = 1$.*

Proof. If G is edgeless, then $p_0(G) = n$, and $P(G; \lambda) = \lambda^n$, by Theorems IX.12, IX.13, and IX.16. So the theorem holds. Assume it true for $|E(G)| < q$, $q > 0$, and consider the case $|E(G)| = q$.

Choose an edge A of G. If it is part of a multiple join, we have $P(G; \lambda) = P(G'_A; \lambda)$, by Theorem IX.19, and $p_0(G) = p_0(G'_A)$, by Theorem I.29. The theorem being true for G'_A, it is therefore true for G.

In the remaining case, G'_A and G''_A are both loopless. By Theorem IX.14, we have

$$P_j(G) = P_j(G'_A) - P_j(G''_A). \qquad (\text{IX.2.12})$$

The theorem holds for G'_A and G''_A. Hence the nonzero values of $P_j(G'_A)$ extend from $j = p_0(G'_A)$ to $j = n$, and those of $-P_j(G''_A)$ from $j = p_0(G''_A)$ to $j = n - 1$. Moreover, if either $P_j(G'_A)$ or $-P_j(G''_A)$ is nonzero, it has the sign of $(-1)^{n-j}$. Hence the theorem holds for G, by Theorems I.29 and II.5. It follows, in general, by induction. □

We write

$$P(G;\lambda) = \lambda^{p_0(G)} U(G;\lambda) \qquad (IX.2.13)$$

for any graph G, where $U(G;\lambda)$ is a polynomial in λ of degree $|V(G)| - p_0(G)$. Let us express $U(G;\lambda)$ as a polynomial in $(\lambda - 1)$. Let us write $U_j(G)$ for the coefficient of $(\lambda - 1)^j$ in the resulting expression.

We shall prove an analogue of Theorem IX.28 for the coefficients $U_j(G)$. In it, $p_0(G)$ is replaced by the block-number $\beta(G)$ of G, defined as the number of blocks of G having at least one edge. Thus $\beta(G)$ is zero for edgeless graphs, and for these only.

Theorem IX.29. *Let A be an edge of G that is neither a loop nor an isthmus. Then*

$$\operatorname{Min}\{\beta(G'_A), \beta(G''_A)\} = \beta(G). \qquad (IX.2.14)$$

Proof. Let B be the block of G that includes A. If B has no other edge and A is not a loop, then A must be an isthmus of G, by Theorems I.45 and III.5. We deduce that B has at least two edges.

Now it is clear that, to within vertex-isomorphism, the blocks of G other than B are blocks of G'_A and G''_A. Hence,

$$\operatorname{Min}\{\beta(G'_A), \beta(G''_A)\} \geq \beta(G),$$

with equality if either B'_A or B''_A is nonseparable. So our theorem follows from Theorem III.33. □

Corollary IX.30. *If A belongs to a multiple join of G, then $\beta(G'_A) = \beta(G)$.*

In the above proof, B''_A now has a loop. It is thus 2-connected only if $|E(B)| = 2$, and in that case B'_A is also 2-connected. □

Theorem IX.31. *Let G be a graph of n vertices. Then $U_j(G)$ is nonzero if and only if $\beta(G) \leq j \leq n - p_0(G)$, and it then has the sign of $(-1)^{n - p_0(G) - j}$. Moreover, the leading coefficient ($j = n - p_0(G)$) is 1.*

Proof. Suppose first that G is a forest. Then

$$P(G;\lambda) = \lambda^{p_0(G)} (\lambda - 1)^{n - p_0(G)},$$

by Theorems IX.13, IX.23, and I.37. Moreover, $\beta(G)$ is now $|E(G)|$, that is, $n - p_0(G)$, and so the theorem holds.

We now carry out an induction on $|E(G)|$ closely resembling that in Theorem IX.28. This time, however, we choose A to be not an isthmus of G, and we use the symbol β instead of p_0. So when A belongs to a multiple join, we appeal to Corollary IX.30 instead of Theorem I.29, and in the remaining case we use Theorem IX.29 instead of Theorems I.29 and II.5. □

The indeterminate λ is often interpreted as a real variable. The results summarized below in Theorem IX.32, easily derived from Theorems IX.28 and IX.31, may then be of interest.

Theorem IX.32. *Let G be a connected loopless graph of n vertices, with at least one edge. Then $P(G; \lambda)$ is nonzero when $-\infty < \lambda < 0$ with the sign of $(-1)^n$, and nonzero when $0 < \lambda < 1$ with the sign of $(-1)^{n-1}$. At $\lambda = 0$ it is zero and its slope is nonzero, with the sign of $(-1)^{n+1}$. If G is 2-connected, then $P(G; \lambda)$ is zero at $\lambda = 1$, and its slope there is nonzero, with the sign of $(-1)^n$.*

It is conjectured that the sequence of absolute values of the coefficients $P_j(G)$ is *unimodal* for each G, meaning that no $P_j(G)$ is less than both $P_{j-1}(G)$ and $P_{j+1}(G)$ in absolute value. An analogous conjecture can be made for the coefficients $U_j(G)$. In each case, no general proof is known and no counterexample has been found.

IX.3. COLORINGS OF GRAPHS

Chromatic polynomials were introduced to count n-colorings of graphs. It may be well therefore to give some further account of the theory of such mappings. Unfortunately, proofs in this theory are rare and usually difficult (if not trivial), especially when, as here, we make no use of planarity. Two examples follow, one trivial and one not.

Theorem IX.33. *Let G be a strict graph in which the maximum valency is k. Then G has a $(k + 1)$-coloring.*

To construct such a coloring, we assign one of the $(k + 1)$ colors to each vertex in turn, giving each vertex a color not already assigned to any adjacent vertex.

Theorem IX.34. (Brooks' Theorem). *Let G be a connected strict graph in which the maximum valency is $k > 2$. Then either G is a $(k + 1)$-clique or it has a k-coloring.*

Proof. If possible, choose G so that the theorem fails and $|V(G)|$ has the least value consistent with this condition. We construct a $(k + 1)$-coloring of G as for Theorem IX.33. We refer to the colors 1 to k in S_{k+1} as the *ordinary* colors, and to $(k + 1)$ as the *critical color*. □

Lemma IX.35. *Let L be an arc in G with ends x and y. Then we can construct a new $(k + 1)$-coloring of G from a given one, changing colors only at vertices of L, so that the critical color occurs at no vertex of L other than x.*

Proof. We write the defining sequence of vertices of L as

$$(a_0, a_1, a_2, \ldots, a_s),$$

where $a_0 = y$ and $a_s = x$. We recolor the vertices of this sequence in succession, according to the following rules:
i) *We assign to a_0 an ordinary color not already appearing at any adjacent vertex other than a_1.*
ii) *Having recolored a_j, where $0 < j < s - 2$, we assign to a_{j+1} an ordinary color not now occurring at any adjacent vertex other than a_{j+2}.*
iii) *Having recolored a_{s-1}, we assign to a_s a color, possibly critical, not now occurring at any adjacent vertex.*

The valency restriction makes this construction possible. It gives us the required new $(k + 1)$-coloring. △

For any vertex x of G, we can, by repeated application of Lemma IX.35, obtain a $(k + 1)$-coloring of G in which only x has the critical color. (By the choice of G, we cannot get rid of the critical color completely.) We call this a *critical coloring* of G, with *critical vertex x*.

Lemma IX.36. *G is regular of valency k.*

Proof. Suppose some vertex x to have valency less than k. Construct a critical coloring of G with critical vertex x. We can now change the color of x to an ordinary color, so as to obtain a k-coloring of G, contrary to the choice of G. △

Lemma IX.37. *G is 2-connected.*

Proof. Suppose the contrary. Then G has a 1-separation (H, K), with a cut-vertex v. Each of H and K is connected, by Theorem III.1, and each has a vertex other than v, by Theorem IV.10. Neither of them can be a $(k + 1)$-clique, since v is incident with at least one edge of each. Evidently the maximum valency in each is k. It follows that both $N(H; k)$ and $N(K; k)$ are nonzero, by the choice of G. Hence $N(G; k)$ is nonzero, by Theorems IX.15 and IX.20, a contradiction. △

Lemma IX.38. *G is 3-connected.*

Proof. Suppose the contrary. Then G has a 2-separation (H, K), with hinges x and y, say. We recall that H and K are connected. (See Theorem IV.20.) Let G_1, H_1, and K_1 be formed from G, H, and K, respectively, by adjoining a new link A with ends x and y. Each of H and K has a vertex not x or y, by Theorem IV.10. Hence each of the connected graphs H_1 and K_1 has maximum valency k.

Suppose, first, that neither H_1 nor K_1 is a $(k+1)$-clique. Then $N(H_1; k)$ and $N(K_1; k)$ are nonzero, by the choice of G. Hence, $N(G_1; k)$ is nonzero, by Theorems IX.20 and IX.27. But a k-coloring of G_1 would be a k-coloring of G.

We may now assume, without loss of generality, that H_1 is a $(k+1)$-clique. But then H has a k-coloring Q_1 in which x and y receive the same color. Moreover, x and y must each have valency 1 in K. Form K_0 from K by identifying x and y. The maximum valency in K_0 is k, the new vertex having valency only 2. Hence K_0 has a k-coloring by the choice of G, and therefore K has a k-coloring Q_2 in which x and y have the same color. We can combine Q_1 and Q_2 to get a k-coloring of G. The Lemma follows. △

Now G has at least $(k+1)$ vertices, by Theorem IX.21. We can find two nonadjacent ones, say t and u, for otherwise G would be a $(k+1)$-clique, by Lemma IX.36. By Lemma IX.35, we can find a critical coloring of G with critical vertex t. Since u and its adjacent vertices receive only ordinary colors, there must be two vertices adjacent to u, say v and w, that receive the same color. Let T and U be the sets of edges incident with t and u, respectively. G is vertically 3-connected, by Lemma IX.38 and Theorem IV.10. Hence the number $\lambda(G; T, U)$, as defined in Sec. II.4, is at least 3. So, by Menger's Theorem (II.34 or II.35), there exist three internally disjoint arcs in G from t to u. Hence there is an arc L in G with ends t and u that does not include v or w. By Lemma IX.35, changing colors only in L, we can convert our critical coloring into a new one in which the critical vertex is u. But the vertices adjacent to u now have only ordinary colors, and at most $(k-1)$ of those. Recoloring u with one of the remaining ordinary colors, we get a k-coloring of G. This contradiction establishes the theorem.
□

For the case $n = 2$ we can replace Theorem IX.34 by the following Theorem IX.39. A 2-colorable graph is, of course, the same as a bipartite graph, as defined in Sec. II.6.

Theorem IX.39. *A graph G is bipartite (2-colorable) if and only if it has no circuit of odd length.*

Proof. A graph G with an odd circuit is clearly nonbipartite.

Conversely, suppose G to be nonbipartite. If it has a loop, it has an odd circuit of length 1. In the remaining case, we can delete links from it, one by one, until we have an edgeless and therefore bipartite graph. At some intermediate stage, we have a nonbipartite graph H and a link A of H such that H'_A is bipartite. Then the two ends x and y of A in H must have the same color in every bipartition of H'_A. This can happen only if they are in the same component of H'_A; the two colors can be interchanged in any component. So x and y are joined by an arc L in H'_A. It is of even length since vertices of the two colors must alternate in it. Adjoining A to L, we obtain an odd circuit of G. □

For completeness, let us add that G is 1-colorable if and only if it is edgeless, and 0-colorable if and only if it is null.

Consider a primitive chain-group N on a set S to a ring R. Let us define a *Hamiltonian* chain of N as a primitive chain h of N whose support has exactly $(|S| - r(N) + 1)$ cells. No primitive chain of N can have more than this number of cells in its support, by Theorem VIII.16.

Now let G be a connected graph. Let Ω be any orientation of G. By Theorem VIII.34, a Hamiltonian chain h of $\Gamma(\Omega, R)$ corresponds to a circuit H of G with $(|E(G)| - p_1(G) + 1)$ edges, that is, $|V(G)|$ edges, by Theorem VIII.37 and Eq. (I.6.1). Thus C is a circuit through all the vertices of G, a *Hamiltonian circuit* of G in the usual terminology.

Consider a Hamiltonian chain h of $\Delta(\Omega; R)$. By Theorem VIII.40, it corresponds to a bond H of G with $(|E(G)| - |V(G)| + 2)$ edges, by Eq. (VIII.5.8). The two end-graphs X and Y of H have between them all the vertices of G, but only $(|V(G)| - 2)$ edges. Hence, each of them must be a tree. (See Theorem I.35.) We are led to the concept of a *Hamiltonian bond* of G, algebraically dual to that of a Hamiltonian circuit. It is a bond of G for which each end-graph is a tree, that is, a bond of exactly $(p_1(G) + 1)$ edges.

Theorem IX.40. *If a connected graph G has a Hamiltonian bond, then it has a 4-coloring.*

Proof. Trees are bipartite, by Theorem IX.39. So we can color one end-graph of the bond with two of the four colors, and the other end-graph with the other two colors. We then have a 4-coloring of G. □

Theorem IX.41 (Grinberg's Theorem). *Let H be a Hamiltonian bond, with end-graphs X and Y, in a connected graph G. Let f_n^X and f_n^Y denote the number of vertices of X and Y, respectively, that have valency n in G ($n = 1, 2, 3 \ldots$). Then*

$$\sum_{n=1}^{\infty} (n-2)(f_n^X - f_n^Y) = 0. \tag{IX.3.1}$$

Proof. Using Theorem I.37, we find

$$\sum_{n=1}^{\infty} f_n^X = |E(X)| + 1. \tag{IX.3.2}$$

It is also clear that

$$\sum_{n=1}^{\infty} n f_n^X = |E(H)| + 2|E(X)|. \tag{IX.3.3}$$

Hence,

$$\sum_{n=1}^{\infty} (n-2) f_n^X = |E(H)| - 2. \tag{IX.3.4}$$

The Flow Polynomial

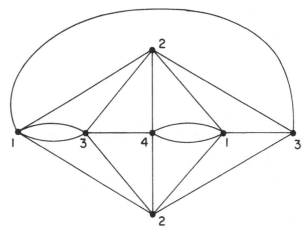

FIGURE IX.3.1

There is an analogous equation with Y replacing X. We subtract it from Eq. (IX.3.4) to obtain Eq. (IX.3.1). □

Grinberg's Theorem can sometimes be used to show that a given graph G has no Hamiltonian bond. Suppose, for example, that a connected graph G has all its vertices of valency congruent to 2 mod 3, with just one exception. Then the left side of Eq. (IX.3.1) cannot be divisible by 3, and therefore no Hamiltonian bond of G can exist. Figure IX.3.1 exhibits a simple example.

In this figure we show a 4-coloring of G, lest it should be thought that Theorem IX.40 has a direct converse.

Much work has been expended in the search for conditions for the existence of an n-coloring in a connected graph G. The search has culminated in the recent immensely long proof of the "Four-Color Theorem" for planar graphs, by Haken and Appel [1,2]. But that proof is thought to be still in need of confirmation. For the graphs with which we are concerned here, and for a general n, only a conjecture can be offered.

Hadwiger's Conjecture asserts that, if G has no n-coloring, then some minor of G is an $(n + 1)$-clique. It has been proved for $n \leq 3$, and the case $n = 4$ has been shown to be equivalent to the Four-Color Theorem. For other values of n, we have no proof and no counterexample.

IX.4. THE FLOW POLYNOMIAL

Like the chromatic polynomial, the *flow polynomial* $F(G; \lambda)$ of a graph G is a graph-function and a polynomial in an indeterminate λ with integral coefficients. We define it by:

$$F(G;\lambda) = (-1)^{|E(G)|+|V(G)|} Q(G; -1, -\lambda). \quad (IX.4.1)$$

By Eq. (IX.1.12), this is equivalent to

$$F(G;\lambda) = (-1)^{|E(G)|} \sum_{S \in E(G)} (-1)^{|S|} \lambda^{p_1(G:S)} \quad (IX.4.2)$$

The flow polynomial has the following properties IX.42, IX.43, and IX.44, deriving from the definition of a V-function, and from Theorem IX.9.

Property IX.42. *If G is null, then $F(G;\lambda) = 1$.*

Property IX.43. *If G is the union of two edge-disjoint subgraphs H and K with at most one vertex in common, then*

$$F(G;\lambda) = F(H;\lambda)F(K;\lambda). \quad (IX.4.3)$$

Property IX.44. *If A is a link of G, then*

$$F(G;\lambda) = F(G_A'';\lambda) - F(G_A';\lambda). \quad (IX.4.4)$$

By Eq. (IX.1.16), we have the following theorems.

Theorem IX.45. $F(X_m;\lambda) = (\lambda - 1)^m$.

Theorem IX.46. *If G has an isthmus, then $F(G;\lambda) = 0$.*

Theorem IX.47. $F(G;\lambda)$ *is topologically invariant.*

The last two results follow from Eq. (IX.4.1) and the fact that the V-function $Q(G; -1, -\lambda)$ takes the value -1 when $G = X_0$, by Eq. (IX.1.16). Now Theorem IX.46 follows from Eq. (IX.1.17), and Theorem IX.47 follows from Theorem IX.7.

We now prove a companion theorem to Theorem IX.22, with nowhere-zero cycles in place of nowhere-zero coboundaries. As before, Ω is an arbitrary orientation of G, with dart-set S, and the ring R of coefficients has exactly n members. (So $n > 0$.) Let us write $M(\Omega, R)$ for the number of nowhere-zero cycles of Ω to R.

By the conventions of Sec. VIII.1, the chain-group $\Gamma(\Omega, R)$ has exactly one chain when S is null, called the zero chain. Paradoxically this zero chain is then nowhere zero. Thus

$$M(\Omega, R) = 1$$

when G is edgeless, and in particular when G is null.

Now let G be a union of disjoint subgraphs H and K. Then Ω induces orientations Ω_H and Ω_K of H and K, respectively, where the darts are the appropriate members of S with the same heads and tails as in Ω. Each cycle of Ω to R is determined by its restrictions to the dart-sets of Ω_H and Ω_K, and, by the definitions of Sec. VIII.4, these are cycles of Ω_H and Ω_K

respectively. Moreover, a cycle of Ω_H and a cycle of Ω_K together determine a unique cycle of Ω, of which they are both restrictions. We infer that

$$M(\Omega, R) = M(\Omega_H, R) M(\Omega_K, R).$$

Now let A be any link of G, corresponding to a dart D of Ω. Then Ω determines orientations Ω'_D and Ω''_D of G'_A and G''_A, respectively, as in Sec. VIII.6. The cycles of Ω''_D to R are the restrictions to $S - \{D\}$ of the cycles of Ω, by Theorem VIII.49. Moreover, a given cycle of Ω''_D is a restriction of only one cycle of Ω. For otherwise, $\{D\}$ would be the support of a cycle of Ω, and so A would have to be a loop.

The nowhere-zero cycles of Ω''_D to R are the restrictions to $S - \{D\}$ of the nowhere-zero cycles of Ω, and of the cycles of Ω with support $S - \{D\}$. But the nowhere-zero cycles of Ω'_D are the restrictions to $S - \{D\}$ of the cycles of Ω with support $S - \{D\}$, by Theorems VIII.47 and VIII.48. From these observations, we infer that

$$M(\Omega, R) = M(\Omega''_D, R) - M(\Omega'_D, R).$$

Consider the case $G = X_m$. Then every chain on S to R is a cycle of Ω. It follows that

$$M(\Omega, R) = (n-1)^m \quad \text{if } G = X_m.$$

The results expressed in the last four equations suffice for an inductive proof that $M(\Omega, R)$ is a graph-function, depending only on G and n. So, by the first three equations, $(-1)^{|E(G)| + |V(G)|} M(\Omega, R)$ is a V-function. By Theorem IX.4 we can identify it with the V-function $Q(G; -1, -n)$. So we have the following theorem.

Theorem IX.48. *The number of nowhere-zero cycles of Ω to R, where Ω is an arbitrary orientation of G and the ring R has n elements, is $F(G; n)$.*

Comparing this result with Theorem IX.22, we see that the graph-functions $U(G, \lambda)$ and $F(G; \lambda)$ are algebraically dual, at least for positive integral values of λ.

Theorem IX.49. $F(G; n) \geq 0$ *for each positive integer n, and for each G. Moreover, if $F(G; n) = 0$ for one such n, then $F(G; m) = 0$ for each positive integer $m < n$.*

Proof. The first part follows from Theorem IX.48. As for the second, suppose $F(G; m) > 0$. Then any orientation Ω of G has a nowhere-zero cycle to I_m, by Theorem IX.48.

Now any cycle of Ω to I becomes one to I_m when we replace each coefficient by its residue mod m. Moreover, each cycle to I_m derives from one to I in this way. This statement is true for the elementary chains of $\Gamma(\Omega, I_m)$ by Theorem VIII.34, and therefore for all chains of $\Gamma(\Omega, I_m)$ by Theorems VIII.18 and VIII.35. It follows from Theorem VIII.29 that there

is a nowhere-zero cycle of Ω to I in which each coefficient is less than m in absolute value. Replacement of its coefficients by their residues mod n gives us a nowhere-zero cycle of Ω to I_n. But then $F(G; n) > 0$, by Theorem IX.48.
□

In the analogous Theorem IX.21, it is obvious that $P(G; n)$, or $U(G; n)$ is nonzero for a sufficiently large integer n. A similar result can be proved here, using the following dual of Theorem IX.18.

Theorem IX.50. *If G has no isthmus, then $F(G; \lambda)$ is a polynomial in λ of degree $p_1(G)$. Hence it is not identically zero.*

This is a consequence of Eq. (IX.4.2) and Theorem I.34, for the only term on the right of Eq. (IX.4.2) in which the index of λ is $\geq p_1(G)$ is that for which $S = E(G)$.

Theorem IX.51. *If G has no isthmus, there exists a positive integer n such that $F(G; n) > 0$.*

Proof. Suppose the contrary. Then the polynomial $F(G; \lambda)$ vanishes for infinitely many values of λ, taken to be a complex variable, and therefore $F(G; \lambda)$ is identically zero, by the Fundamental Theorem of Algebra. This is contrary to Theorem IX.50.
□

There have been some noteworthy improvements upon this result. For each positive integer n, let us define an *n-flow* of G as a nowhere-zero cycle on some orientation Ω of G to I in which each coefficient has absolute value less than n. Given such an n-flow, we can always reorient so as to transform it into an n-flow in which each coefficient is positive. As in the proof of Theorem IX.49, we can show that G has an n-flow if and only if Ω has a nowhere-zero cycle to I_n, that is, if and only if $F(G; n) > 0$, by Theorem IX.48. The Five-Flow Conjecture asserts that every isthmusless graph G has a 5-flow; that is, it satisfies $F(G; 5) > 0$. F. Jaeger showed in [6] that $F(G; 8) > 0$ for each such G, and P. Seymour bettered this result in [8], showing that $F(G; 6) > 0$.

On the other hand, as we shall see in the next section, there are isthmusless graphs G for which $F(G; 4) = 0$.

IX.5. TAIT COLORINGS

Let G be any cubic graph. Consider a set $T = \{\alpha, \beta, \gamma\}$ of three distinct elements called "colors." A *Tait* coloring of G is a mapping t of $E(G)$ into T so that the edges incident with any vertex x are mapped onto three distinct colors. This definition implies that, if G has a loop, it has no Tait coloring.

There is a ring R_4 of four elements, which we can write as $0, 1, \omega, \omega^2$. It satisfies the multipicative rule $\omega^3 = 1$ and the following additive rules:

$$z + z = 0 \quad \text{for each } z \in R_4,$$

and

$$1 + \omega + \omega^2 = 0.$$

Since each element of R_4 is its own negative, cycles and coboundaries to R_4 for G can be regarded as chains on $E(G)$, with no reference to any particular orientation of G, as explained in Sec. VIII.4. The Tait colorings of G can be regarded as its nowhere-zero cycles to R_4, the nonzero elements of R_4 being taken as the three colors.

Theorem IX.52. *A cubic graph G has a Tait coloring if and only if $F(G; 4) > 0$, by Theorem IX.48.*

Theorem IX.53. *Let B be a bond of G, and t a Tait coloring of G. Let n_α, n_β, and n_γ be the numbers of edges of B colored α, β, and γ, respectively, in t. Then n_α, n_β, and n_γ are either all odd or all even.*

Proof. B is the support of a coboundary of G to R_4 in which each coefficient is 1, by Theorem VIII.40. This coboundary is orthogonal to the cycle t, by Theorem VIII.42. Hence

$$n_\alpha \alpha + n_\beta \beta + n_\gamma \gamma = 0.$$

The theorem now follows, by the rules of addition in R_4. □

In particular, G can have no Tait coloring if it has an isthmus, as is required also by Theorem IX.46.

Now let G be connected, and let B be a bond of G with end-graphs H and K. Suppose first that $|B| = 2$. From H we can form a new cubic graph H_1 by adjoining a new edge A_1 whose ends are the ends in H of the members of B. We call H_1 the *residual graph* of B containing H, and we refer to A_1 as its *completing edge*. There is, of course, a second residual graph of B, containing K.

Suppose next that $|B| = 3$. From G we can form a new cubic graph H_1 by contracting K to a single new vertex k, incident with the three edges of B. Similarly, we form K_1 by contracting H. We now call H_1 and K_1 the *residual* graphs of B containing H and K, respectively.

Theorem IX.54. *Let G be a connected cubic graph having a bond B such that $|B| = 2$ or 3. Suppose each of the two residual graphs of B in G to have a Tait coloring. Then G has a Tait coloring.*

Proof. Let the residual graphs of B be H_1 and K_1, with Tait colorings s and t, respectively.

In the case $|B| = 2$, we permute the three colors in t, if necessary, so as to arrange that the completing edges of H_1 and K_1 have the same color, α say. We can now obtain a Tait coloring of G by assigning to each edge of B the color α, and to each other edge of G its color in s or t.

In the case $|B| = 3$, we permute in t, if necessary, so as to give each edge of B the same color in t as in s. We then form a Tait coloring of G by assigning to each edge its color in s or t. □

Theorem IX.55. *The Petersen graph has no Tait coloring.*

Proof. The Petersen graph, shown in Fig. IV.2.8, is described in Sec. IV.2 as consisting of two disjoint 5-circuits $a_1 a_2 a_3 a_4 a_5$ and $b_1 b_3 b_5 b_2 b_4$, and five edges $a_i b_i$, one for each suffix i. These five edges make up a bond B having the two 5-circuits as its end-graphs.

Suppose the graph to have a Tait coloring t. By Theorem IX.53, we may assume without loss of generality that three edges of B are colored α, one β, and one γ. Let us assume that the ends in $a_1 a_2 a_3 a_4 a_5$ of the three edges of B colored α are not three consecutive vertices of that pentagon. Taking advantage of obvious automorphisms, we can suppose these ends to be a_1, a_3, and a_5, the vertices a_2 and a_4 being ends of edges of B of colors β and γ, respectively. But now the edges $a_1 a_2$ and $a_2 a_3$ are both forced to have color γ in t, and we have a contradiction.

We infer that the edges of B colored α have ends at three consecutive vertices of the first pentagon, and similarly at three consecutive vertices of the second. But these two requirements are incompatible. □

The Petersen graph satisfies $F(G; 4) = 0$, by Theorems IX.52 and IX.55. Consequently it has no 4-flow. The term "snark" is now used for a connected cubic graph that has no Tait coloring and no bond of fewer than four edges. The Petersen graph has been found to be the simplest example of a snark. A noncolorable graph with two or three edges in some bond B is easily reduced to a smaller one, a residual graph of B in the terminology of Theorem IX.54. So this case is regarded as trivial, and is excluded. The case of a bond of one edge is trivial by Theorem IX.46.

For examples of snarks, some in infinite families, reference should be make to a paper of R. Isaacs, published in 1975 [5].

It has been conjectured that every snark has a minor that is a Petersen graph.

We define a *Tait subgraph* of a cubic graph G as a spanning subgraph of G in which each component is a circuit of even length. If H is such a subgraph, then the corresponding Tait cycle of G is the chain on $E(G)$ to I_2 whose support is $E(H)$. It is a cycle of $\Gamma(G; I_2)$.

Consider any Tait coloring t of G. If δ is any one of the three colors, then the edges of the other two colors define a Tait subgraph $H_\delta(t)$ of G, and we denote the corresponding Tait cycle by $J_\delta(t)$. In this way each Tait coloring t gives rise to three Tait cycles $J_\alpha(t)$, $J_\beta(t)$, an $J_\gamma(t)$. Evidently they satisfy the equation

$$J_\alpha(t) + J_\beta(t) + J_\gamma(t) = 0. \qquad \text{(IX.5.1)}$$

Let us consider a converse problem. Let J be a Tait cycle of G, corresponding to a Tait subgraph H. We ask how many distinct Tait colorings of G have J as an associated Tait cycle $J_\delta(t)$? Here we count two Tait cycles as distinct if we cannot change one into the other merely by a permutation of the three colors. In view of this definition, we can suppose the color γ given to the edges of G not in the support of J, and to these edges only. We can then arrange for the colors α and β to alternate in each component of H in $2^{p_0(H)}$ ways. Allowing for an interchange of α with β, we arrive at the following result.

Theorem IX.56. *The Tait cycle determined by a Tait subgraph H of G is associated with exactly*

$$2^{p_0(H)-1}$$

distinct Tait colorings of G.

From this result we can derive information about the Hamiltonian circuits of a cubic graph G. We note that a Hamiltonian circuit of G can be characterized as a connected Tait subgraph.

Let us sum Eq. (IX.5.1) over a complete set of distinct Tait colorings t of G. Since our coefficients are residues mod 2, only the Tait cycles corresponding to Hamiltonian circuits, that is, those for which $p_0(H) = 1$, survive in the sum. Our result is that the sum of all such Tait cycles is zero. It can be stated as follows.

Theorem IX.57. *(Smith's Theorem). Let A be any edge of a cubic graph G. Then the number of Hamiltonian circuits of G that include A is even.*

(Let us not forget that zero is an even number). As a simple but striking consequence of Smith's Theorem, we have:

Theorem IX.58. *If a cubic graph has a Hamiltonian circuit, then it has at least three.*

IX.6. THE DICHROMATE OF A GRAPH

The *dichromate* $\chi(G) = \chi(G; x, y)$ of a graph G is a slightly modified dichromatic polynomial. It is a graph-function that is a polynomial in two independent indeterminates x and y. It is symmetrically related to the two dual functions $U(G; \lambda)$ and $F(G; \lambda)$. We define it as follows:

$$\chi(G; x, y) = (x-1)^{-p_0(G)} Q(G; x-1, y-1). \qquad \text{(IX.6.1)}$$

By Eq. (IX.1.12), this is equivalent to

$$\chi(G;x,y) = \sum_{S \subseteq E(G)} (x-1)^{p_0(G:S)-p_0(G)} \times (y-1)^{p_1(G:S)}$$

(IX.6.2)

It is clear that $p_0(G)$ never exceeds $p_0(G:S)$. So Eqs. (IX.6.1) and (IX.6.2) do indeed define a polynomial in x and y. From Eqs. (IX.2.1) and (IX.4.1), we have

$$U(G;\lambda) = (-1)^{\rho(G)} \chi(G;1-\lambda,0),$$ (IX.6.3)

$$F(G;\lambda) = (-1)^{p_1(G)} \chi(G;0,1-\lambda),$$ (IX.6.4)

where $\rho(G)$ is the coboundary rank of G.

The dichromate has the two following properties, derived from the definition of a V-function and from Eq. (IX.1.13).

Theorem IX.59. $\chi(G;x,y) = 1$ if G is null.

Theorem IX.60. If G is the union of two edge-disjoint subgraphs H and K, with at most one common vertex, then

$$\chi(G) = \chi(H)\chi(K).$$ (IX.6.5)

Formula (IX.1.2) survives only in the following modified form.

Theorem IX.61. Let A be an edge of G that is neither a loop nor an isthmus. Then

$$\chi(G) = \chi(G'_A) + \chi(G''_A).$$ (IX.6.6)

Proof. If A is not an isthmus of G, we have

$$p_0(G) = p_0(G'_A) = p_0(G''_A),$$

by Theorems I.29 and II.5. Hence Eq. (IX.6.6) follows from Eq. (IX.6.1) and the properties of V-functions. □

Let us evaluate $\chi(G)$ in some simple cases. It follows from Eq. (IX.6.2) that $\chi(G)$ is 1 for a vertex-graph, x for a link-graph, and y for a loop-graph. Hence, by Theorem IX.60, we have the following rules.

Theorem IX.62. If G is a forest, then $\chi(G) = x^{|E(G)|}$ and if G is linkless, then $\chi(G) = y^{|E(G)|}$.

Let us now write C_k and L_k for a k-circuit and k-linkage, respectively. The case $k = 1$ is included in Theorem IX.62. For $k > 1$ we find, using Theorem IX.61, that

$$\chi(C_k) = \chi(N) + \chi(C_{k-1})$$

and

$$\chi(L_k) = \chi(L_{k-1}) + \chi(X_{k-1}),$$

The Dichromate of a Graph

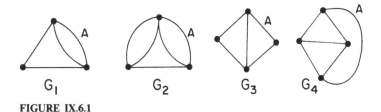

FIGURE IX.6.1

where N is an arc of $(k-1)$ edges. From Theorem IX.62, by a simple induction, we find

$$\chi(C_k) = y + x + x^2 + \cdots + x^{k-1}, \quad \text{(IX.6.7)}$$
$$\chi(L_k) = x + y + y^2 + \cdots + y^{k-1}. \quad \text{(IX.6.8)}$$

We go on to the graphs of Fig. IX.6.1, applying Theorems IX.60 and IX.61, with the edge A as indicated.

$$\chi(G_1) = \chi(C_3) + \chi(C_2)\chi(C_1)$$
$$= (y + x + x^2) + (y + x)y$$
$$= x^2 + xy + y^2 + x + y.$$
$$\chi(G_2) = \chi(G_1) + \chi(L_3)\chi(C_1)$$
$$= \chi(G_1) + (x + y + y^2)y$$
$$= y^3 + x^2 + 2xy + 2y^2 + x + y.$$
$$\chi(G_3) = \chi(C_3)\chi(L_1) + \chi(G_1)$$
$$= x^3 + 2x^2 + 2xy + y^2 + x + y.$$
$$\chi(G_4) = \chi(G_3) + \chi(G_2)$$
$$= x^3 + y^3 + 3x^2 + 4xy + 3y^2 + 2x + 2y.$$

We write $\chi_{ij}(G)$ for the coefficient of $x^i y^j$ in $\chi(G)$.

Theorem IX.63. $\chi(G)$ *has degree* $\rho(G)$ *in* x *and* $p_1(G)$ *in* y.

Proof. This is equivalent to the assertion that $\chi(G)$, regarded as a polynomial in $x-1$ and $y-1$, has degree $\rho(G)$ in $x-1$ and $p_1(G)$ in $y-1$. But that follows from Eq. (IX.6.2). In each product on the right of Eq. (IX.6.2), the index of $x-1$ is at most $|V(G)| - p_0(G) = \rho(G)$, and the index of $y-1$ is at most $p_1(G)$, by Theorem I.34. Moreover, these maximal values do occur, the first for a null S and the second for $S = E(G)$. □

We define the χ-*matrix* of G as a matrix of $(\rho(G)+1)$ rows and $(p_1(G)+1)$ columns, the entry in the $(i+1)$th row and $(j+1)$th column being $\chi_{ij}(G)$ $(i, j = 0, 1, 2, \ldots)$. By Theorem IX.63, all the nonzero coefficients in $\chi(G)$ are elements of the χ-matrix. When we write the χ-matrix as a table of the coefficients in $\chi(G)$, we usually omit the zeros, or most of them. Thus in the case of the 3-wheel, G_4 of Fig. IX.1.1, we tabulate the coefficients as follows:

$$\begin{bmatrix} 0 & 2 & 3 & 1 \\ 2 & 4 & & \\ 3 & & & \\ 1 & & & \end{bmatrix}$$

Theorem IX.64. *The coefficients $\chi_{ij}(G)$ are all nonnegative.*

Proof. For graphs with at most one edge, this follows from Theorems IX.59 and IX.62. Assume it true whenever $|E(G)|$ is less than some positive integer q and consider the case $|E(G)| = q$.

If every edge of G is an isthmus, the theorem is still true, by Theorem IX.62. In the remaining case, some edge A is not an isthmus. If A is a loop, we have $\chi(G) = y\chi(G'_A)$, by Theorems IX.60 and IX.62. The theorem holds for G'_A by the inductive hypothesis, and therefore it holds for G. If A is not a loop, we have $\chi(G) = \chi(G'_A) + \chi(G''_A)$, by Theorem IX.61. The theorem holds for G'_A and G''_A, by the inductive hypothesis, and therefore it holds for G. So the theorem is true for $|E(G)| = q$, and the induction succeeds. □

It is conjectured that each row and column of a χ-matrix is unimodal, i.e., that no element in it is less than both its neighbors.

Another regularity provable by induction is that $\chi_{00}(G) = 0$ whenever G has an edge. This is just the first member of a family of linear relations between coefficients in $\chi(G)$. The second member asserts that $\chi_{01}(G) = \chi_{10}(G)$, and the third that

$$\chi_{20}(G) + \chi_{02}(G) = \chi_{11}(G) + \chi_{10}(G).$$

The second member is valid when $|E(G)| > 2$ and the third when $|E(G)| > 3$.

These equations are related to the identity

$$\chi(G; 1 + z^{-1}, 1 + z) = z^{-\rho(G)}(1 + z)^{|E(G)|}, \qquad \text{(IX.6.9)}$$

which is easily proved by substitution in Eq. (IX.6.2). To get a member of our family of equations, we differentiate

$$\sum_{i,j} \chi_{ij}(G)(1 + z^{-1})^i (1 + z)^j$$

n times with respect to z, substitute $z = -1$, and equate the result to zero. The equation will be valid when $|E(G)| > n$.

For a connected graph G, the dichromate is closely related to the set of spanning trees of G. We can, for example, show that

$$\chi(G; 1, 1) = T(G), \qquad \text{(IX.6.10)}$$

by substitution in Eq. (IX.6.2), for the surviving products on the right of Eq. (IX.6.2), each having the value 1, correspond to those edge-sets S for which $p_0(G:S) = 1$ and $p_1(G:S) = 0$, that is, for which $G:S$ is a spanning tree of G.

We go on to prove a more general theorem, after some preliminary definitions.

Let the edges of a connected graph G be enumerated as $A_1, A_2, \ldots, A_n$, and let T be any spanning tree of G. Now let A_k be any edge of T, and let T_1 and T_2 be its end-graphs in T. It may happen that every edge A_j of G that has one end in T_1 and one in T_2 satisfies $j \geq k$. If so, we say that A_k is *internally active* in T. We write $a(T)$ for the number of edges of T that are internally active in it.

Now consider an edge A_k of G that is not in T. If A_k is a link, its ends are joined by a unique arc L_k in T, by Theorem I.44. If A_k is a loop, we write L_k for the vertex-graph defined by its incident vertex. If every edge A_j of G that belongs to L_k satisfies $j > k$ we say that A_k is *externally active* in T. We write $b(T)$ for the number of edges of G, all outside T, that are externally active in T.

According to these definitions, an isthmus of G, which must belong to every spanning tree of G, is internally active in every such tree and externally active in none. But a loop of G belongs to no spanning tree. It is externally active in every spanning tree of G and internally active in none.

Theorem IX.65. *If G is a connected graph, then*

$$\chi(G; x, y) = \sum_T x^{a(T)} y^{b(T)}, \qquad (IX.6.11)$$

for an arbitrary enumeration of the edges of G.

Proof. Let us denote the sum on the right of Eq. (IX.6.11), defined in terms of an arbitrary enumeration $E = (A_1, A_2, \ldots, A_n)$ of the edges, by $W(G)$.

If $|E(G)| = 0$, then G is a vertex-graph. It has only one spanning tree, G itself, and $a(G)$ and $b(G)$ are both zero. Hence, $W(G) = 1$, and the theorem holds, by Theorem IX.62.

Assume as an inductive hypothesis that the theorem is true whenever $|E(G)|$ is less than some positive integer q, and consider the case $|E(G)| = q$.

Suppose first that each edge of G is a loop or an isthmus, there being i isthmuses and ℓ loops. Then there is just one spanning tree T, its edges being the isthmuses. By our preliminary observations, $a(T) = i$ and $b(T) = \ell$. Accordingly, $W(G) = x^i y^\ell$. Hence $W(G) = \chi(G; x, y)$, by Theorems IX.60 and IX.62.

In the remaining case, let A_m be the edge of greatest suffix that is neither a loop nor an isthmus of G. We write also $A_m = A$. To each of the graphs G'_A and G''_A, we assign the enumeration E' of edges obtained from E by deleting A.

As in Sec. II.2, the spanning trees of G'_A are those spanning trees of G that do not include A, and the spanning trees of G''_A are obtained from those spanning trees of G that do contain A, by contracting that edge.

By the choice of m, the edge A can be neither externally nor internally active in any spanning tree of G. Hence, still by the choice of m, $a(T)$ and $b(T)$ have the same values in G'_A as in G, for any spanning tree T of G that does not include A. Furthermore, if T is any spanning tree of G that does include A, then the numbers $a(T)$ and $b(T)$ for G and E are the numbers $a(T''_A)$ and $b(T''_A)$, respectively, for G''_A and E'. We deduce that

$$W(G) = W(G'_A) + W(G''_A)$$

with respect to the enumerations E and E'. Hence,

$W(G) = \chi(G'_A) + \chi(G''_A)$ (by the inductive hypothesis)

$= \chi(G)$ (by Theorem IX.61).

We conclude that the theorem is true also when $|E(G)| = q$, and that the induction succeeds. □

By Eq. (IX.6.11), the coefficient $\chi_{ij}(G)$, for a connected G, is the number of spanning trees T of G such that $a(T) = i$ and $b(T) = j$.

IX.7. SOME REMARKS ON RECONSTRUCTION

Theorem IX.66. *The dichromate of a graph is reconstructible.*

Proof. Referring to Eq. (IX.6.2), we see that we can reconstruct $\chi(G; x, y)$ if we can reconstruct the number of spanning subgraphs $G: S$ of G with given values of $p_0(G: S)$ and $p_1(G: S)$. But by Theorem V.11, we can reconstruct the number of disconnected spanning subgraphs of this kind. This leaves over spanning subgraphs $G: S$ such that $p_0(G: S) = 1$ and therefore $p_1(G: S) = |S| - |V(G)| + 1$. So we now require to be able to find the number of connected spanning subgraphs of G with a given number of edges. But the number of spanning subgraphs with k edges is known; it is the number of ways of choosing k things out of $|E(G)|$. The number of disconnected ones is reconstructible by Theorem V.11, and so the number of connected ones can be found by subtraction. We conclude that the sum on the right side of Eq. (IX.6.2) is reconstructible. □

The reconstructibility of the dichromate implies that of the dichromatic polynomial, the chromatic polynomial, and the flow polynomial, by

Eqs. (IX.6.1), (IX.2.1), and (IX.4.1). In particular, the chromatic number and the property of having or not having a 5-flow are reconstructible.

By a continuation of the argument for Theorem IX.66, we can prove the following theorem.

Theorem IX.67. *The number of Hamiltonian circuits of a graph G is reconstructible.*

Proof. We saw, in the proof of Theorem IX.66, that the number of connected spanning subgraphs of G with k edges is reconstructible. But by Theorem V.15, the number of such subgraphs with no spanning block is reconstructible. So, by subtraction, we can find the number of connected spanning subgraphs of G with k edges and with a spanning block. But for $k = |V(G)|$, these are the Hamiltonian circuits of G. (See Eq. (I.1.1) and Theorem I.27.) □

We should perhaps add here a comment on the reconstructibility of the *characteristic polynomial* of a graph G, denoted here by $J(G; \lambda)$. It is usually discussed in terms of strict graphs.

Let G be a strict graph, and let its vertices be enumerated as $v_1, v_2, \ldots, v_m$. To avoid trivialities, we suppose $m \geq 3$. The adjacency matrix $A(G)$ of G is an $m \times m$ matrix for which the entry in the ith row and jth column is the number of edges, 1 or 0, joining v_i and v_j. The characteristic polynomial of G is defined as follows:

$$J(G; \lambda) = \det(\lambda I - A(G)), \qquad (IX.7.1)$$

where I is the unit matrix of order m. The zeros of $J(G; \lambda)$, the eigenvalues of $A(G)$, have been much studied. See, for example, [3].

We can expand $J(G; \lambda)$ in the form

$$J(G; \lambda) = \sum_{j=0}^{m} B_j \lambda^j, \qquad (IX.7.2)$$

where the B_j are integers. Apart from multiplication by $(-1)^{m-j}$, the number B_j is the sum of the determinants of the adjacency matrices of the induced subgraphs of G with $(m - j)$ vertices. Except for $B_0 = \det A(G)$, the reconstructibility of each B_j follows directly from Kelly's Lemma.

The reconstructibility of $J(G; \lambda)$ now depends on that of $\det A(G)$. Considering the expansion of this determinant we see that each term corresponds to a spanning subgraph of G in which each component is either a circuit or a link-graph. We can indeed evaluate the determinant if we know the number of spanning subgraphs of G with k link-graphs and l circuits as components, for each pair (k, ℓ). But for $k + \ell > 1$, this number is reconstructible, by Theorem V.11. For $k + \ell = 1$, it is the number of Hamiltonian circuits, since $m > 3$, and this is reconstructible, by Theorem IX.67. We thus have another theorem.

Theorem IX.68. *The characteristic polynomial of a strict graph G is reconstructible.*

Graph theorists sometimes express the hope that a reconstructible polynomial may be found whose coefficients determine the corresponding graph G uniquely. Then the reconstructibility of the polynomial will imply that of G. But if such a polynomial exists, it is not the dichromate and not the characteristic polynomial, for all forests of k edges have the same dichromate, by Theorem IX.62. Moreover, a construction is explained in [14] for "codichromatic" pairs of graphs with connectivities up to 5. A well-known theorem of A. J. Schwenk asserts that, for almost every tree, there is another tree with the same characteristic polynomial [7].

IX.8. NOTES

IX.8.1. V-functions and 1-Factors

It is shown in [10] and [15] that there is a V-function such that for a cubic graph of $2n$ vertices, $(-1)^n V(G)$ is the number of 1-factors. We have

$$V(X_k) = \tfrac{1}{2}(3^k + 1) \cdot (-1)^{k+1}.$$

IX.8.2. The Beraha Numbers

Chromatic polynomials are studied most often in connection with triangulations of the sphere. For these, some curious special theorems were found for the value $\lambda = \tfrac{1}{2}(3 + \sqrt{5})$. This is one member of an infinite sequence of numbers, pointed out by S. Beraha, that seem to be of special importance in the theory, but their significance is not yet properly understood. See [12], [13], and [16].

IX.8.3. Grinberg's Theorem

E. Grinberg [4] presented his theorem as a result about planar graphs. In this chapter, I have taken the liberty of dualizing it, the dual form being valid for all graphs.

IX.8.4. Rotors

Reversing a rotor with at most 5-fold rotational symmetry does not alter the chromatic or dichromatic polynomial of a graph. (See [14].)

EXERCISES

1. If a V-function satisfies $V(X_k) = \frac{1}{2}(3^k + 1)(-1)^{k+1}$, find its values for the graphs of Figs. I.7.4 and I.7.5.
2. Find the dichromatic polynomials of the graphs of Figs. II.2.1. and II.2.2.
3. Find the chromatic polynomial of (i) the wheel of order n, (ii) the vertex-join of this wheel, and (iii) the bipyramid B_n obtained from two such wheels by identifying their rims.
4. Find 3-colorings of the cube-graph and the Petersen graph.
5. A cubic graph G is obtained from a cubic graph H by expanding one vertex into a 3-circuit (see Sec. IV.2). Show that G and H have the same number of Hamiltonian circuits.
6. A graph G is defined by two circuits $a_1 a_2 \ldots a_9$ and $b_1 b_5 b_9 b_4 b_8 b_3 b_7 b_2 b_6$, together with the joins $a_j b_j$. Show that G has exactly three Hamiltonian circuits, and no Tait coloring except the one given by these.
7. Find a third member of the family of linear relations for coefficients in χ discussed in Sec. IX.6. Can you find a general formula for such linear relations from Eq. (IX.6.9)?

REFERENCES

[1] Appel, K., and W. Haken, Every planar map is four colorable; I. Discharging, *Illinois J. Math.*, **21** (1977), 429-490.
[2] Appel, K., W. Haken, and J. Koch, Every planar map is four-colorable. II. Reducibility, *Illinois J. Math.*, **21** (1977), 491-567.
[3] Biggs, N., *Algebraic graph theory*, Cambridge Univ. Press, 1974.
[4] Grinberg, E., Plane homogeneous graphs of degree three without Hamiltonian circuits. *Latvian Math. Yearbook* **4**, (1968), 51-58 (Russian). Izdat. "Zinatne" Riga.
[5] Isaacs, R., Infinite families of nontrivial trivalent graphs which are not Tait-colorable, *Amer. Math. Monthly*, **82** (1975) 221-239.
[6] Jaeger, F., On nowhere-zero flows in multigraphs, *Congressus Numerantium* **XV**, *Utilitas Math.* Winnipeg, Man. (1976) 373-378.
[7] Schwenk, A. J., "Almost all trees are cospectral," in *New Directions in the Theory of Graphs*, Academic Press, New York, 1973, pp. 275-307.
[8] Seymour, P. D., Nowhere-zero 6-flows, *J. Comb. Theory*, (B) **30** (1981), 130-135.
[9] Stanley, R. P., Acyclic orientations of graphs, *Discrete Math.* **5** (1973), 171-178.
[10] Tutte, W. T., A ring in graph theory, *Proc. Cambridge Phil. Soc.*, **43** (1947), 26-40.
[11] ———, The factorization of linear graphs, *J. London Math. Soc.*, **22** (1947), 107-111.

[12] ———, On chromatic polynomials and the golden ratio, *J. Comb. Theory*, **9** (1970), 289–296.
[13] ———, The golden ratio in the theory of chromatic polynomials, *Annals New York Acad. Sci.*, **175** (1970), 391–402.
[14] ———, Codichromatic graphs, *J. Comb. Theory*, (B), **16** (1974) 168–174.
[15] ———, 1-factors and polynomials, *Europ. J. Combinatorics*, **1** (1980), 77–87.
[16] ———, Chromatic solutions, *Can. J. Math.*, **34** (1982), 741–758.
[17] Whitney, H., A logical expansion in mathematics, *Bull. Amer. Math. Soc.*, **38**, (1932), 572–579.

Chapter X

Combinatorial Maps

X.1. DEFINITIONS AND PRELIMINARY THEOREMS

In recent years, some purely combinatorial theories of maps have been presented [3, 4, 6, 7]. The following account is based on [9].

We postulate a finite set S of elements called *crosses*. The number of elements of S is to be divisible by 4. We write $S = 4n$. The set S is partitioned into n disjoint quartets called *cells*. If X is any cross, then the other members of its cell are denoted by θX, ϕX, and $\theta\phi X$.

The symbols θ and ϕ can be regarded as denoting two permutations of S. These permutations are required to satisfy the two following axioms.

Axiom X.1. $\theta^2 = \phi^2 = I$ and $\theta\phi = \phi\theta$.

Axiom X.2. If X is any cross, then the four crosses X, θX, ϕX, and $\theta\phi X$ are all distinct.

Here I denotes the identical permutation of S. We observe that each of the permutations θ and ϕ leaves each cell invariant, except for a permutation of its members among themselves. Each of θ and ϕ is an *involution*; it partitions S into disjoint pairs of distinct elements, and it interchanges the members of each such pair. We refer to θ as the *first involution* and to ϕ as the *second involution*.

253

Now let P be any permutation of S. It arranges the crosses in disjoint cyclic sequences called the *orbits* of P. Each orbit is said to be *through* each of its member crosses. We can write the orbit through a cross X as

$$J(P, X) = (X, PX, P^2X, \ldots, P^{m-1}X), \quad (X.1.1)$$

where m is the least positive integer such that

$$P^m X = X. \quad (X.1.2)$$

Since an orbit is cyclic, it does not matter which of its crosses we write first; the orbit through X is also the orbit through PX and the orbit through P^2X, and so on.

The involutions θ and ϕ have their orbits. In their case $m = 2$ for each X.

We say that the permutations θ, ϕ, and P define a *premap* $L = L(\theta, \phi, P)$ on S if the following two axioms are satisfied.

Axiom X.3. $P\theta = \theta P^{-1}$.

Axiom X.4. *For each cross X, the orbits of P through X and θX are distinct.*

We then say that the premap L has θ and ϕ as its *first* and *second involutions*, respectively, and P as its *basic permutation*.

We refer to the number of members of an orbit $J(Q, X)$ as the *length* of that orbit.

Let us compare the orbits through X and θX in the case of a premap $L = L(\theta, \phi, P)$, for an arbitrary cross X. The two orbits are distinct by Axiom X.4. They also have the same length. For, if m is the length of $J(P, X)$, we have

$$P^{-m}\theta X = \theta X \quad \theta X = P^m \theta X, \quad (X.1.3)$$

by Eq. (X.1.2) and Axiom X.3. Hence the length of $J(P, \theta X)$ is not greater than that of $J(P, X)$. Similarly the length of $J(P, \theta^2 X)$, that is, $J(P, X)$, is not greater than that of $J(P, \theta X)$.

With $J(P, X)$ given by Eq. (X.1.1), we can now write:

$$J(P, \theta X) = (\theta X, P\theta X, P^2\theta X, \ldots, P^{m-1}\theta X),$$
$$J(P, \theta X) = (\theta X, \theta P^{-1}X, \theta P^{-2}X, \ldots, \theta P^{-m+1}X),$$
$$J(P, \theta X) = (\theta X, \theta P^{m-1}X, \theta P^{m-2}X, \ldots, \theta PX), \quad (X.1.4)$$

by Axiom X.3 and Eq. (X.1.2).

Thus $J(P, \theta X)$ is formed from $J(P, X)$ by reversing the cyclic order of the elements and then premultiplying each by θ. We express this relation by saying that $J(P, \theta X)$ is the *conjugate* orbit to $J(P, X)$ in L. Conjugacy of orbits is evidently a symmetrical relation.

Definitions and Preliminary Theorems

Theorem X.5. *In any premap $L = L(\theta, \phi, P)$, the orbits of P through X and θX are conjugate for any cross X. A cross Y then belongs to one of them if and only if θY belongs to the other.*

We define a *vertex* of the premap $L = L(\theta, \phi, P)$ as a pair v of conjugate orbits of P. Each of the two orbits is an *oriented vertex* of L, and an *oriented form of v*.

A cell $A = (X, \theta X, \phi X, \theta\phi X)$ can be called an *edge* of L. Then the unordered pairs $\{X, \theta X\}$ and $\{\phi X, \theta\phi X\}$ are its two *half-edges* in L. For each half-edge $\{Y, \theta Y\}$, there is a unique vertex v of L such that Y belongs to one of the oriented forms of v, and θY to the other. We express this relation by saying that v *absorbs* the half-edge, and that A is *incident* with v in L. The two half-edges of A may be absorbed by the same vertex or by two different vertices. In the first case, A is a *loop* of L, and in the second case a *link*.

The definitions above do not exclude the possibility that S is null. In that case, we postulate a unique permutation I on S. There is then a unique premap $L(I, I, I)$ on S. We call this a null premap. It has no vertex and no edge.

With the above definition of incidence, the vertices and edges of L are the vertices and edges, respectively, of a graph $G(L)$, called the *graph* of L. Evidently $G(L)$ has no isolated vertex. The loops and links of L are the loops and links, respectively, of $G(L)$.

We shall say that a cross X *belongs* to any edge or oriented vertex of L of which it is an element, that it belongs to a vertex of L if it belongs to an oriented form of that vertex, and that it belongs to a subgraph H of $G(L)$ if it belongs to some edge or vertex of H.

Given a graph G with no isolated vertex, we may ask for a premap L, on some set S, such that G and $G(L)$ are isomorphic. Such a premap can be found by *Edmonds' Construction* [4]. Corresponding to each edge A of G we define a set $A' = (X, \theta X, \phi X, \theta\phi X)$ of four elements called crosses, making the eight crosses on any two distinct edges all distinct. We then have a set S of $4|E(G)|$ crosses, and the notation determines first and second involutions θ and ϕ satisfying Axioms X.1 and X.2. We split each set A' into its two pairs $\{X, \theta X\}$ and $\{\phi X, \theta\phi X\}$, assigning one of these pairs to each end of A. Both pairs are assigned to the same vertex if and only if A is a loop of G. For each vertex of G, it is now easy to arrange the members of its assigned pairs in two conjugate orbits so as to satisfy Axioms X.3 and X.4. We now have our required premap $L(\theta, \phi, P)$.

We return to our fundamental set S of crosses. A nonnull set K of permutations of S generates a group Ψ_K of such permutations. It partitions S, if nonnull, into nonnull equivalence classes such that two crosses belong to the same equivalence class if and only if some member of Ψ_K maps one onto the other. In the case of a premap $L(\theta, \phi, P)$, we shall be interested in the cases $K = \{\theta, \phi, P\}$ and $K = \{\theta\phi, P\}$.

Consider the case $K = \{\theta, \phi, P\}$. We then call the equivalence classes of Ψ_K the *connection-sets* of L. If L has exactly one connection-set, necessarily S, we say that L is *connected*, or equivalently, that L is a map on S.

Let T be any connection-set of L. The effect on T of a member Q of Ψ_K is to permute its elements among themselves. Thus each Q has a unique restriction Q_T to T, that is a permutation of T having the same effect on each member of T as does Q itself. The restrictions to T of θ, ϕ, and P evidently satisfy the four axioms for a premap $L_T = L(\theta_T, \phi_T, P_T)$ on T. Moreover, T is a connection-set of L_T and therefore L_T is a map on T. The maps L_T corresponding to the connection-sets of L are the *components* of L. These components clearly satisfy the rules presented in the following theorem.

Theorem X.6. *Let L be a premap on a nonnull set S. Then each cross of S belongs to exactly one component of L, and each edge, oriented vertex, and vertex of L is an edge, oriented vertex, and vertex, respectively, of some component of L. Moreover, the edges and vertices of a component L_T of L are those edges and vertices of L whose crosses are all in the corresponding connection-set T of L.*

A null premap has no connection-sets and no components.

Theorem X.7. *Let L be a premap on a nonnull set S. Then the graphs of the components of L are the components of the graph of L.*

Proof. Let X and Y be crosses. If they belong to the same link-graph or loop-graph of $G(L)$, then X can be transformed into Y by a sequence of permutations each of which is θ, ϕ or P, that is, by a member of Ψ_K where $K = \{\theta, \phi, P\}$. This follows from the definitions of an edge and vertex of L. It now follows, from Theorem I.43, that X is transformed into Y by some member of Ψ_K whenever X and Y belong to the same component of $G(L)$.

But if X and Y belong to different components of $G(L)$ neither θ, ϕ, nor P can transform X into Y: Each of those operations preserves either an edge or a vertex of $G(L)$. We infer that no member of Ψ_K can then transform X into Y.

We conclude that the crosses belonging to any given component of $G(L)$ determine a connection-set of L. The theorem follows, by Theorem X.6. □

In what follows we shall be interested in maps rather than premaps. We shall usually use the symbol M to denote a map. If a premap $L(\theta, \phi, P)$ is a map we shall denote it also by $M(\theta, \phi, P)$. By definition, a map has exactly one connection-set, and therefore the null premap is not a map. We have seen that:

Theorem X.8. *The components of a premap are maps.*

As a special case of Theorem X.7, we have:

Theorem X.9. *The graph of a map is connected.*

It follows from Theorem X.7 that whenever Edmonds' Construction is applied to a connected graph (with no isolated vertex), the resulting premap is a map.

X.2. ORIENTABILITY

Let $M = M(\theta, \phi, P)$ be a map on a set S. We write $H = \{\theta, \phi, P\}$ and $K = \{\theta\phi, P\}$.

Theorem X.10. *Any element h of Ψ_H can be expressed in the form $k\theta^i$, where $k \in \Psi_K$ and the index i is either 0 or 1.*

Proof. The rule $\theta f = f^{-1}\theta$ holds for $f = \theta\phi$ and for $f = P$, by Axioms X.1 and X.3. Since $\theta^2 = I$, it follows that for each $f \in \Psi_K$ there exists $f' \in \Psi_K$ such that $\theta f = f'\theta$. But an arbitrary element h of Ψ_H can be written

$$h = k_1 \theta k_2 \theta k_3 \theta \cdots k_{m-1} \theta k_m,$$

where m is some positive integer and the k_i are elements of Ψ_K. In some cases, k_i may be the unit element of Ψ_K. Hence

$$h = k_1 k_2' k_3 k_4' \cdots k_{m-1}' k_m \quad \text{or} \quad h = k_1 k_2' k_3 k_4' \cdots k_{m-1} k_m' \theta,$$

according as m is odd or even. □

Theorem X.11. *Let X and Y be crosses of M. Then there is an element ψ of Ψ_K that transforms X into either Y or θY.*

Proof. By the definition of a map, there is an element h of Ψ_H such that $X = hY$. By Theorem X.10, there exists $k \in \Psi_K$ such that $X = k\theta^i Y$, where $i = 0$ or 1. We can take k^{-1} as the required element ψ. □

Let us refer to the equivalence classes of Ψ_K as the *orientation classes* of M. It follows from Theorem X.11 that the number of such classes is either 1 or 2. Moreover, if there are two of them and the cross X is in one, then the crosses θX and $\phi X = (\theta\phi)\theta X$ must be in the other. We say M is *orientable* or *unorientable*, according as the number of its orientation classes is 2 or 1. Thus:

Theorem X.12. *A map $M = M(\theta, \phi, P)$ is unorientable if and only if some cross X of M is transformed into θX (or into ϕX) by some operation of Ψ_K.*

We consider the orientable case, writing the two orientation classes as U and W. The permutation θ reduces to a 1–1 correspondence between U and W, as does ϕ. Any orbit of P is in either U or W. Moreover, one member of each pair of conjugate orbits is in U and the other is in W. We note, too, that one member of each half-edge of M is in U, and the other in W.

Consider a particular orientation class U. The permutations $\theta\phi$ and P have restrictions $(\theta\phi)_U = \theta_U \phi_U$ and P_U, respectively, to U, and these two restrictions generate a permutation group $\Psi(U)$ acting on U. The members of Ψ_K and $\psi(U)$ effect the same transformations in U. Hence, $\psi(U)$ is transitive on U, that is, has only one equivalence class.

It may be convenient to have an independent definition of an "oriented map." An *oriented map* $Q = Q(\xi, P)$ is defined by a set T of an even number of elements, and two permutations ξ and P on T. The permutation ξ is restricted like θ and ϕ in Axioms X.1 and X.2, that is, $\xi^2 = I$ and ξX is distinct from X for each $X \in T$. The permutation group $\Psi(T)$ generated by ξ and P is required to be transitive on T, but otherwise P is unrestricted.

An orbit $(X, \xi X)$ of ξ is an *edge*, and an orbit of P a *vertex* of Q. An edge and vertex of Q are *incident* if some member of T belongs to both. An edge $(X, \xi X)$ is a *loop* of Q if X and ξX belong to the same vertex, and a *link* of Q otherwise. With this definition of incidence, the edges, vertices, loops, and links of Q are the edges, vertices, loops, and links, respectively, of a graph $G(Q)$, called the *graph* of Q.

Consider an orientable map $M = M(\theta, \phi, P)$, with its two orientation classes U and W. We observe that M determines an oriented map $Q_U = Q((\theta\phi)_U, P_U)$ on U, and an oriented map $Q_W = Q((\theta\phi)_W, P_W)$ on W. These two oriented maps are the two orientations or oriented forms of M. The graph of each is isomorphic to $G(M)$. Conversely, an orientable map M can be derived from any given oriented map by using the following theorem:

Theorem X.13. *Let a set S of $4n > 0$ elements be partitioned into two sets U and W of $2n$ elements each. Let Q be any oriented map on U. Then we can construct an orientable map M on S having Q as one of its oriented forms.*

Proof. Write $Q = Q(\xi', P')$. We set up a 1–1 mapping θ of U into W. We interpret θ as a permutation of S by taking it as its own inverse. We extend ξ' and P' as permutations ξ and P, respectively, of S by assigning to them in W orbits conjugate to those in U. Thus $\xi\theta = \theta\xi$ and $\xi^2 = \theta^2 = I$ in S. We now write $\phi = \xi\theta$, observing that $\phi^2 = I$ and $\theta\phi = \phi\theta$ in S. The permutations θ, ϕ, and P of S now satisfy the four axioms of a premap $L = L(\theta, \phi, P)$ on S. Moreover, L is a map $M = M(\theta, \phi, P)$. This is because U must be contained in some connection-set T of L, by the definition of an oriented map, and T must also contain W by the operation of θ. But then M

is orientable by Theorem X.12, with orientation classes U and W, and Q is its oriented form on U. □

A premap L is said to be *orientable* if each of its components is an orientable map, and *unorientable* if any one of its components is unorientable.

X.3. DUALITY

Consider a premap $L = L(\theta, \phi, P)$ on a set S of crosses. There are constructions for deriving other premaps from it. For example, we can deduce from Axioms X.3 and X.4 that $P^2\theta = \theta P^{-2}$ and that the orbits of P^2 through X and θX are distinct, for each $X \in S$. We can therefore assert the existence of a premap $L(\theta, \phi, P^2)$ on S. Another premap on S is $L(\theta, \phi, P^{-1})$, derived from L by reversing the cyclic order for each orbit of P.

Let us study the permutation

$$P^* = P\theta\phi. \quad (X.3.1)$$

Theorem X.14. $P^*\phi = \phi(P^*)^{-1}$.

To prove Theorem X.14, we observe that

$$P^*\phi = P\theta = \theta P^{-1} = \phi(\phi\theta P^{-1}) = \phi(P^*)^{-1}, \quad \text{by Axiom X.3}$$

Theorem X.15. *For each cross X the orbits of P^* through X and ϕX are distinct.*

Proof. If the theorem is false, there is an X such that

$$(P^*)^m \phi X = X \quad (X.3.2)$$

for some nonnegative integer m. Choose such an X and m so that m has the least possible value. Then $m > 0$ by Axiom X.2. If $m = 1$, we have $(P\theta\phi)\phi X = X$, that is, $P(\theta X) = X$, and this is contrary to Axiom X.4. So $m \geq 2$. But for $m > 2$ we have

$$(P^*)^{m-2} P\theta X = (P^*)^{-1} X = \theta\phi P^{-1} X = \phi P\theta X,$$

by X.3.2 and Axioms X.1 and X.3. Equivalently, we write

$$(P^*)^{m-2} \phi(\phi P\theta X) = \phi P\theta X.$$

This contradicts the choice of X and m; we can reduce m by 2 if we replace X by the cross $\phi P\theta X$. The theorem follows. □

By Theorems X.14 and X.15, together with Axioms X.1 and X.2, we see that the permutations ϕ, θ, and P^* satisfy the four axioms for a premap,

$$L^* = L(\phi, \theta, P^*)$$

on S. Let it be emphasized that, in the passage from L to L^*, first and second involutions are interchanged. We call L^* the *dual* premap of L. The definition of the dual implies the following:

Theorem X.16. $(L^*)^* = L$, *for any premap L.*

The oriented vertices and vertices of L^* are called the *oriented faces* and *faces* of L, respectively. Then, in accordance with Theorem X.16, the oriented faces and faces of L^* are the oriented vertices and vertices of L, respectively.

The edges of L remain as the edges of L^*, but they are partitioned differently into half-edges in the two premaps. Thus, an edge $A = (X, \theta X, \phi X, \theta\phi X)$ has the half-edges $\{X, \theta X\}$ and $\{\phi X, \theta\phi X\}$ in L and the half-edges $\{X, \phi X\}$ and $\{\theta X, \theta\phi X\}$ in L^*.

We write $\alpha_0(L)$, $\alpha_1(L)$, and $\alpha_2(L)$ for the numbers of vertices, edges, and faces of L, respectively. By the preceding definitions

$$\alpha_0(L^*) = \alpha_2(L), \qquad \alpha_1(L^*) = \alpha_1(L), \qquad \alpha_2(L^*) = \alpha_0(L). \tag{X.3.3}$$

The combination

$$N(L) = \alpha_0(L) - \alpha_1(L) + \alpha_2(L) \tag{X.3.4}$$

is called the *Euler characteristic* of the premap L.

Theorem X.17. *Dual premaps have equal Euler characteristics, by Eqs.* (X.3.3) *and* (X.3.4).

It is clear that the permutations ϕ, θ, and $P^* = P\theta\phi$ generate the same permutation group as do θ, ϕ, and P. Hence, L^* has the same connection-sets as L. Within any one connection-set T, we observe that

$$(P^*)_T = (P\theta\phi)_T = P_T \theta_T \phi_T,$$

by the definition of restrictions to T. We may therefore assert the following:

Theorem X.18. *The components of L^* are the duals of the components of L.*

Theorem X.19. *The dual of a map is a map* (by Theorem X.18).

By Theorem X.18, and Theorem X.6 applied to both L and L^*, we have:

Theorem X.20. *The vertices, edges, and faces of a premap L are the vertices, edges, and faces, respectively, of the components of L, each appearing in exactly one component.*

By Theorem X.20, we can write

$$\alpha_i(L) = \sum_{j=1}^{k} \alpha_i(M_j) \qquad (i = 1, 2, 0), \tag{X.3.5}$$

where the maps M_j are the components of L. Hence, by Eq. (X.3.4), we have the following theorem.

Theorem X.21. *The Euler characteristic of a premap is the sum of the Euler characteristics of its components.*

Now the permutations $\phi\theta$ and P^* generate the same permutation group as do $\theta\phi$ and P. We thus have

Theorem X.22. *If M is an orientable map, then M^* is orientable. Moreover, M and M^* have the same orientation classes.*

We can define the *dual* of an oriented map $Q = Q(\xi, P)$ as the oriented map $Q^* = Q(\xi, P\xi)$. Then the oriented forms of M^* in Theorem X.22 become the duals of the oriented forms of M.

X.4. ISOMORPHISM

Consider two premaps $L_1 = L(\theta_1, \phi_1, P_1)$ and $L_2 = L(\theta_2, \phi_2, P_2)$ on sets S_1 and S_2, respectively. An *isomorphism* of L_1 onto L_2 is a 1–1 mapping σ of S_1 onto S_2 with the following property: If $\sigma X_1 = X_2$, where X_1 and X_2 are crosses of S_1 and S_2, respectively, then also

$$\sigma(\theta_1 X_1) = \theta_2 X_2, \qquad \sigma(\phi_1 X_1) = \phi_2 X_2, \qquad \text{and} \qquad \sigma(P_1 X_1) = P_2 X_2.$$

If such an isomorphism exists, we say that L_1 and L_2 are *isomorphic*. We can express the relation between L_1 and L_2 given by an isomorphism σ of L_1 onto L_2 by the equation

$$L_2 = L\left(\sigma\theta\sigma^{-1}, \sigma\phi\sigma^{-1}, \sigma P\sigma^{-1}\right). \tag{X.4.1}$$

The identical permutation of S_1 is an isomorphism of L_1 onto itself. If σ is an isomorphism of L_1 onto L_2, then σ^{-1} is an isomorphism of L_2 onto L_1. Moreover, if μ is an isomorphism of L_2 onto a premap L_3, then the mapping-product $\mu\sigma$ is an isomorphism of L_1 onto L_3. Isomorphism of premaps is an equivalence relation. It partitions the class of all premaps into disjoint nonnull *isomorphism classes* so that two premaps belong to the same class if and only if they are isomorphic.

An isomorphism σ of L_1 onto L_2 transforms the orbits of θ_1, ϕ_1, and P_1 into those of θ_2, ϕ_2, and P_2, respectively. It therefore maps the oriented vertices, vertices, and edges of L_1 into those of L_2. It transforms the group of permutations generated by θ_1, ϕ_1, and P_1 into that generated by θ_2, ϕ_2, and P_2. It therefore preserves connection-sets, establishes a 1–1 correspondence between the components of L_1 and those of L_2, and induces an isomorphism of each component of L_1 onto the corresponding component of L_2. In particular, if L_1 is a map, then L_2 is a map. Since the isomorphism transforms $P_1^* = P_1\theta_1\phi_1$ into $P_2^* = P_2\theta_2\phi_2$, it determines an isomorphism of L_1^* onto L_2^*.

Let L_1 and L_2 be maps. Then σ transforms the group of permutations generated by $\theta_1\phi_1$ and P_1 into that generated by $\theta_2\phi_2$ and P_2. Hence, L_2 is orientable if L_1 is, and σ maps the orientation classes of L_1 onto those of L_2.

Briefly, two isomorphic premaps are copies of one another. An isomorphism is the operation of making a copy.

In order to have some simple examples, we now make a complete list of nonisomorphic maps of one edge only. Without loss of generality, we can write our set S as

$$\{X, \theta X, \phi X, \theta\phi X\},$$

the involutions θ and ϕ being supposed given. S is our single edge; all premaps on S are maps. Since orbits of P come in conjugate pairs, P is completely determined by its effect on X. Write $M = M(\theta, \phi, P)$ for a map on S.

Suppose, first that $PX = X$. Then the oriented vertices of M are the four orbits (X), (θX), (ϕX), and $(\theta\phi X)$. The vertices of M are the two pairs

$$\{(X),(\theta X)\} \quad \text{and} \quad \{(\phi X),(\theta\phi X)\}.$$

Each of these absorbs one of the half-edges, so that $G(M)$ is a link-graph. We call M the *link-map*.

Suppose next that $PX = \theta\phi X$. Then the oriented vertices of M are $(X, \theta\phi X)$ and $(\theta X, \phi X)$. Together these constitute a single vertex, so that $G(M)$ is a loop-graph. We now call M the *loop-map*.

Evidently the link-map and loop-map described above are duals. So, for the link-map, we have

$$\alpha_0(M) = 2, \quad \alpha_1(M) = 1, \quad \alpha_2(M) = 1, \quad N(M) = 2.$$

For a loop-map M, in accordance with Eqs. (X.3.3) and (X.3.4),

$$\alpha_0(M) = 1, \quad \alpha_1(M) = 1, \quad \alpha_2(M) = 2, \quad N(M) = 2.$$

The case $PX = \theta X$ is excluded by Axiom X.4, so there remains only the case $PX = \phi X$. The oriented vertices of M are now $(X, \phi X)$ and $(\theta X, \theta\phi X)$. Again, $G(M)$ is a loop-graph. For the orbits of $P\theta\phi$, that is, the oriented vertices of M^*, we find $(X, \theta X)$ and $(\phi X, \theta\phi X)$. There is thus an isomorphism of M onto M^*. This isomorphism leaves the crosses X and $\theta\phi X$ unaltered, but it interchanges θX with ϕX. A map or premap isomorphic with its own dual is said to be *self-dual*. The particular self-dual map under consideration is the *canonical projective map*. For it we have

$$\alpha_0(M) = 1, \quad \alpha_1(M) = 1 \quad \alpha_2(M) = 1, \quad N(M) = 1.$$

Our list of maps of one edge is now complete.

We note that the link-map and loop-map are orientable, with orientation classes $\{X, \theta\phi X\}$ and $\{\theta X, \phi X\}$ in each case. But the canonical projective map is unorientable, for the group generated by $\theta\phi$ and $P = \phi$ is transitive on S.

An isomorphism of a premap L onto itself is an *automorphism* of L. The automorphisms of L are the elements of a permutation group $A(L)$ called the *automorphism group* of L. Its unit element is the identical permutation on the set of crosses. We conclude this section with two theorems about the automorphisms of a map M.

Theorem X.23. *Let $M = M(\theta, \phi, P)$ be a map on a set S. Then if an automorphism σ of M leaves any cross invariant, it is the identical automorphism.*

Proof. If $\sigma X = X$ for some cross X of S, then also

$$\sigma(\theta X) = \theta X, \quad \sigma(\phi X) = \phi X, \quad \text{and} \quad \sigma(PX) = PX,$$

by the definition of an isomorphism. Hence $\sigma(\psi X) = \psi X$ for any element ψ of the group of permutations generated by θ, ϕ, P. But this group is transitive on S, since M is a map. □

Theorem X.24. *Let $M = M(\theta, \phi, P)$ be a map on a set S. Then the order of its automorphism group is a divisor of $|S|$* [5].

Proof. $A(M)$ partitions the set S into disjoint nonnull equivalence classes so that two crosses belong to the same equivalence class if and only if each can be transformed into the other by the operation of some element of $A(M)$.

Let C be one of these equivalence classes, and X one of its crosses. Then the elements of C are the crosses ψX, $\psi \in A(M)$. But if ψ_1 and ψ_2 are elements of $A(M)$ such that $\psi_1 X = \psi_2 X$, then we have $\psi_2^{-1}\psi_1 X = X$, whence $\psi_1 = \psi_2$, by Theorem X.23. We deduce that each equivalence class has exactly $|A(M)|$ elements. Hence, $|S| = q|A(M)|$, where q is the number of equivalence classes. □

In enumerative map theory, use is made of the concept of a *rooted map*. A rooted map is a map M in which some cross X is distinguished as the *root*. The dual rooted map is then M^* with the same cross as root. Theorem X.23 is basic in enumerative theory. It asserts that a rooted map has no nontrivial automorphism.

X.5. DRAWINGS OF MAPS

We consider how to represent a map M in a diagram. We can start with a drawing of the connected graph $G(M)$. It may happen that we can draw

$G(M)$ in the plane without crossings of edges. It is then found that the drawing partitions the rest of the plane into disjoint connected regions. These are called the *faces* of the diagram. Sometimes these faces can be made to represent the faces of M, according to the scheme described below.

Consider an edge A in the drawing of $G(M)$, with incident vertices x and y. An associated cross X of M is represented by an arrow drawn along the edge, not actually in the edge but displaced a little to one side. We make the rules that θX has an arrow drawn along A in the same direction as for X but on the opposite side, and that ϕX is to be drawn on the same side as X but in the opposite direction. So $\theta\phi X$ has to be drawn on the opposite side to X and in the opposite direction. The four crosses associated with A are thus shown in Fig. X.5.1, which shows only a portion of a drawing of some map. (See Note 2.)

It is well to regard A as cut at some internal point into two half-edges, one with end x and one with end y. These are to represent the half-edges of A in M. The crosses X and θX, constituting one half-edge of A in M, are to be drawn against the corresponding half-edge in the diagram, directed away from its end-vertex. Figure X.5.1 shows the case in which A is a link. Figure X.5.3 shows an edge A that is a loop. It is not necessary to show an actual point of subdivision of A into half-edges, provided that each arrow is drawn sufficiently near the vertex from which it is directed. Even this precaution is necessary only in the case of a loop. This assignment of crosses to the edges of the diagram specifies the two involutions θ and ϕ and makes it clear that they satisfy Axioms X.1 and X.2.

Consider all the half-edges radiating from a vertex x. The diagram imposes a definite cyclic order on them. We count the diagram as representing M if the permutation P rotates each cross X one place in the cyclic order at the end of its half-edge, to the side of that half-edge opposite X. Figure X.5.1 shows the operation of this rule at the two ends of A. We note that it requires $\theta P^{-1} X$ to be identical with $P\theta X$ for each X, that is, it necessitates

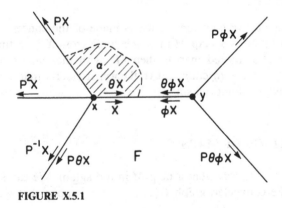

FIGURE X.5.1

Axiom X.3. The two oriented vertices of M associated with the vertex x of the diagram are rotations around it in two opposite directions, each corresponding to the diagrammatic cyclic order of half-edges at x. The members of one oriented vertex of M occupy the right sides of the half-edges from x, and those of the other the left sides. This observation implies Axiom X.4. We have now imposed enough conditions on our diagram to make it represent some map M and to be a drawing, in the sense of Chapter I, of the connected graph $G(M)$.

Let us now interpret the faces of the diagram. Each has two bounding sequences, which we define informally below. (We consider that no formal definitions or proofs are needed in the present section.) One stands just inside a face F and walks round the boundary, keeping one hand on that boundary. One thus traces out a cyclic sequence of boundary edges, each on a definite side and in a definite direction. In other words, one traces out a cyclic sequence of crosses. This is a *bounding sequence* of F. Inspection of a diagram such as Fig. X.5.1 or Fig. X.5.3 shows that in any such bounding sequence a cross X must always be succeeded by $P\theta\phi X$. The bounding sequences of the faces of the diagram, therefore, are simply the orbits of $P\theta\phi$, the oriented faces of M. So the faces of the diagram represent the faces of M. Each face of the diagram has two oppositely directed bounding sequences, and these constitute the corresponding face of M.

In works on map theory, special attention is given to the "proper maps." Such a map is represented by a diagram in which each face is bounded by a simple closed curve, corresponding to a circuit in the graph. Figure X.5.2 shows an example.

We can give the oriented vertices of the map of Fig. X.5.2, showing one member of each conjugate pair, as

$$(X, Y, Z), \quad (U, \phi Y, \theta\phi W), \quad (V, \phi X, \theta\phi U), \quad (W, \phi Z, \theta\phi V).$$

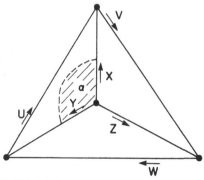

FIGURE X.5.2

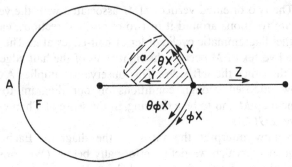

FIGURE X.5.3

Likewise we can give the oriented faces as

$$(U,V,W), \quad (X,\theta V,\theta\phi Z), \quad (Y,\theta U,\theta\phi X), \quad (Z,\theta W,\theta\phi Y).$$

In Fig. X.5.3 we have a diagram of an improper map. We can give the oriented vertices as

$$(\phi Y), \quad (X,Y,\theta\phi X,\theta Z), \quad \text{and} \quad (\phi Z),$$

and the oriented faces as

$$(Y,\theta\phi Y,\theta\phi X) \quad \text{and} \quad (X,\theta Z,\phi Z).$$

Figure X.5.4 gives diagrams of the loop-map and link-map. Their orbits are as found in Sec. X.4.

So far our diagrams have been in the Euclidean plane. We can describe them as on a sphere, spherical diagrams being transformed into planar ones by the device of stereographic projection. Not all maps can be drawn on the sphere, but it is known that each can be drawn on some closed surface. Some can be drawn in the real projective plane.

One such is the canonical projective map of Sec. X.4. A diagram of it appears below in Fig. X.5.5.

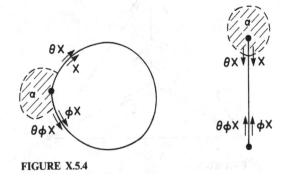

FIGURE X.5.4

Drawings of Maps   267

FIGURE X.5.5

In terms of the neighborhood of the vertex x, we have represented X and ϕX on opposite sides of the line A, and this may seem paradoxical. However, A is meant to be a complete projective line, broken into two half-edges by its point at infinity. We cannot distinguish two sides of a projective line. However, if we exclude a small neighborhood of x, we can distinguish two sides of the rest of the line, in its own neighborhood. Then X and ϕX, as shown in the diagram, are on the same side of the residual line. So we consider that the four crosses are properly drawn as shown. Our rule for P, which concerns only the neighborhood of x, gives us $(X, \phi X)$ and $(\theta X, \theta \phi X)$ as the two conjugate oriented vertices. The diagram has only one face, the connected complement of A in the projective plane. To construct a bounding sequence, we can start with X, proceed to the point at infinity on A, and continue along the other half-edge back to x. This brings us back below the line, as shown in the figure, in opposition to ϕX. We continue to trace out the boundary by proceeding along θX. We take this to mean that θX succeeds X in the bounding sequence. Likewise, X succeeds θX. We conclude that the two conjugate bounding sequences are $(X, \theta X)$ and $(\phi X, \theta \phi X)$. These are the oriented faces of M as found in Sec. X.4.

As another example of a surface, we take the torus. This is usually represented by a rectangle $ABCD$, with the rule that the directed side AB is to be identified with DC, and AD with BC. Figure X.5.6 shows a map on

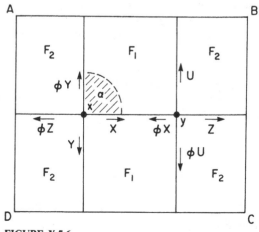

FIGURE X.5.6

this torus. It has, in the diagram, two vertices x and y and two faces F_1 and F_2. There are four edges.

Here the rule for P gives the oriented vertices, one member of each conjugate pair, as $(X, \theta\phi Y, \theta\phi Z, Y)$ and $(U, \theta\phi X, \theta\phi U, Z)$. We trace out the bounding sequence $(X, \theta\phi U, \theta\phi X, \theta\phi Y)$ for F_1, and $(Y, \theta\phi Z, U, Z)$ for F_2. It can be verified that these bounding sequences are orbits of $P\theta\phi$.

Topologically, a surface can be constructed to fit any given map. We start with the graph of the map and adjoin topological discs to it so that they have the orbits of $P\theta\phi$ as bounding sequences. But our object in this chapter is to emphasize that there is a purely combinatorial map theory, independent of topology. We therefore avoid topological concepts and constructions, unless our diagrams are to be counted as such.

X.6. ANGLES

Let $M = M(\theta, \phi, P)$ be a map on a set S. For any $X \in S$ we describe the ordered pair (X, PX) as a *turn* of M at the oriented vertex through X. There is a *conjugate* turn $(\theta PX, \theta X)$, that is, $(P^{-1}\theta X, \theta X)$, at the conjugate oriented vertex. The unordered pair

$$\{(X, PX), (P^{-1}\theta X, \theta X)\}$$

of conjugate turns is an *angle* at the corresponding vertex of M. This angle, denoted by α, is represented by a small shaded region in each diagram of Sec. X.5. We can think of this as the region across which P rotates X and $P^{-1}\theta X$.

The *dual turn* of (X, PX) is the turn $(\phi PX, \theta X)$, that is $(P^{*-1}\theta X, \theta X)$, of M^*. Its conjugate turn in M^*, allowing for the interchange of θ and ϕ, is $(\theta\phi X, PX)$. These two turns constitute an angle α^* at the corresponding vertex of M^*. We call α^* the *dual angle* of α. It can be represented by the same shaded region as α in a diagram, being called a "corner" of the face containing that region.

A vertex and a face of M are said to be incident if some angle at the vertex has its dual angle at the face. If they are related in this way by m pairs of dual angles, they are said to be *m-tuply incident*. Thus, in Fig. X.5.3, x and F are doubly incident.

X.7. OPERATIONS ON MAPS

Let $M = M(\theta, \phi, P)$ be a map on a set S. As in Sec. X.3, we consider the formation of other maps from M. But now we allow the introduction of a new cell or the suppression of an old one.

Let a set S_1 be derived from S by adjoining a new cell $A = (Z, \theta Z, \phi Z, \theta \phi Z)$. We extend the involutions θ and ϕ from S to S_1, as indicated by the notation. Thus extended, they still satisfy Axioms X.1 and X.2. Let us now consider possible ways of converting P into a permutation P_1 on S_1.

Choose $X \in S$. We write the conjugate orbits of P through X and θX as follows:
$$u = (X, PX, P^2X, \ldots, P^{m-1}X),$$
$$u' = (\theta X, \theta P^{m-1}X, \theta P^{m-2}X, \ldots, \theta PX),$$

m being some positive integer. We now choose an integer j such that $0 \le j < m$, and we partition u into the following two linear sequences:
$$\ell_1 = (X, PX, \ldots, P^{j-1}X),$$
$$\ell_2 = (P^jX, P^{j+1}X, \ldots, P^{m-1}X).$$

We also make a corresponding partition of u' into linear sequences:
$$\ell_1' = (\theta P^{j-1}X, \ldots, \theta PX, \theta X),$$
$$\ell_2' = (\theta P^{m-1}X, \ldots, \theta P^{j+1}X, \theta P^jX).$$

If $j = 0$, we interpret ℓ_1 as a null sequence. Then ℓ_2 is written just like u, but is regarded as linear, not cyclic.

We now adjoin one of the new crosses to each of our four linear sequences, thereby converting them into the following cyclic sequences:
$$w_1 = (Z, X, PX, \ldots, P^{j-1}X),$$
$$w_2 = (\theta \phi Z, P^jX, P^{j+1}X, \ldots, P^{m-1}X),$$
$$w_1' = (\theta Z, \theta P^{j-1}X, \ldots, \theta PX, \theta X),$$
$$w_2' = (\phi Z, \theta P^{m-1}X, \ldots, \theta P^{j+1}X, \theta P^jX).$$

If $j = 0$, then w_1 and w_1' are to be interpreted as the cyclic sequences (Z) and (θZ), of length 1, respectively. Then X is to be found in w_2, not w_1. Correspondingly, θX is in w_2', not w_1'.

We take the orbits of P other than u and u', together with the four new cyclic sequences w_1, w_1', w_2, and w_2'. We then have the orbits of a permutation P_1 on S_1 that satisfies Axioms X.3 and X.4. One pair of conjugate orbits is $\{w_1, w_1'\}$, and another is $\{w_2, w_2'\}$. We have a premap $L_1 = L(\theta, \phi, P_1)$ on S_1.

We observe that the permutations P and P_1 have the same effect on any cross of S, except for

$$P^{j-1}X, \quad P^{m-1}X, \quad \theta X, \quad \text{and} \quad \theta P^jX \quad \text{if } j > 0$$

and for

$$P^{m-1}X \quad \text{and} \quad \theta X \quad \text{if } j = 0.$$

These exceptional crosses are transformed by P_1 into members of A. Of course, the effect of θ or ϕ on any cross of S is the same in L_1 as in M.

The premap L_1 is a map. For suppose not. Then it has a connection-set T that contains no member of A. Then T does not contain $P^{m-1}X$ or θX, and if $j > 0$ it does not contain $P^{j-1}X$ or $\theta P^j X$. Hence the operations θ, ϕ, and P_1 of L_1 have the same effects on each cross of T as do θ, ϕ, and P, respectively. This implies that T is a connection-set of the map M, that is, $T = S$. But this contradicts the definition of T, for the new orbit w_2 includes both a member of S and a member of A. Accordingly, we replace the symbol L_1 by M_1, and write $M_1 = M(\theta, \phi, P_1)$.

We relate the operation to the two angles

$$\alpha = (P^{m-1}X, X), (\theta X, \theta P^{m-1}X)$$

and

$$\beta = (P^{j-1}X, P^j X), (\theta P^j X, \theta P^{j-1}X)$$

at the vertex z of M defined by the orbits u and u'. (If $j = 0$, then $\alpha = \beta$.) We say that M_1 is formed from M by *splitting* the vertex z between these two angles. Figure X.7.1 illustrates the operation in the case $j = 0$, and Fig. X.7.2 in the case $j > 0$.

In Fig. X.7.1 we note the replacement of z by two new vertices z_1 and z_2. The latter is monovalent in $G(M_1)$. It corresponds to the new orbits (Z) and (θZ).

It is clear from the construction that

$$\alpha_0(M_1) = \alpha_0(M) + 1, \qquad \alpha_1(M_1) = \alpha_1(M) + 1. \qquad (X.7.1)$$

To find the change in $\alpha_2(M)$, we must study the dual maps

$$M^* = M(\phi, \theta, P\theta\phi) \quad \text{and} \quad M_1^* = M(\phi, \theta, P_1\theta\phi).$$

From the definition of P_1, we see that $P_1\theta\phi$ has the same effect on each cross of S as does $P\theta\phi$, except for ϕX and $\theta\phi P^{m-1}X$, and for $\theta\phi P^{j-1}X$

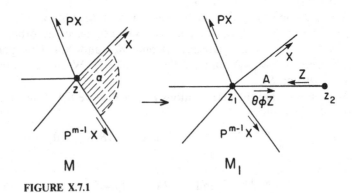

FIGURE X.7.1

Operations on Maps

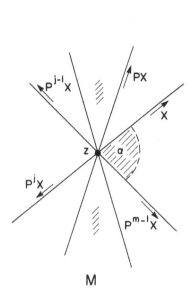
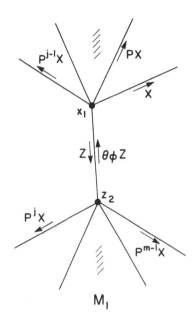

FIGURE X.7.2

and $\phi P^j X$ if $j > 0$. But if $j = 0$, we have

$$P_1^*\phi X = P_1\theta X = \phi Z,$$
$$P_1^*\phi Z = P_1\theta Z = \theta Z,$$
$$P_1^*\theta Z = P_1\phi Z = \theta P^{m-1}X = P\theta X = P^*\phi X.$$

By conjugacy in M_1^* we have also

$$P_1^*\theta\phi P^{m-1}X = \theta\phi Z, \quad P_1^*\theta\phi Z = Z, \quad \text{and} \quad P_1^* Z = X = P^*\theta\phi P^{m-1}X.$$

We conclude that the effect on the orbits of P^*, in the change to P_1^*, is to lengthen the members of one pair of conjugate orbits. This occurs by the insertion of ϕZ and θZ, in that order, in the turn $(\phi X, P^*\phi X)$, and correspondingly $\theta\phi Z$ and Z into the conjugate turn $(\theta\phi P^{m-1}X, X)$. Thus the number of orbits is the same for P_1^* as for P^*.

Now take the case $j > 0$. We have

$$P_1^*\phi X = P_1\theta X = \theta Z, \quad P_1^*\theta Z = P_1\phi Z = \theta P^{m-1}X = P^*\phi X,$$
$$P_1^*\theta\phi P^{m-1}X = P_1 P^{m-1}X = \theta\phi Z, \quad P_1^*\theta\phi Z = P_1 Z = X = P^*\theta\phi P^{m-1}X,$$
$$P_1^*\theta\phi P^{j-1}X = P_1 P^{j-1}X = Z, \quad P_1^* Z = P_1\theta\phi Z = P^j X = P^*\theta\phi P^{j-1}X,$$
$$P_1^*\phi P^j X = P_1\theta P^j X = \phi Z, \quad P_1^*\phi Z = P_1\theta Z = \theta P^{j-1}X = P^*\phi P^j X.$$

Thus the change from P^* to P_1^* is effected by inserting θZ into the turn $(\phi X, P^*\phi X)$, Z into the turn $(\theta\phi P^{j-1}X, P^*\theta\phi P^{j-1}X)$, and making the two conjugate changes. So whether j is zero or nonzero, we can supplement Eq. (X.7.1) by writing

$$\alpha_2(M_1) = \alpha_2(M). \qquad (X.7.2)$$

Theorem X.25. *Let a map M_1 be formed from a map M by splitting a vertex z. Then $N(M_1) = N(M)$. Moreover, M_1 is orientable if and only if M is orientable.*

Proof. That $N(M) = N(M)$ follows from Eqs. (X.7.1) and (X.7.2), together with the definition in Eq. (X.3.4).

Suppose M is orientable, with orientation classes U and W. We may suppose $X \in U$ and $\theta X \in W$. Form U' by adjoining Z and $\theta\phi Z$ to U, and form W' by adjoining θZ and ϕZ to W. The effect of P_1, and of $\theta\phi$ on U' is simply to permute its elements among themselves. Hence M_1 is orientable, with orientation classes U' and W'.

Conversely, suppose M_1 orientable, with orientation classes U' and W'. We may suppose U' to include X, and therefore Z, $\theta\phi Z$, and $P^i X$ for each i, by our formulae for the new orbits of P_1. Form U from U' by deleting Z and $\theta\phi Z$; form W from W' by deleting θZ and ϕZ. Then the effect on U of the permutations P and $\theta\phi$ of M is simply to permute its elements among themselves. Hence, M is orientable, with orientation classes U and W. □

An operation that transforms orientable maps into orientable ones, and unorientable maps into unorientable ones, is said to preserve *orientability-character*. An operation that preserves both the Euler characteristic and the orientability-character of a map will be described in what follows as *superficial*. Thus Theorem X.25 asserts that vertex-splitting is a superficial operation.

We go on to some operations in which the number of edges is diminished by 1. The first is that of *contracting a link*.

Suppose a map $M = M(\theta, \phi, P)$ on S to have a link A incident with vertices x and y. We write the crosses of A as Z, θZ, ϕZ, and $\theta\phi Z$. These must belong to four distinct orbits of P, in two conjugate pairs corresponding to the vertices x and y. Let us write the orbits through Z and $\theta\phi Z$ as

$$u = (Z, PZ, P^2 Z, \ldots, P^j Z)$$

and

$$v = (\theta\phi Z, P\theta\phi Z, \ldots, P^k \theta\phi Z),$$

where j is a nonnegative integer and k a positive one. We do not define the operation for the case $j = k = 0$, that is, we do not define it for the link-map.

From the orbits u and v we construct the single cyclic sequences

$$w = (PZ, P^2 Z, \ldots, P^j Z, P\theta\phi Z, P^2\theta\phi Z, \ldots, P^k \theta\phi Z).$$

From the conjugate orbits u' and v' of u and v, respectively, we likewise construct the cyclic sequence

$$w' = (\theta P^k \theta \phi Z, \ldots, \theta P \theta \phi Z, \theta P^j Z, \ldots, \theta P^2 Z, \theta P Z).$$

If $j = 0$, we interpret w as

$$(P \theta \phi Z, P^2 \theta \phi Z, \ldots, P^k \theta \phi Z),$$

and w' as

$$(\theta P^k \theta \phi Z, \ldots, \theta P^2 \theta \phi Z, \theta P \theta \phi Z).$$

The restriction on k ensures that w and w' do not reduce to null cyclic sequences.

We take the orbits of P other than u, u', v, and v', together with the new cyclic sequences w and w'. We then have the orbits of a permutation P_0 on $S - A$. Let us denote the restrictions of θ and ϕ to $S - A$ by the same symbols θ and ϕ. Then the permutations θ, ϕ, and P_0 on $S - A$ satisfy Axioms X.1–X.4. We have a premap $L_0 = L(\theta, \phi, P_0)$ on $S - A$, w and w' being conjugate oriented vertices, corresponding to a vertex z.

The premap L_0 is actually a map. One way to prove this is to observe that $G(L_0)$ can be identified with $(G(M))''_A$. Hence, $G(L_0)$ is connected, by Theorem II.5, and so L_0 is a map, by Theorem X.7. So we have a map $M_0 = M(\theta, \phi, P_0)$ on $S - A$.

We say that M_0 is derived from M by *contracting the link* A. We can verify that M is derived from M_0 by splitting the vertex z between the angles corresponding to the turns $(P^j Z, P \theta \phi Z)$ and $(P^k \theta \phi Z, PZ)$ if $j > 0$ and to the turn $(P^k \theta \phi Z, P \theta \phi Z)$ if $j = 0$. So by Theorem X.25 we have the following:

Theorem X.26. *Let a map M_0 be formed from a map M by contracting a link A. Then $N(M_0) = N(M)$. Moreover, M_0 is orientable if and only if M is orientable.*

Contracting a link is a superficial operation.

The next operation is that of *deleting an edge*. As before, we consider a map $M = M(\theta, \phi, P)$ on S, with an edge $A = \{Z, \theta Z, \phi Z, \theta \phi Z\}$. The operation simply deletes from each orbit of M the crosses belonging to A, while retaining the cyclic order of the other crosses. We apply it only when each orbit of M includes a cross not belonging to A; then no orbit reduces to a null cyclic sequence. The operation changes the orbits of P into the orbits of a permutation P_0 on $S - A$. We denote the restrictions of θ and ϕ to $S - A$ by the same symbols θ and ϕ, respectively. We can now observe that the permutations θ, ϕ, and P_0 on $S - A$ satisfy Axioms X.1–X.4. We thus have a premap $L_0 = L(\theta, \phi, P_0)$ on $S - A$. We say that it is derived from M by *deleting* A. We note that

$$\alpha_0(L_0) = \alpha_0(M), \quad \alpha_1(L_0) = \alpha_1(M) - 1. \quad \text{(X.7.3)}$$

Let us now study the effect of the operation on M^*. For $X \in S - A$, we have

$$P_0^* X = P_0 \theta \phi X = P \theta \phi X = P^* X$$

unless $P\theta\phi X$, that is, P^*X, belongs to A. One consequence of this observation is that those orbits of P^* that include no cross of A are also orbits of P_0^*. As a further simple consequence, of some utility in the theory of planar maps, we have the following theorem.

Theorem X.27. *Let a premap L_0 be derived from a map M by deleting an edge A, and let A be a loop of M^*. Then $N(L_0) > N(M)$.*

Proof. There are $\alpha_2(M)-1$ conjugate pairs of orbits of P^* that involve no cross of A, and all these orbits are orbits of P_0^*. But P_0^* must have at least one other orbit. For otherwise A would be a connection-set of M^* and therefore identical with S, by Theorem X.19. But then the operation of edge-deletion would not be applicable to M. We deduce that

$$\alpha_2(L_0) > \alpha_2(M).$$

Hence,

$$N(L_0) > \alpha_2 M) - \alpha_1(M) + 1 + \alpha_0(M), \quad \text{(by Eqs. (X.3.4) and (X.7.3))}$$
$$> N(M) \quad \text{(by Eq. (X.3.4)).} \qquad \square$$

The case in which A is a link of M^* is of more immediate interest. Then the four crosses on A belong to four distinct orbits of P^*, in two conjugate pairs. Let the ones through Z and $\theta\phi Z$ be

$$(Z, P^*Z, P^{*2}Z, \ldots, P^{*j}Z),$$
$$(\theta\phi Z, P^*\theta\phi Z, \ldots, P^{*k}\theta\phi Z).$$

It is not necessary to write down their conjugates in M^*. We can specify j simply as a nonnegative integer, but then we must require k to be positive. For if $j = k = 0$, then M has only one edge by the connection of M^*, and the operation of edge-deletion is not applicable. The apparent asymmetry between j and k involves no loss of generality. If $j = 0$, the first orbit is simply (Z).

Suppose $j = 0$. By the preceding observations we have $P_0^* X = P^* X$ for each cross X of $S - A$ in the above two orbits, save only for $P^{*k}\theta\phi Z$. But

$$P(\theta\phi P^{*k}\theta\phi Z) = \theta\phi Z, \quad P(\theta\phi Z) = P^*Z = Z, \quad \text{and} \quad P(Z) = P^*\theta\phi Z.$$

Hence $P_0(\theta\phi P^{*k}\theta\phi Z) = P^*\theta\phi Z$, that is,

$$P_0^*(P^{*k}\theta\phi Z) = P^*\theta\phi Z.$$

The effect of the operation on M^* is thus to suppress the orbit (Z) and its conjugate (ϕZ), and to delete $\theta\phi Z$ and θZ from the orbits through these crosses. It is equivalent to contracting the link A in M^*.

When $j > 0$, the exceptional crosses in the two orbits are $P^{*j}Z$ and $P^{*k}\theta\phi Z$. But now $P(\theta\phi P^{*j}Z) = Z$ and $P(Z) = P^*\theta\phi Z$. Hence $P_0(\theta\phi P^{*j}Z) = P^*\theta\phi Z$. Similarly, $P_0(\theta\phi P^{*k}\theta\phi Z) = P^*Z$. So we have

$$P_0^* P^{*j}Z = P^*\theta\phi Z, \qquad P_0^* P^{*k}\theta\phi Z = P^*Z.$$

The effect of the operation on M^* is thus to replace the above two orbits of P^* by the single orbit

$$\left(P^*Z, \ldots, P^{*j}Z, P^*\theta\phi Z, \ldots, P^{*k}\theta\phi Z\right)$$

and to make the corresponding change of conjugate orbits. Again it is equivalent to contracting the link A in M^*.

Making use of Theorem X.26, we summarize and extend the above results as follows.

Theorem X.28. *Let a premap L_0 be derived from a map M by deleting an edge A that is a link of M^*. Then L_0^* is derived from M^* by contracting the link A in that map. Hence L_0^* is a map. Accordingly, L_0 is a map $M_0 = M(\theta, \phi, P_0)$ on $S - \{A\}$, by Theorem X.19. Moreover, by Theorems X.26, X.17, and X.22, $N(M_0) = N(M)$ and M_0 is orientable if and only if M is orientable.*

X.8. COMBINATORIAL SURFACES

We define a *surface* as the class of all maps with a given Euler characteristic and a given orientability-character, provided that the class is nonnull. "A given orientability-character" means either all orientable or all unorientable. A map belonging to a combinatorial surface is usually said to be *on* that surface.

Topologically, the maps of such a class can be defined as those realizable in a given 2-manifold. It is an important theorem for the classification of 2-manifolds that the transformations noted in Sec. X.7 as "superficial" can be effected by constructions in the 2-manifold. Another important theorem is that if M_1 and M_2 are maps in the same 2-manifold, then M_1 can be transformed into an isomorph of M_2 by a finite sequence of superficial transformations.

In what follows we prove an adaptation of the second theorem to the theory of combinatorial surfaces. The superficial transformations used will be vertex-splitting, contracting a link, and replacing a map by its dual. The last operation is superficial by Theorems X.17 and X.22.

We make use of the following definitions: A map M is *unitary* if $\alpha_0(M) = \alpha_2(M) = 1$. The canonical projective map of Sec. X.5 is an example. The *combinatorial sphere* is the class of orientable maps with Euler characteristic 2, and maps on it are called *planar*. Thus the link-map and loop-map, studied in Sec. X.5, are planar.

Theorem X.29. *If M is a unitary map, then $N(M) < 2$.*

This follows from Eq. (X.3.4) and the fact that each map has at least one edge.

By Theorem X.29 there is no unitary map on the combinatorial sphere. However, a special convention is sometimes made to introduce such a map. It is the "vertex-map," with one vertex, one face, and no edges. In a diagram, one point of the plane is distinguished as the vertex and the rest of the plane represents the face. The "map" is regarded as self-dual. But we shall not recognize this vertex-map in the present section.

We suppose given a surface U.

Theorem X.30. *Let M be a map on U. Then by contracting links in M, we can transform it into either a link-map or a map with just one vertex.*

Proof. We may suppose M to have at least two vertices. By connection, it has a link A. If M is not a link-map, then one vertex incident with A is incident with a second edge, and the operation of link-contraction is applicable to A. We apply it. We repeat this process, diminishing the number of edges at each step, until it terminates. We must then have either a link-map or a map with a single vertex. □

Theorem X.31. *Let M be a map on U. Then by a finite sequence of link-contractions and dual replacements, we can transform M into a link-map, a loop-map, or a unitary map.*

Proof. If $\alpha_0(M) > 1$, then either we have a link-map or we can apply link-contraction, as in Theorem X.30. If $\alpha_2(M) > 1$, then either M^* is a link-map, that is, M is a loop-map, or we can use link-contraction in M^*. We apply link-contractions to M and M^* until the operation is no longer possible. M must then have been transformed into a loop-map, a link-map, or a unitary map. □

Theorem X.32. *Let M be a planar map. Then, by a finite sequence of link-contractions and dual replacements, we can transform M into a link-map.*

Since a link-map is the dual of a loop-map, this theorem follows from Theorems X.29 and X.31.

Theorem X.33. *There is no map whose Euler characteristic exceeds 2, and no unorientable map whose Euler characteristic exceeds 1.*

This is a consequence of Theorems X.29 and X.31, since the link-map and loop-map are both orientable.

Let us now study the case in which U is not the combinatorial sphere. Starting with an arbitrary map on U, we change it by superficial transformations, as in Theorem X.31, into a unitary map M on U.

We write $M = M(\theta, \phi, P)$. We specify the structure of M by giving an orbit u of P. It is not necessary to state the conjugate orbit explicitly.

Let X be a cross of u. Then u must include also either ϕX or $\theta\phi X$, but not both of them. The cross θX is excluded by Axiom X.4. It is convenient to write u as

$$(X, R_1, \phi X, R_2) \quad \text{or} \quad (X, R_1, \theta\phi X, R_2),$$

where the symbols R_1 and R_2 stand for certain linear sequences of crosses, possibly null. Given such a sequence R_i, we write R'_i for the linear sequence obtained from it by reversing the order of the terms and then premultiplying each by θ. So with u written as above, we would write the conjugate orbit u' of P as

$$(\theta X, R'_2, \theta\phi X, R'_1) \quad \text{or} \quad (\theta X, R'_2, \phi X, R'_1).$$

If X and ϕX occur in u, we say that the pair $\{X, \phi X\}$ is a *cross-cap* of u. It is an *assembled* cross-cap of u if X and ϕX are consecutive crosses of u. If X and $\theta\phi X$ occur in u, they constitute a *conforming pair* of u, *assembled* if they are consecutive. If $\{X, \theta\phi X\}$ is one conforming pair of u, there may be a second, $\{Y, \theta\phi Y\}$, such that y can be written as

$$(X, R_1, Y, R_2, \theta\phi X, R_3, \theta\phi Y, R_4).$$

We then say that the conforming pair $\{X, \theta\phi X\}$ (and similarly $\{Y, \theta\phi Y\}$) is *bound* in u, and that the four crosses X, Y, $\theta\phi X$, and $\theta\phi Y$ constitute a *handle* of u. Such a handle is *assembled* if its four crosses occur consecutively.

Theorem X.34. *By a finite sequence of vertex-splittings and link-contractions, we can convert a given unitary map M into another unitary map M_0 for which all the cross-caps are assembled.*

Proof. Write $M = M(\theta, \phi, P)$. Suppose an orbit u of P to have an unassembled cross-cap $\{X, \phi X\}$. Then we can write u as $(X, R_1, \phi X, R_2)$.

By suitably splitting the vertex of M, we convert M into a map M_1, of two vertices, in which the orbits of the basic permutation are

$$(X, R_1, Z), \quad (\theta\phi Z, \phi X, R_2),$$

and their two conjugates. Here Z and $\theta\phi Z$ are crosses of the new edge. We can equally well specify the structure of M_1 by the two orbits

$$(X, R_1, Z) \quad (\theta\phi X, \phi Z, R'_2).$$

We have replaced one orbit by its conjugate. Contraction of the link with crosses X and $\theta\phi X$ gives us a new map M_2, of one vertex, in which these two orbits are replaced by the single orbit

$$(R_1, Z, \phi Z, R'_2).$$

We note first that M_2 is unitary. For
$$\alpha_0(M_2) = \alpha_0(M) = 1, \quad \alpha_1(M_2) = \alpha_1(M), \quad \text{and} \quad N(M_2) = N(M_1),$$
since we have used only superficial transformations. Hence, $\alpha_2(M_2) = \alpha_2(M) = 1$, by Eq. (X.3.4). Next we observe that, in the passage from M to M_2, the unassembled cross-cap $\{X, \phi X\}$ has been replaced by the assembled one $\{Z, \phi Z\}$. Moreover, no assembled cross-cap in M has been lost; such a cross-cap $\{Y, \phi Y\}$ persists either in R_1 or, in the form $\{\theta\phi Y, \theta Y\}$, in R'_2.

By this procedure we can reduce to the case in which every cross-cap of u is assembled. Then clearly every cross-cap in u' is correspondingly assembled. □

Theorem X.35. *Let $M = M(\theta, \phi, P)$ be a unitary map on U. Then no orbit of P has an unbound conforming pair.*

Proof. Suppose the orbit u of P to have an unbound conforming pair $\{X, \theta\phi X\}$. We can write u as $(X, R_1, \theta\phi X, R_2)$, where R_1 may be null. Let R_3 be the linear sequence $(R_1, \theta\phi X)$.

Let Y be any cross of R_3. If $Y = \theta\phi X$, then $P^*Y = PX$, whence $P^*Y \in R_3$. If Y is in R_1, then $P^*Y = P\theta\phi Y$. But $\theta\phi Y$ is in R_1 since $\{X, \theta\phi X\}$ is unbound. So again, $P^*Y \in R_3$.

We deduce that one orbit of P^* consists solely of crosses from R_3. But then P^* must have a second orbit through X. This contradicts the hypothesis that M is unitary. □

Consider our unitary map in which all the cross-caps are assembled. Then Theorem X.35 tells us that either the orbits of the basic permutation consist solely of cross-caps or each of them has a handle. The next theorem deals with the assembly of handles.

Theorem X.36. *By a finite sequence of vertex-splittings and link-contractions, we can convert a given unitary map into one in which the orbits of the basic permutation consist solely of assembled cross-caps and assembled handles.*

Proof. Suppose our unitary map M to have a handle. One of the orbits of its basic permutation P can be written as
$$u = (X, R_1, Y, R_2, \theta\phi X, R_3, \theta\phi Y, R_4).$$
By an alternating sequence of vertex-splittings and link-contractions, we transform u as follows.
$$u \to (R_4, X, R_1, Z), (\theta\phi Z, Y, R_2, \theta\phi X, R_3, \theta\phi Y)$$
$$\to (R_1, Z, R_4, R_3, \theta\phi Y, \theta\phi Z, Y, R_2)$$
$$\to (Y, R_2, R_1, T), (\theta\phi T, Z, R_4, R_3, \theta\phi Y, \theta\phi Z)$$
$$\to (R_2, R_1, T, \theta\phi Z, \theta\phi T, Z, R_4, R_3),$$
Z and T being new crosses.

Combinatorial Surfaces 279

This process transforms M into another unitary map, by the argument in Theorem X.34. The handle represented by X, Y, $\theta\phi X$, and $\theta\phi y$ has been replaced by the assembled handle $(T, \theta\phi Z, \theta\phi T, Z)$. Any assembled cross-cap or other assembled handle of u appears as a subsequence of an R_i, and is preserved.

By repetition of the operation just described, we can replace unassembled handles by assembled ones until we have a unitary map of the kind required. (No crosses will be left over, by Theorem X.35.) □

Theorem X.37. *By a finite sequence of vertex-splittings and link-contractions we can convert a given unitary map into one in which the orbits of the basic permutation consist either entirely of assembled cross-caps or entirely of assembled handles.*

Proof. We may suppose our unitary map $M = M(\theta, \phi, P)$ already in the form specified in Theorem X.36. Suppose it to have at least one assembled cross-cap and at least one assembled handle. Then an orbit u of P can be written as follows:

$$u = (Y, Z, \theta\phi Y, \theta\phi Z, X, \phi X, R_1).$$

This form displays a handle immediately followed by a cross-cap. In the following formulae each line, as in Theorem X.36, describes a map by listing one oriented vertex from each conjugate pair. The transformation from one line to the next may represent a vertex-splitting, a link-contraction, or merely the replacement of an orbit by its conjugate. The symbols X_1 and X_2 stand for new crosses introduced by vertex-splitting:

$$u \to (X_1, \theta\phi Y, \theta\phi Z, X), (\theta\phi X_1, \phi X, R_1, Y, Z)$$
$$\to (X, X_1, \theta\phi Y, \theta\phi Z), (\theta\phi X, \phi X_1, \theta Z, \theta Y, R_1')$$
$$\to (X_1, \theta\phi Y, \theta\phi Z, \phi X_1, \theta Z, \theta Y, R_1')$$
$$\to (X_2, \theta\phi Z, \phi X_1, \theta Z, \theta Y), (\theta\phi X_2, R_1', X_1, \theta\phi Y)$$
$$\to (\theta Y, X_2, \theta\phi Z, \phi X_1, \theta Z), (\phi Y, \theta X_1, R_1, \phi X_2)$$
$$\to (X_2, \theta\phi Z, \phi X_1, \theta Z, \theta X_1, R_1, \phi X_2).$$

At this stage we still have a unitary map M_1, by the argument in Theorem X.34. We have an assembled cross-cap in the form $(\phi X_2, X_2)$ and an unassembled one, $\{\theta\phi Z, \theta Z\}$. The original assembled handles and cross-caps, other than $\{X, \phi X\}$, appear as subsequences of R_1 and are preserved.

Our next step is to replace the cross-cap $\{\theta\phi Z, \theta Z\}$ by an assembled one, as in Theorem X.34. The assembled cross-caps and handles of M_1 are, in the notation of Theorem X.34, represented by subsequences of R_1 and R_2 and are therefore preserved, perhaps in conjugate form. The remaining pair of crosses, $\{\phi X_1, \theta X_1\}$, can only be converted into another cross-cap, by Theorem X.35. Finally, we assemble this cross-cap, too. We thus obtain a unitary map, still of the form specified in Theorem X.36, but with one handle fewer and two more cross-caps. Repetition of the above procedure leads us to a unitary map in which the orbits of the basic permutation consist entirely of assembled cross-caps.

Only if the orbits of P consist entirely of assembled handles is the above procedure inapplicable. The theorem follows. □

Let us define a *canonical* map as either a link-map or a unitary map $M = M(\theta, \phi, P)$ in which the orbits of P consist either entirely of assembled cross-caps or entirely of assembled handles. In the unitary case the number of cross-caps or handles is the *genus* of the canonical map. The genus of the link-map is taken to be zero.

The link-map is orientable, and so is a unitary canonical map of handles, an orientation class being the set of crosses belonging to one orbit of P. But a canonical map of cross-caps is unorientable, by Theorem X.12. From now on, we distinguish the two kinds of unitary canonical map as "orientable" and "unorientable."

We observe that an orientable canonical map can be constructed with an arbitrary nonnegative integer as genus, and that an unorientable one can be constructed with an arbitrary positive integer as genus. There are no other possibilities for the genus. Since a canonical map is completely determined by one orbit of its permutation we have

Theorem X.38. *To within an isomorphism, there is at most one canonical map with a given genus and orientability-character.*

Theorem X.39. *An orientable canonical map of genus k has Euler characteristic $2-2k$. An unorientable canonical map of genus k has Euler characteristic $2-k$.*

Theorem X.39 follows from the fact that the number of edges is $2k$ in the first case and k in the second.

Theorem X.40. *To within an isomorphism, there is exactly one canonical map on each combinatorial surface.*

Proof. First we show that there is a canonical map on any given surface. For the combinatorial sphere, we have the link-map. For any other surface, the required result follows from Theorems X.31 and X.37. We then observe that there is at most one canonical map on the surface, by Theorems X.38 and X.39. □

Theorem X.41. *Let M_1 and M_2 be maps on the same surface. Then M_1 can be transformed into M_2 by a finite sequence of vertex-splittings, link-contractions, and dual replacements.*

Proof. Let C be the canonical map on the surface. Then, to within an isomorphism, each of M_1 and M_2 can be transformed into C by a sequence of the kind specified, by Theorems X.32 and X.37. Since vertex-splitting and link-contraction are inverse operations, as pointed out in Sec. X.7, there is another such sequence transforming C into M_2. The theorem follows. □

The genus of a canonical map is also called the *genus* of its surface, and of any other map on that surface. A surface is called *orientable* or *unorientable* according as the maps on it are orientable or unorientable. We can now say that there is just one orientable surface whose genus is a given nonnegative integer, and just one unorientable surface whose genus is a given positive integer, and moreover there are no other surfaces. The unorientable surface of genus 1 is the combinatorial *projective plane*, and the one of genus 2 is the *Klein bottle*. The canonical map on the former is the "canonical projective map" described in Sec. X.5. The orientable surface of genus 0 is the combinatorial sphere. That of genus 1 is the combinatorial *torus*, on which we have the map of Fig. X.5.6.

X.9. CYCLES AND COBOUNDARIES

Let M and M^* be dual maps on a set S, and on a surface V. Then $G(M)$ and $G(M^*)$, or isomorphs thereof, are said to be *dual graphs* with respect to V.

In this section we discuss the case in which V is orientable. We write $M = M(\theta, \phi, P)$, and denote the orientation classes of M by U and V. Let Q_U and Q_U^* be the oriented forms of M and M^*, respectively, on U. (See Theorem X.22.) Instead of $G(M)$ and $G(M^*)$, we can study their isomorphs $G(Q_U)$ and $G(Q_U^*)$, respectively. In the notation of Sec. X.2, Q_U is the oriented map $Q(\theta\phi, P_U)$ and Q_U^* is $Q(\theta\phi, (P\theta\phi)_U)$. The edges of each of the graphs $G(Q_U)$ and $G(Q_U^*)$ are the orbits $(X, \theta\phi X)$ of $\theta\phi$, the vertices of Q_U are the orbits of P_U, and the vertices of Q_U^* are the orbits of $(P^*)_U = (P\theta\phi)_U = P_U\theta\phi = (P_U)^*$. We have used the symbols θ and ϕ for the restrictions of these involutions to U. We say that Q_U and Q_U^* are *dual oriented maps*.

Consider an orientation Ω of the graph $G(Q_U)$. We can identify the darts of $G(Q_U)$ on an edge A with the two ordered pairs $(X, \theta\phi X)$ and $(\theta\phi X, X)$ of crosses of A. The head of the dart $(X, \theta\phi X)$ is the vertex through X, and the tail is the vertex through $\theta\phi X$. One, but not both, of these two darts appears in Ω. We denote the set of darts of Ω by T. Given a cross X of U, we write

$$\beta(\Omega, X) = 1 \quad \text{or} \quad -1,$$

according as the dart $(X, \theta\phi X)$ or the dart $(\theta\phi X, X)$ belongs to T. We note

that
$$\beta(\Omega, X) = -\beta(\Omega, \theta\phi X) \qquad (X.9.1)$$
for each cross X of U. We write $D(\Omega, X)$ for the dart of Ω corresponding to the cross X.

The numbers $\beta(\Omega, X)$ can easily be related to the incidence numbers $\eta(D, v)$ defined in Sec. VIII.4. Let us take v to be the vertex through X. If the edge A of $G(Q_U)$ corresponding to X is a link of $G(Q_U)$, the definitions imply
$$\eta(D(\Omega, X), v) = \beta(\Omega, X). \qquad (X.9.2)$$
If A is a loop of $G(Q_U)$, that is, if both X and $\theta\phi X$ are crosses of v, we can write instead
$$\eta(D(\Omega, X), v) = 0 = \beta(\Omega, X) + \beta(\Omega, \theta\phi X), \qquad (X.9.3)$$
by Eq. (X.9.1).

Having chosen an orientation Ω of $G(Q_U)$, we can recognize a corresponding orientation Ω^* of $G(Q_U^*)$ having the same dart-set T. We say that Ω and Ω^* are *dual digraphs* with respect to the orientable surface V. We note that $(\Omega^*)^* = \Omega$, and that
$$\beta(\Omega^*, X) = \beta(\Omega, X) \qquad (X.9.4)$$
for each $X \in U$.

We consider chains on T to a ring R, as described in Sec. VIII.1. We can call them the 1-chains of either Ω or Ω^*. Let f be such a 1-chain and let ∂f be its boundary in Ω. We discuss the coefficient in ∂f of a vertex
$$v = (X_1, X_2, \ldots, X_k)$$
of Ω. By Sec. VIII.4, the coefficient is the sum of the products of the incidence numbers $\eta(D, v)$ with the corresponding coefficients $f(D)$, taken over the darts D corresponding to the crosses of v. So, by Eqs. (X.9.2) and (X.9.3), we can write
$$(\partial f)(v) = \sum_{j=1}^{k} \beta(\Omega, X_j) f(D(\Omega, X_j)). \qquad (X.9.5)$$

We next study the vertex-coboundary δv in Ω. By Sec. VIII.5, the coefficient of $D(\Omega, X_j)$ in it is $\beta(\Omega, X_j)$, where $1 \le j \le k$. We investigate the boundary $\partial^*(\delta v)$ of this 1-chain in Ω^*.

Let α be any vertex of Ω^*. In the orbit α of P_U, crosses of the form X_j or $\theta\phi X_j$ with X_j belonging to v, occur as disjoint sequences of the form $(\theta\phi X_i, X_{i+1})$. (We put $X_{k+1} = X_1$.) Applying Eq. (X.9.5) in Ω^* to the 1-chain δv, and using Eq. (X.9.4), we find that the contribution of such a pair to the expression on the right of Eq. (X.9.5) is
$$\beta(\Omega, X_i)\beta(\Omega, \theta\phi X_i) + \beta(\Omega, X_{i+1})\beta(\Omega, X_{i+1}) = (-1) + 1 = 0.$$

It follows that the coefficient of α in $\partial^*(\delta v)$ is zero. Since α can be chosen arbitrarily, we can assert that any vertex-coboundary of Ω is a cycle of Ω^*. Hence by Theorem VIII.44, we have the following theorem.

Theorem X.42. *Let Ω and Ω^* be dual digraphs with respect to an orientable surface V. Then the coboundaries to R of either one are cycles to R of the other.*

In the usual terminology a cycle of Ω is said to be *bounding* if it is a coboundary of Ω^*, and *nonbounding* otherwise.

Let C be a cell-base of $\Delta(\Omega, R)$. Then C is also a cell-base of $\Gamma(\Omega^*, R) \cdot C$, by Theorem X.42. Hence, by Theorem VII.20, there is a cell-base C^* of $\Gamma(\Omega^*, R)$ such that $C \subseteq C^*$. We have

$$|C| = \alpha_0(M) - 1 \quad \text{(by Eq. (VIII.5.8))}$$
$$|C^*| = \alpha_1(M^*) - \alpha_0(M^*) + 1 \quad \text{(by Theorem VIII.37)}.$$

Hence,

$$|C^*| - |C| = \alpha_1(M) - \alpha_2(M) - \alpha_0(M) + 2,$$
$$|C^*| - |C| = 2 - N(M). \tag{X.9.6}$$

Considering the chain-base of $\Gamma(\Omega^*, R)$ corresponding to C^*, we see that it must include nonbounding cycles of Ω^* unless $N(M) = 2$, that is, unless V is the combinatorial sphere. But when $N(M) = 2$, the sets C and C^* are identical, and the chain-base of $\Delta(\Omega, R)$ corresponding to C is also the chain-base of $\Gamma(\Omega^*, R)$ corresponding to C^*.

Theorem X.43. *Let Ω and Ω^* be dual planar digraphs. Then*

$$\Delta(\Omega, R) = \Gamma(\Omega^*, R) \quad \text{and} \quad \Gamma(\Omega, R) = \Delta(\Omega^*, R).$$

For any pair of dual planar graphs, Theorem X.43 tells us that the properties of one, insofar as they can be expressed in terms of cycles and coboundaries, are algebraically dual to the properties of the other.

X.10. NOTES

X.10.1. Rooted Maps

There are many papers concerned with the enumeration of rooted maps of various kinds. Usually the maps are planar, as in [8]. But maps on other surfaces have been considered [2].

X.10.2. Crosses

One way of representing a cross X on an edge A in a diagram is by two crossed arrows. One arrow is drawn along A and one across it. The involution θ reverses the arrow crossing A, and ϕ reverses the arrow along A. The name "cross" was suggested by this representation.

EXERCISES

1. If the map of Fig. X.5.2 is denoted by $M = M(\theta, \phi, P)$, determine the vertices, faces, and edges, and the components, of the premaps $L(\theta, \phi, P^2)$ and $L(\theta, \phi, P^3)$.
2. Show that there is a map M on the projective plane whose graph is a 5-clique. Find the graph of the dual map.
3. Consider the following variation on the operation of vertex-splitting defined at the beginning of Sec. X.7. Instead of introducing a new edge, we replace the two conjugate orbits u and u' by the four sequences ℓ_1, ℓ_2, ℓ'_1, and ℓ'_2, each made cyclic by taking the first term as the successor of the last. Under what conditions does this replacement leave the Euler characteristic unchanged?
4. Can you construct a map on the 5-clique whose Euler characteristic is not 1? If so, determine its orientability character and Euler characteristic.
5. Find a map on the projective plane having exactly three faces, each hexagonal, whose graph is the Thomsen graph. Find all the Hamiltonian circuits, and distinguish the bounding ones from the nonbounding.

REFERENCES

[1] Brahana, H. R., Systems of circuits on two-dimensional manifolds, *Ann. Math.* (2), **23** (1921), 144–168.
[2] Brown, W. G., On the enumeration of nonplanar maps, *Memoirs Amer. Math. Soc.*, **65** (1966).
[3] Cori, R., *Graphes planaires et systèmes de paranthèses*, Centre National de la Recherche Scientifique, Institut Blaise Pascal (1969).
[4] Edmonds, J. R., A combinatorial representation for polyhedral surfaces, *Notices Amer. Math. Soc.*, **7** (1960), 646.
[5] Harary, F., and W. T. Tutte, On the order of the group of a planar map, *J. Comb. Theory*, **1** (1966), 394–395.
[6] Jacques, A., Sur le genre d'une paire de substitutions, *C. R. Acad. Sci. Paris*, **267** (1968), 625–627.
[7] ———, "Constellations et graphes topologiques," in *Combinatorial Theory and its Applications*, Budapest, 1970.
[8] Mullin R. C., and P. J. Schellenberg, The enumeration of c-nets via quadrangulations, *J. Comb. Theory*, **4** (1968), 259–276.
[9] Tutte, W. T., "What is a map?" in *New Directions in the Theory of Graphs*, Acad. Press, New York 1973, pp. 309–325.

Chapter XI

Planarity

XI.1. PLANAR GRAPHS

We defined a planar map in Chapter X as a map on the combinatorial sphere. So we can characterize it as a map (necessarily orientable) of Euler characteristic 2.

In this section we acknowledge the vertex-map, as described in Sec. X.8. Its graph is the vertex-graph.

We extend the definition of a premap by allowing it to have vertex-maps as components. A premap L is defined as in Sec. X.1, but with the adjunction of $k \geq 0$ vertex-maps as extra components. The dual L^* of L is to be defined with the same number k of adjoined vertex-maps. Moreover, the vertices and faces of the vertex-maps of L^* are to be the faces and vertices, respectively, of the vertex-maps of L. By this device, the validity of Theorem X.18 is preserved.

A premap is called *planar* if its components are all planar maps. A graph is called *planar* if it is isomorphic to the graph of a planar premap, that is, if each of its components is the graph of a planar map. (See Theorems X.7 and X.8.)

Two planar graphs G and G^* are said to be *dual* to one another if there is a planar premap L such that

$$G \cong G(L) \quad \text{and} \quad G^* \cong G(L^*).$$

We should not ignore the possibility that there may be two or more nonisomorphic premaps L such that $G \cong G(L)$, and that therefore the isomorphism class of G^* may not be uniquely determined by G. However, we can assert that G^* is connected if and only if G is connected, by Theorems X.7 and X.18 and our conventions about vertex-maps. Indeed, this assertion can be generalized as follows:

Theorem XI.1. *Let G and G^* be dual planar graphs. Then they are either both connected or both disconnected. If they are connected, they have equal connectivities.*

Proof. The first part is already proved. If G is a vertex-graph, then G^*, edgeless and connected, is also a vertex-graph. In the remaining case, there is a planar map M, not a vertex-map, such that $G = G(M)$ and $G^* = G(M^*)$. Let Ω be an orientation of $G(M)$, and Ω^* the corresponding orientation of $G(M^*)$. Then

$$\kappa(G) = \kappa(G(M)) = \kappa(\Gamma(\Omega, I)) = \kappa(\Delta(\Omega^* \cdot I)) = \kappa(G(M^*)) = \kappa(G^*),$$

by Theorems VIII.66 and X.42. Here I is the ring of integers. □

Two connected dual planar graphs G and G^* can be drawn together, with their diagrams simply related. We put each vertex of G^* in the corresponding face of the diagram of G, and draw each edge of G^* across the corresponding edge of G. We then find that each vertex of G is represented within the corresponding face of G^*. The next two figures give simple examples. Figure XI.1.1 shows the duality between the link-graph and the loop-graph. The symbol X represents a cross of a corresponding planar map. In all such diagrams a cross is to be represented twice, by arrows attached to corresponding edges of G and G^*. It can be arranged that the arrow attached to G representing a cross X can be transformed, by a quarter-turn anticlockwise, into the X-arrow attached to G^*. The quarter-turn is about the point of intersection of the curves for G and G^* representing the edge concerned. We make this arrangement in the following diagrams.

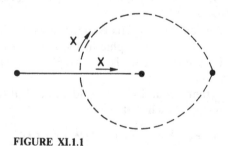

FIGURE XI.1.1

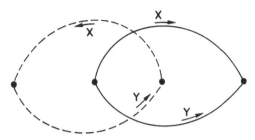

FIGURE XI.1.2

The full lines in Fig. XI.1.2 represent a planar map M whose graph is a 2-circuit. The two vertices of one oriented form of M are the cyclic sequences (X, Y) and $(\theta\phi X, \theta\phi Y)$. Hence the faces of this form are $(X, \theta\phi Y)$ and $(Y, \theta\phi X)$. The broken lines represent the dual map.

We observe that the maps M and M^* of Fig. XI.1.2 are isomorphic, and so the graphs $G(M)$ and $G(M^*)$ are isomorphic. We therefore say that the 2-circuit is a self-dual planar graph.

Theorem XI.2. *A map whose graph is a tree is planar and has exactly one face. Moreover, any planar map with a single face has a tree as its graph.*

Proof. If the graph $G(M)$ of a map M is a tree, then $\alpha_0(M) = \alpha_1(M) + 1$, and therefore $N(M) = \alpha_2(M) + 1$. Since $\alpha_2(M)$ is always positive, it follows, by Theorem X.33, that $\alpha_2(M) = 1$ and $N(M) = 2$.

On the other hand, if M is planar and $\alpha_2(M) = 1$, we have $\alpha_0(M) = \alpha_1(M) + 1$. Since $G(M)$ is connected, by Theorem X.7, it follows that $p_1(G(M)) = 0$, and that $G(M)$ is a tree, by Theorem I.35. □

Theorem XI.3. *To within an isomorphism, there is exactly one orientable and exactly one unorientable map whose graph is an n-circuit, for each positive integer n. The orientable map is planar, and the unorientable one is on the combinatorial projective plane.*

Proof. Let $M = M(\theta, \phi, P)$ be a map such that $G(M)$ is an n-circuit. Let the defining sequence of vertices of the circuit be $(a_1, a_2, \ldots, a_n)$ and the defining sequence of edges $(A_1, A_2, \ldots, A_n)$, as in Sec. I.1. We write the four crosses of A_j as X_j, θX_j, ϕX_j, and $\theta\phi X_j$.

Now a_j is incident only with A_j and A_{j+1} ($A_{n+1} = A_1$). We can adjust the notation so that $PX_j = \theta\phi X_{j+1}$ whenever $1 \le j < n$. Then, as an oriented form of a_n, we must have either $(X_n, \theta\phi X_1)$ or $(X_n, \phi X_1)$. For if $n = 1$, the equalities $PX_1 = X_1$ and $PX_1 = \theta X_1$ are ruled out by Axiom X.4 and the fact that a_n is not monovalent. In the case $n > 1$, the crosses X_1 and θX_1 belong to oriented forms of the different vertex a_1.

Consider the first possibility $PX_n = \theta\phi X_1$. Then the orbits of P are the n cyclic sequences $(X_j, \theta\phi X_{j+1})$ and their conjugates in M. We can verify that the four axioms for a premap are indeed satisfied, and that this premap has the correct graph. In this case we call the map a *circuit-map* of order n. The orbits of P^* in this circuit-map M are found to be

$$(X_n, X_{n-1}, \ldots, X_1), \quad (\theta X_n, \theta X_{n-1}, \ldots, \theta X_1)$$

and their conjugates in M^*. So we have

$$\alpha_0(M) = \alpha_1(M) = n, \quad \alpha_2(M) = 2, \quad \text{and} \quad N(M) = 2.$$

Hence the circuit-map is planar and therefore orientable.

In the second case the orbits of P are the cyclic sequences $(X_j, \theta\phi X_{j+1})$ and $(\theta X_j, \phi X_{j+1})$ for $j < n$, together with the cyclic sequences $(X_n, \phi X_1)$ and $(\theta X_n, \theta\phi X_1)$. The four axioms are thus satisfied, and a corresponding map M exists. (See, Theorem X.7.) The map is unorientable, by Theorem X.12. It satisfies $N(M) = \alpha_2(M) \geq 1$. Hence, $N(M) = 1$, by Theorem X.33. □

XI.2. SPANNING SUBGRAPHS

In any map M we define the *valency* of a vertex or face as the number of crosses in each of its oriented forms. Thus the valency of a vertex of M is its valency in $G(M)$.

Theorem XI.4. *In any planar map M, there is either a vertex of valency less than 3 or a face of valency less than 6.*

Proof. Suppose the theorem to fail for some planar map M. Then

$$3\alpha_0(M) \leq 2\alpha_1(M), \quad 6\alpha_2(M) \leq 2\alpha_1(M).$$

Hence,

$$N(M) \leq \frac{2\alpha_1(M) - 3\alpha_1(M) + \alpha_1(M)}{3} = 0,$$

contrary to the definition of a planar map. □

Consider two dual planar graphs G and G^*, and a planar map M such that $G \cong G(M)$ and $G^* \cong G(M^*)$. Then M determines a 1–1 mapping f of $E(G)$ onto $E(G^*)$. The rule is that $A \in E(G)$ and $fA \in E(G^*)$ correspond to the same edge of M under the respective graph isomorphisms. We call f a *duality-mapping*.

The next theorem is concerned with the cycle-ranks and coboundary-ranks of the spanning subgraphs of G and G^*. As in Sec. VIII.5, the cycle-rank of a graph H is $p_1(H)$ and the coboundary-rank is $|V(H)| - p_0(H)$, abbreviated here as $\rho(H)$.

Theorem XI.5. *Let G and G^* be dual planar graphs. Let f be a duality-mapping for them of $E(G)$ onto $E(G^*)$. Let T be any subset of $E(G)$.*

Spanning Subgraphs

Write $T^* = E(G^*) - f(T)$. *Then*

$$p_1(G:T) + \rho(G^*:T^*) = \rho(G^*), \quad (XI.2.1)$$
$$p_1(G^*:T^*) + \rho(G:T) = \rho(G). \quad (XI.2.2)$$

Proof. Let M be the planar map defining f, by the isomorphisms $G = G(M)$ and $G^* = G(M)$. Let Ω and Ω^* be corresponding orientations of $G(M)$ and $G(M^*)$, with dart-set W. Let U and U^* be the sets of darts of W on the edges of $G(M)$ and $G(M^*)$ corresponding to the edges of T and T^* in G and G^*, respectively. Thus $U^* = W - U$. With chain-groups to the ring I we have:

$p_1(G:T) + \rho(G^*:T^*)$

$= r(\Gamma(\Omega:U)) + r(\Delta(\Omega^*:U^*))$ (by Theorem VIII.37 and Eq. (VIII.5.8))

$= r(\Gamma(\Omega) \times U) + r(\Delta(\Omega^*) \cdot U^*)$ (by Theorems VIII.47–49)

$= r(\Delta(\Omega^*) \times U) + r(\Delta(\Omega^*) \cdot U^*)$ (by Theorem X.43)

$= r(\Delta(\Omega^*))$ (by Eq. (VIII.2.4))

$= \rho(G^*)$ (by Eq. (VIII.5.8)).

We thus have Eqs. (XI.2.1) and (XI.2.2) by interchanges of G with G^* and T with T^*. □

In particular,

$$p_1(G) = \rho(G^*), \quad p_1(G^*) = \rho(G). \quad (XI.2.3)$$

For another consequence of Theorem XI.5, suppose $G:T$ is a spanning tree of G, that is, $p_1(G:T) = 0$ and $G:T$ is connected. Since G is connected, being isomorphic to the graph of a map, the second condition is equivalent to $\rho(G:T) = \rho(G)$. So, from Theorem XI.5, we can infer the following:

Theorem XI.6. $G:T$ *is a spanning tree of* G *if and only if* $G^*:T^*$ *is a spanning tree of* G^*.

We note that dual planar graphs have equal tree-numbers.

We now use Theorem XI.5 to prove a duality theorem for the dichromate.

Theorem XI.7. *Let* G *and* G^* *be dual planar graphs. Then*

$$\chi(G; x, y) = \chi(G^*; y, x). \quad (XI.2.4)$$

Proof. Equation (IX.6.2) is equivalent to

$$\chi(G; x, y) = \sum_{S \subseteq E(G)} \{(x-1)^{\rho(G) - \rho(G:S)} \times (y-1)^{p_1(G:S)}\}.$$

(XI.2.5)

By Theorem XI.5 we can rewrite the expression on the right as

$$\sum_{T \subseteq E(G^*)} \left\{ (x-1)^{p_1(G^*:T)} (y-1)^{\rho(G^*) - \rho(G^*:T)} \right\}.$$

But this is $\chi(G^*; y, x)$ by Eq. (XI.2.5), applied to G^*. □

We can now infer from Eqs. (IX.6.3) and (IX.6.4) that

$$U(G^*; \lambda) = F(G; \lambda). \qquad (XI.2.6)$$

For a positive integral λ, it now follows from the results of Chapter IX that G has a λ-coloring if and only if G^* has a λ-flow.

Theorem XI.8. *Let G and G^* be dual planar graphs, with a duality-mapping f of $E(G)$ onto $E(G^*)$. Then an edge A of G is an isthmus of G if and only if fA is a loop of G^*.*

Proof. A is an isthmus of G if and only if $\rho(G'_A) = \rho(G) - 1$, by Theorem I.29, that is, if and only if $p_1(G^*:\{fA\}) = 1$, by Theorem XI.5, that is, if and only if fA is a loop of G^*. □

The Four-Color Theorem, as usually stated asserts that every planar graph without a loop has a 4-coloring. An equivalent assertion, by Eq. (XI.2.6) and Theorem XI.8 is that every planar graph without an isthmus has a 4-flow. Analogously, the 5-Flow Conjecture is an attempted generalization of the Five-Color Theorem from planar graphs to general graphs.

XI.3. JORDAN'S THEOREM

We consider a planar map $M = M(\theta, \phi, P)$ and the associated graphs $G(M)$ and $G(M^*)$. We denote the sets of vertices, edges, and faces of M by $V(M)$, $E(M)$, and $F(M)$, respectively.

Let U be an orientation-class of M, and let Q_U and Q_U^* be the oriented forms of M and M^*, respectively, on U. The graph $G(Q_U)$ is essentially the same as $G(M)$. In passing from $G(M)$ to $G(Q_U)$, we replace each vertex by its oriented form in U and each edge by its pair of crosses in U, but the incidences remain as before. The same observation applies to $G(M^*)$ and $G(Q_U^*)$.

Let Ω and Ω^* be corresponding orientations of $G(Q_U)$ and $G(Q_U^*)$, respectively. The elementary chains of $\Gamma(\Omega, I)$ are those of $\Delta(\Omega^*, I)$, by Theorem X.43. So by Theorems VIII.34 and VIII.40, we have the following theorem.

Theorem XI.9. *Let T be any nonnull subset of $E(M)$. Then T is the set of edges of a circuit of $G(M)$ if and only if it is a bond of $G(M^*)$.*

We offer Theorem XI.9 as a combinatorial analogue of Jordan's topological theorem about simple closed curves on the sphere. Let C be any circuit of $G(M)$. Then $E(C)$ is a bond of $G(M^*)$ by Theorem XI.9. Let H and K be the end-graphs of this bond. They are subgraphs of $G(M^*)$. We call them the *residual graphs* of C, with respect to M, considering them to be analogous to the two residual domains of Jordan's curve.

The *trace* $T(u)$ in $G(M)$ of a vertex or face u of M is the set of edges incident with u in M. The *trace-graph* of u in $G(M)$ is the reduction $G(M) \cdot T(u)$. Sometimes the trace-graph is known to be a circuit, and we speak of a *trace-circuit*.

Theorem XI.10. *Let f be any face of M. Let $(X_1, X_2, \ldots, X_m)$ be one of its oriented forms. For each suffix j, let A_j be the edge to which X_j belongs, so that the edges A_j constitute $T(f)$ in $G(M)$. Then for each j, there is a vertex v_j of $G(M)$ incident with both A_j and A_{j+1} ($A_{m+1} = A_1$). Consequently, the trace-graph of f in $G(M)$ is connected.*

Proof. We have only to demonstrate the existence of v_j. But $X_{j+1} = P\theta\phi X_j$ ($X_{m+1} = X_1$). Hence, there is a vertex of $G(M)$ with an oriented form through $\theta\phi X_j$ and X_{j+1}. This is incident with both A_j and A_{j+1}, and can be taken as v_j. □

We return to our circuit C and its two residual graphs H and K.

An edge A of M is an edge either of C or of one of its residual graphs. If A is an edge of H, we say also that it is *enclosed* by H. A vertex v of $G(M)$ is *enclosed* by H if all the crosses belonging to v belong also to H.

Theorem XI.11. *A vertex v of $G(M)$ is either (i) a vertex of C, (ii) enclosed by H, or (iii) enclosed by K, and these three possibilities are mutually exclusive.*

Proof. v is a vertex of C if and only if some cross belonging to v belongs also to an edge of C.

In the remaining case, all crosses belonging to v belong also to $H \cup K$. Hence the trace-graph T_v of v in $G(M^*)$ is a subgraph of $H \cup K$. But T_v is connected, by Theorem XI.10 applied to M^*. Hence, T_V is a subgraph either of H or of K, by Theorem I.24. □

Let us write the defining sequence of vertices of C as $(a_1, a_2, \ldots, a_n)$ and the defining sequence of edges as $(A_1, A_2, \ldots, A_n)$, as in Sec. I.1. Then A_j is incident with a_{j-1} and a_j ($a_0 = a_n$). On the edge A_j, we have the four crosses X_j, θX_j, ϕX_j, and $\theta\phi X_j$, for each j. We adjust the notation so that X_j and $\theta\phi X_j$ are in U, so that the half-edge $\{X_j, \theta X_j\}$ is absorbed by a_{j-1}, and so that the half-edge $\{X_j, \theta\phi X_j\}$ is absorbed by a_j.

We suppose Ω specified so that the dart D_j of Ω corresponding to A_j is the ordered pair $(X_j, \theta\phi X_j)$ for each j. Then, by the theory of Sec. X.9, there is an elementary chain of $\Gamma(\Omega, I)$, with support $E(C)$, in which the coefficient of each of these darts is 1. But this chain also corresponds to a bond $E(C)$ of $G(M^*)$, with end-graphs H and K. By Theorem VIII.40 and Sec. X.9, we can adjust the notation so that the n crosses X_j belong to H and the n crosses $\theta\phi X_j$ to K. Then the crosses ϕX_j must belong to H and the crosses θX_j to K.

Consider a vertex a_j of C, and its oriented form through X_j. That cyclic sequence must take in the cross $\theta\phi X_{j+1}$ from A_j, not ϕX_{j+1}, because that is not in U. Since a_j is incident only with A_j and A_{j+1} in C, the cyclic sequence involves only these two crosses from C. We can therefore write the oriented form of a_j through X_j as:

$$(R_1, X_j, R_2, \theta\phi X_{j+1}). \tag{XI.3.1}$$

As in Sec. X.8 we write R_i for a linear sequence of crosses, possibly null. In the notation of that section, we would write the conjugate of the cyclic sequence (XI.3.1) as $(\phi X_{j+1}, R'_2, \theta X_j, R'_1)$.

Theorem XI.12. *Let a_j be a vertex of C. Let its oriented form through X_j be as in Eq. (XI.3.1). Then the crosses of R_1 belong to H, and those of R_2 to K.*

Proof. In the linear sequence (R_1, X_j), any two consecutive crosses belong to a common vertex of M^*, by Theorem XI.10. Hence if the sequence R_1 is a nonnull, the edges to which its crosses belong define a connected reduction J_1 of $H \cup K$. Moreover, this reduction includes the vertex with cross X_j, and this vertex belongs to H. Hence J_1 is a subgraph of the component H of $H \cup K$. Similarly, if R_2 is nonnull, its crosses define a connected reduction J_2 of $H \cup K$, J_2 involves the vertex with cross $\theta\phi X_{j+1}$, and J_2 is a subgraph of K. □

Let S_H be the set of crosses of M belonging to H or C. We proceed to define a permutation P_H of S_H by giving the orbits of $P_H\theta\phi$.

If a cross X of S_H belongs to a vertex of H, we put

$$P_H\theta\phi X = P\theta\phi X.$$

We thus assign to $P_H\theta\phi$ all orbits of $P^* = P\theta\phi$ that correspond to vertices of H. This leaves only the crosses of C that are denoted by $\theta\phi X_j$ and θX_j. For these, we write

$$P_H\theta\phi(\theta\phi X_j) = \theta\phi X_{j+1}, \qquad P_H\theta\phi(\theta X_j) = \theta X_{j-1},$$

thus giving to $P_H\theta\phi$ two new conjugate orbits,

$$\text{Orb}(1) = (\theta\phi X_1, \theta\phi X_2, \ldots, \theta\phi X_n) \tag{XI.3.2}$$

and

$$\text{Orb}(2) = (\theta X_n, \theta X_{n-1}, \ldots, \theta X_1). \tag{XI.3.3}$$

Jordan's Theorem

The definition of $P_H\theta\phi$, and therefore of P_H, as a permutation of S_H is now complete.

With ϕ and θ, restricted to S_H, as first and second involutions, respectively, the permutation $P_H\theta\phi$ satisfies the four axioms for a premap $L = L(\phi, \theta, P_H\theta\phi)$ on S_H. But $G(L)$ is derived from H by adjoining a new vertex x_H, corresponding to Orb(1) and Orb(2), and then adjoining the edges of C. Each of these is to be incident with x_H and with the vertex of H that is one of its ends in $G(M^*)$. In other words, $G(L)$ is isomorphic with the contraction of $G(M^*)$ to $E(C) \cup E(H)$, with invariance of edges: K contracts into x_H. We note that $G(L)$ is connected, by the connection of H. Accordingly, L is a map on S_H. We call it the *pinch-map*

$$\text{Pin}(M, C, H) = M(\phi, \theta, P_H\theta\phi)$$

of C in M *covering* H. There is likewise a pinch-map $\text{Pin}(M, C, K)$ of C in M covering K. We call the graph of $\text{Pin}(M, C, H)$ the *pinch-graph* of C in M covering H.

In naming the pinch-maps, we have in mind a diagram of M and M^* on the sphere, with the circuit C along a great circle. We think of C as a string that can be drawn tight, thus pinching the sphere into two topological spheres and squeezing M^* into the two pinch-maps.

Let us determine the faces of $\text{Pin}(M, C, H)$. First we make note of a lemma.

Lemma XI.13. *If* $X \in S_H$, *then* $P_H X = PX$, *unless* $X = X_j$ *or* ϕX_j *for some suffix j.*

Proof. We have $P_H X = P_H \theta\phi(\theta\phi X)$. By the definition of $P_H\theta\phi$, this is $P\theta\phi(\theta\phi X) = PX$ unless $\theta\phi X$ is one of the crosses $\theta\phi X_j$ or θX_j belonging to x_H. △

We now consider the orbits of P in M, which define the vertices of M and $G(M)$. For a vertex of M enclosed by K, no cross of S_H is involved. For one enclosed by H, the orbits of P are orbits of P_H, by Lemma XI.13. Any remaining vertex is one of the vertices a_j of C, by Theorem XI.11.3.3. In that case, we can represent one of its oriented forms by Formula (X.3.1), so that Theorem XI.12 holds. We then find that the cyclic sequence

$$(R_1, X_j, \theta\phi X_{j+1}) \qquad (XI.3.4)$$

is an orbit of P_H. For $P_H X_j = P_H \theta\phi(\theta\phi X_j) = \theta\phi X_{j+1}$, by Eq. (XI.3.2) and for any other cross X appearing in Eq. (XI.3.4), $P_H X$ is the same as PX, by Lemma XI.13.

Similarly, the cyclic sequence $(R_2, \theta\phi X_{j+1}, X_j)$ is an orbit of P_K. We refer to the orbits $(R_1, X_j, \theta\phi X_{j+1})$ and $(R_2, \theta\phi X_{j+1}, X_j)$ of P_H as the *H-tightening* and *K-tightening*, respectively, of the orbit (XI.3.1) of P. Likewise, their conjugates $(\theta X_j, \phi X_{j+1}, R'_2)$ and $(\phi X_{j+1}, \theta X_j, R'_1)$ are the K-tightening and H-tightening, respectively, of the orbit $(\phi X_{j+1}, R'_2, \theta X_j, R'_1)$ of P.

We see that an orbit of P persists as an orbit of P_H or P_K, but not of both, unless it corresponds to a vertex of C. In that case, it splits into exactly one orbit of P_H and one of P_K, the corresponding H-tightening and K-tightening. We deduce that:

$$\alpha_2(\text{Pin}(M,C,H)) + \alpha_2(\text{Pin}(M,C,K)) = \alpha_2(M^*) + n. \quad \text{(XI.3.5)}$$

But from the definitions we have

$$\alpha_0(\text{Pin}(M,C,H)) + \alpha_0(\text{Pin}(M,C,K)) = \alpha_0(M^*) + 2, \quad \text{(XI.3.6)}$$

$$\alpha_1(\text{Pin}(M,C,H)) + \alpha_1(\text{Pin}(M,C,K)) = \alpha_1(M^*) + n. \quad \text{(XI.3.7)}$$

Hence,

$$N(\text{Pin}(M,C,H)) + N(\text{Pin}(M,C,K)) = N(M^*) + 2$$
$$= 4. \quad \text{(XI.3.8)}$$

Applying Theorem X.33, we deduce that

$$N(\text{Pin}(M,C,H)) = N(\text{Pin}(M,C,K)) = 2. \quad \text{(XI.3.9)}$$

Theorem XI.14. *The pinch-maps of C in M are planar* (by Eq. (XI.3.9)).

We call the dual of $\text{Pin}(M,C,H)$ the *cap-map* $\text{Cap}(M,C,H)$ of C in M covering H. There is also the cap-map $\text{Cap}(M,C,K) = (\text{Pin}(M,C,K))^*$ of C in M covering K.

From our determination of the structure of $\text{Pin}(M,C,H)$, we see that the graph of $\text{Cap}(M,C,H)$ is vertex-isomorphic with the subgraph $\text{Cap}(G(M),C,H)$ of $G(M)$ specified as follows: Its edges are the edges of C and H, and its vertices are those of C together with the vertices of M enclosed by H. Because this subgraph is isomorphic to the graph of a map, we have the following:

Theorem XI.15. *The graphs* $\text{Cap}(G(M),C,H)$ *and* $\text{Cap}(G(M), C,K)$ *are connected.*

In the change from $\text{Cap}(G(M),C,H)$ to the graph of $\text{Cap}(M,C,H)$, we merely replace each oriented vertex belonging to C by its H-tightening: that is, we delete from it those of its crosses that belong to K.

We call $\text{Cap}(G(M),C,H)$ the *cap-graph* of C in M covering H. The two cap-graphs, by their definitions, have the following properties:

$$\text{Cap}(G(M),C,H) \cup \text{Cap}(G(M),C,K) = G(M), \quad \text{(XI.3.10)}$$

$$\text{Cap}(G(M),C,H) \cap \text{Cap}(G(M),C,K) = C. \quad \text{(XI.3.11)}$$

Jordan's Theorem

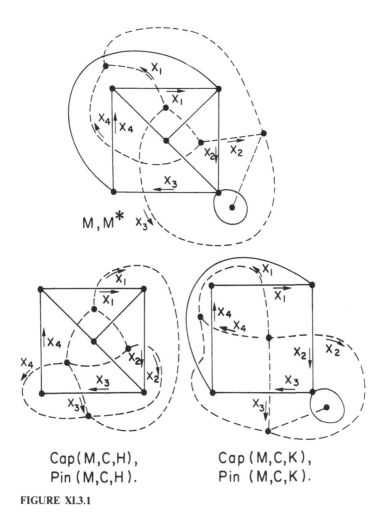

Cap(M,C,H), Cap(M,C,K),
Pin(M,C,H).   Pin(M,C,K).

FIGURE XI.3.1

Figure XI.3.1 shows M and C, with their pinch-maps and cap-maps, in a particular case. The edges of M and the cap-maps are shown with full lines, those of M^* and the pinch-maps with broken ones.

It is convenient to introduce here the notion of a *subpremap* of M. Let G_0 be a nonnull subgraph of $G(M)$ without isolated vertices. Let S_0 be the set of all crosses of M belonging to G_0. We form a permutation P_0 of S_0 by taking the orbits of P corresponding to the vertices of G_0 and deleting from them all crosses not in S_0. We retain the cyclic order of the remaining crosses, of which there must be at least one. Another way of putting this is to say that for each $X \in S_0$, the cross $P_0 X$ is the first member of the sequence $PX, P^2X, P^3X,\ldots$ that is in S_0. Then θ and ϕ, restricted to S_0, together with P_0, satisfy the axioms for a premap $L = L(\theta, \phi, P_0)$ on S_0, and the graph $G(L)$ of L is vertex-isomorphic with G_0. We call L the *subpremap of M* on G_0. If G_0 is connected, then L must be a map. We then call it the *submap* of M on G_0. These definitions can be applied to any map M, planar or not. They imply the rule:

Theorem XI.16. *Any subpremap of a subpremap of M is a subpremap of M.*

As an example we can take the submap M_C of our planar map M on the circuit C. Its oriented vertices are the cyclic sequences $(X_j, \theta\phi X_{j+1})$ and their conjugates $(\theta X_j, \phi X_{j+1})$. Hence, it follows that the oriented faces of M_C are the orbits Orb(1) and Orb(2) of Eqs. (XI.3.2) and (XI.3.3), together with

$$\text{Orb}(3) = (X_n, X_{n-1}, \ldots, X_1) \qquad (XI.3.12)$$

and

$$\text{Orb}(4) = (\phi X_1, \phi X_2, \ldots, \phi X_n). \qquad (XI.3.13)$$

Orb(1) and Orb(2) are the oriented forms of a face f_H of M_C. Orb(3) and Orb(4) are the oriented forms of another face f_K. We call f_H and f_K the faces of M_C *opposite* H and K, respectively. We note that M_C is a circuit-map, as defined in the proof of Theorem XI.3. It is planar.

We can now characterize Pin(M, C, H) as a map whose vertices are those of H, together with the face of M_C opposite H. The face of M_C opposite K appears as a vertex of Pin(M, C, H) only when H is a vertex-graph. In that case, Pin(M, C, K) is identical with M^*.

We can characterize Cap(M, C, H) as the submap of M on Cap($G(M), C, H$). If H is a vertex-graph, then Cap(M, C, H) is identical with M. Otherwise we can characterize the faces of Cap(M, C, H) as the face of M_C opposite H, together with all faces of M having crosses belonging to H. By Theorem XI.16, M_C is a submap of Cap(M, C, H) and of Cap(M, C, K).

XI.4. CONNECTIVITY IN PLANAR MAPS

We call a map M k-connected if $G(M)$ is k-connected. Thus, for planar maps, M^* is k-connected if and only if M is k-connected, by Theorem XI.1.

Let x be a vertex of a graph G. We denote the set of edges incident with x by E_x. We call x a bond-vertex of G if E_x is a bond of G. In that case, E_x is nonnull, x is incident only with links, and one end-graph of E_x is the vertex-graph at x.

Theorem XI.17. *Let G be a connected graph that is neither a vertex-graph nor a loop-graph. Then a vertex x of G is a bond-vertex if and only if it is not a cut-vertex.*

Proof. Suppose, first, that x is incident with a loop A. Then E_x is not a bond of G. On the other hand, x is incident with at least one other edge, since G is not a loop-graph, and therefore x is a cut-vertex of G. From now on, we may suppose that x is incident only with links.

Let H be the graph obtained from G by deleting x and its incident edges. If H is not connected, then x is a cut-vertex and E_x is not a bond. If H is connected, then E_x is a bond and x cannot be a cut-vertex. To complete the proof, we observe that H is nonnull because G is not a vertex-graph. □

Theorem XI.18. *Let M be a planar map that is neither a vertex-map nor a link-map. Then, in order that the trace-graph in $G(M)$ of each face of M shall be a circuit, it is necessary and sufficient that M shall be 2-connected.*

Proof. By Jordan's Theorem (XI.9) the trace-graphs are all circuits if and only if the vertices of M^* are all bond-vertices of $G(M^*)$, that is if and only if $G(M^*)$ is 2-connected, by Theorem XI.17. □

We now present some constructions for planar maps and graphs.

Theorem XI.19. *Every subgraph of a planar graph is planar.*

Proof. We first show that any connected subgraph H of a connected planar graph G is planar. We can identify G with the graph $G(M)$ of some planar map M.

If $|E(G)| \leq 1$, the result is trivial, M being then the vertex-map, the link-map, or the loop-map. Assume that the theorem, in the connected case, holds whenever $|E(G)|$ is less than some integer $q \geq 2$, and consider the case $|E(G)| = q$.

If $H = G$, the theorem holds. We may therefore suppose G to have an edge A not in $E(H)$. We can also assume that H is not a vertex-graph, that case being trivial.

If A is not an isthmus of G, we can delete it from M to get a new planar map M_1 (by Theorem X.28). By the description of this operation in Sec. X.7, the graph of M_1 is vertex-isomorphic to G'_A. Since H is a subgraph of G'_A, it is planar, by the inductive hypothesis.

If A is an isthmus of G, then H is a subgraph of an end-graph K of A in G. Moreover, we can contract the link A in M to get a new planar map M_2 (by Theorem X.26). As observed in Sec. X.7, the graph of M_2 can be identified with $(G(M))''_A$, that is, with G''_A. But K is vertex-isomorphic to a subgraph of G''_A, whence H is planar, by the inductive hypothesis.

The induction has succeeded: The theorem holds in the connected case. Now consider a general subgraph H of a general planar graph G. Each component of H is a subgraph of a component of G, and each component of G is planar. Hence, each component of H is planar, by the result just proved. Thus H is a planar graph. □

We go on to discuss the construction of a map M on a given connected graph G (with at least one edge). The construction assigns four crosses X, θX, ϕX, and $\theta \phi X$ to each edge A, a set S of $4|E(G)|$ distinct crosses in all. We associate X and θX with one end of A, and ϕX and $\theta \phi X$ with the other, albeit the two may coincide. For each vertex v, we arrange the associated crosses in two conjugate orbits, to obtain a permutation P of S. Our map is $M = M(\theta, \phi, P)$, the two involutions being determined by the notation. It is clear that $G(M)$ and G are isomorphic. There is no harm in identifying them, saying that A is the edge of $G(M)$ determined by the given four crosses. This is Edmonds' Construction, as described in Section X.1. The following theorem gives an example of its use.

Theorem XI.20. *Let a connected graph G have a 1-separation (H, K), with cut-vertex v. Let H and K be planar. Then G is planar.*

Proof. We can identify H and K with the graphs of planar maps $M_H = M(\theta_H, \phi_M, P_H)$ and $M_K = M(\theta_K, \phi_K, P_K)$, respectively. We construct a map $M = M(\theta, \phi, P)$ on G as follows.

The crosses of M are those of M_H and M_K, and have the same associations with the vertices as in those maps. For each cross X, we put $\theta X = \theta_H X$ or $\theta_K X$ according as X belongs to H or to K, and the same rule holds for ϕ. The orbits of P are those of P_H and P_K, except for orbits associated with v.

Let $(X_1, X_2, \ldots, X_p)$ and $(Y_1, Y_2, \ldots, Y_q)$ be oriented forms of v in M_H and M_K, respectively. From these we construct the cyclic sequence

$$(X_1, \ldots, X_p, Y_1, \ldots, Y_q). \tag{XI.4.1}$$

We adopt this and its conjugate as orbits of P. The permutation P and the map M are now completely determined. We have:

$$\alpha_1(M) = \alpha_1(M_H) + \alpha_1(M_K), \tag{XI.4.2}$$

$$\alpha_0(M) = \alpha_0(M_H) + \alpha_0(M_K) - 1. \tag{XI.4.3}$$

By our construction, PX is equal to $P_H X$ or $P_K X$ except when X is X_p, Y_q, θX_1, or θY_1. Hence P^*X is $P_H^* X$ or $P_K^* X$ unless X is $\theta \phi X_p$, $\theta \phi Y_q$, ϕX_1, or ϕY_1. In these cases, P^*X is found to be Y_1, X_1, θY_q, or θX_p, respectively. We infer that the face of M_H through $\theta \phi X_p$ and X_1 and the face of M_K through $\theta \phi Y_q$ and Y_1 are combined into a single new face of M. The other faces of M_H and M_K persist as faces of M. We infer that

$$\alpha_2(M) = \alpha_2(M_H) + \alpha_2(M_K) - 1. \tag{XI.4.4}$$

From Eqs. (XI.4.2)–(XI.4.4), we find that $N(M) = N(M_H) + N(M_K) - 2 = 2$. Thus M is a planar map and G is a planar graph. □

Theorem XI.21. *A connected graph G is a planar if and only if its blocks are planar.*

Proof. If the blocks of G are planar we can, by repeated use of the construction of Theorem XI.19, construct a planar map on G. (See Theorem III.22.) On the other hand, if G is planar then its blocks are planar, by Theorem XI.19 □

In the theory of planar graphs we are much concerned with tests for planarity. Our definitions tell us that we need such a test only for connected graphs, a graph being planar if and only if its components are planar. Similarly, by Theorem XI.21, we need a test only for 2-connected graphs. For other connected graphs, we resolve them into their blocks and test each block in turn. For 2-connected graphs, we describe a test due to S. MacLane [5].

A *planar mesh* in a graph G is a family

$$F = (C_1, C_2, \ldots, C_n)$$

of n circuits C_j of G, not necessarily all distinct, which satisfy the three following conditions:
i) The order n of F is $p_1(G)+1$.
ii) For each edge A of G, there are exactly two suffixes j such that $A \in E(C_j)$.
iii) For each nonnull proper subset X of $\{1,2,\ldots,n\}$ there is an edge A of G such that $A \in E(C_j)$ for exactly one $j \in X$.

Condition (i) shows that the order of F is at least 1, and then (ii) shows that $n \geq 2$. If $n = 2$, then $C_1 = C_2$, by (ii). This can happen only when $G \cdot E(G)$ is a circuit. If $n > 2$, then the n circuits C_j must all be distinct. If, for example, $C_1 = C_2$, then (iii) fails when $X = \{1,2\}$.

For $n \geq 3$, we can express Conditions (ii) and (iii) by saying that the only linear relation mod 2 between the n sets $E(C_j)$ is

$$\sum_{j=1}^{n} E(C_j) = 0. \tag{XI.4.5}$$

Here we identify a subset of $E(G)$ with the chain on $E(G)$ to I_2 of which it is the support. (See Sec. VIII.4.) Every proper subset of F is linearly independent. We shall have occasion to use this mod 2 algebra of subsets in what follows.

Theorem XI.22. *If a graph G without isolated vertices has a planar mesh F, then G is 2-connected.*

Proof. By (iii) we can enumerate the members of F as $C_1, C_2, \ldots, C_n$, so that each, except C_1, has an edge in common with one of its predecessors. Using Theorems III.5 and III.9, we then find, for each integer k in succession from 1 to n, that the union of the first k members of the sequence is 2-connected. □

Theorem XI.23. *Let M be a 2-connected planar map that is neither a vertex-map nor a link-map. Then $G(M)$ has a planar mesh whose member-circuits are the trace-graphs in $G(M)$ of the faces of M.*

Proof. By Theorem XI.18 the trace-graphs of the faces of M constitute a family $F = (C_1, C_2, \ldots, C_n)$ of $n = \alpha_2(M)$ circuits of $G(M)$.

Now for any planar map M, 2-connected or not, we have

$$\alpha_2(M) = N(M) + \alpha_1(M) - \alpha_0(M)$$
$$= 2 + |E(G(M))| - |V(G(M))|,$$
$$\alpha_2(M) = p_1(G(M)) + 1. \qquad \text{(XI.4.6)}$$

So our family F satisfies Condition (i).

If Condition (ii) fails, then some edge A of M is incident with only one face. Then A is a loop of $G(M^*)$, and an isthmus of $G(M)$, by Theorem XI.8. Since $G(M)$ is 2-connected and not a link-graph, this is impossible, by Theorem III.3.

Condition (iii) follows from the connection of $G(M^*)$. □

Theorem XI.24. *Let G be a 2-connected graph with a planar mesh $F = (C_1, C_2, \ldots, C_n)$. Then G is planar.*

Proof. We can assume that G is not a circuit, for circuits are planar, by Theorem XI.3. So, by Theorem III.3, we may suppose G loopless.

We attempt to construct an appropriate planar map on G. As in Edmonds' Construction, we assign four crosses X, θX, ϕX, and $\theta \phi X$ to each edge A, associating X and θX with one end x, and ϕX and $\theta \phi X$ with the other end y. But A belongs to two members of F, with distinct suffices. We make a second partition, assigning X and $\theta \phi X$ to one of these members of F, and θX and ϕX to the other.

Let S be the complete set of $4|E(G)|$ crosses. The notation defines involutions θ and ϕ on S. We now define another permutation P^* of S as follows: For an arbitrary member C_i of F, we write the defining sequence of vertices as $(a_1, a_2, \ldots, a_m)$ and the defining sequence of edges as $(A_1, A_2, \ldots, A_m)$ as in Sec. I.1. Then the ends of A_j are a_j and a_{j+1} ($a_{m+1} = a_1$, etc.). We write X_j for the cross on A_j that is associated with both a_j and C_i. We now assign to P^* the orbit $(X_1, X_2, \ldots, X_m)$, and its conjugate $(\phi X_m, \ldots, \phi X_1)$. This being done for each member C_i of F, we have a permutation P^* of S, and it satisfies the conditions for a map $M^* = M(\phi, \theta, P^*)$. This has a dual map $M = M(\theta, \phi, P)$, where $P = P^*\theta\phi$.

Consider the orbits of P. Our cross X_j is associated with C_i and a_j. Hence $\theta\phi X_j$ is associated with a_{j+1}, and therefore PX_j, that is, $P^*\theta\phi X_j$, is associated with a_j. We conclude that, in each orbit or pair of conjugate orbits of P, all the crosses are associated with the same vertex of G.

We can now assert that there is a mapping f of $V(M)$ onto $V(G)$ such that all the crosses of a vertex v of M are associated with the vertex

$f(v)$ of G. If f is a 1-1 correspondence, it follows that G and $G(M)$ are vertex-isomorphic, and we are entitled to identify them and say that M is a map on G. We can prove that f is one to one by showing that $|V(M)| = |V(G)|$, in the following way:

$$|V(M)| = N(M) + \alpha_1(M) - \alpha_2(M)$$
$$\leq 2 + |E(G)| - n \quad \text{(by construction)}$$
$$\leq 1 + |E(G)| - p_1(G) \quad \text{(by (i))}$$
$$\leq |V(G)|.$$

But $|V(M)|$ is certainly not less than $|V(G)|$. Hence, strict equality holds; $|V(M)| = |V(G)|$ and $N(M) = 2$. Hence, M is a planar map on G, and G is a planar graph. □

The construction of Theorem XI.24 differs from that of Edmonds, in that orbits are chosen to fit circuits, required to be trace-graphs of faces, rather than vertices. It is best regarded as an example of the older construction used by H. R. Brahana in his work on the topological classification of surfaces [1].

Theorem XI.24 gives us MacLane's Test for planarity: a 2-connected graph, other than a vertex-graph or link-graph, is planar if and only if it has a planar mesh. (See Theorem XI.23 [5].)

Theorem XI.25. *Let G be a connected graph with a planar mesh $F = (C_1, C_2, \ldots, C_n)$. Then there is essentially only one planar map M on G whose n faces have the C_i as their trace-circuits.*

This is clear from the construction of Theorem XI.24. There each cross must be associated with just one C_i and just one vertex. Each is uniquely distinguished by its two associations. In two constructions for M, crosses with different symbols could be assigned, but their pairs of associations would be the same. These pairs determine the orbits of M and M^*. So two maps M satisfying the requirements of Theorem XI.25 would be related by an isomorphism preserving each edge and vertex of G.

We use MacLane's Test in the following proofs.

Theorem XI.26. *Let G be the union of a graph, without isolated vertices, and an arc L that avoids H but has its two ends x and y in H. Let $F = (C_1, C_2, \ldots, C_n)$ be a planar mesh of H such that x and y are in $V(C_n)$, separating C_n into two internally disjoint arcs L_1 and L_2. Then G is planar, having a planar mesh F' derived from F by replacing C_n by the two new member-circuits $L \cup L_1$ and $L \cup L_2$.*

Proof. Since H is connected, by Theorem XI.22, it follows from Eq. (I.6.1) that $p_1(G) = p_1(H) + 1$. Thus F' satisfies Condition (i). It satisfies (ii) by construction.

Suppose some proper subfamily of F' sums, in its edge-sets, to zero mod 2. Then it involves both $L \cup L_1$ and $L \cup L_2$, or neither of them. Replacing the two new circuits by C_n if necessary, we derive a proper subfamily of F that sums to zero. Hence, if Condition (iii) failed for F', it would fail also for F. □

Theorem XI.27. *Let a graph G be the union of two graphs H and K, without isolated vertices, whose intersection is a circuit C. Let H and K have planar meshes $F_H = (C_1, C_2, \ldots, C_h)$ and $F_K = (D_1, D_2, \ldots, D_k)$, respectively, such that $C_1 = D_1 = C$. Then G is planar, having the planar mesh $F = (C_2, \ldots, C_h, D_2, \ldots, D_k)$.*

Proof. Using Theorem XI.22 and Eq. (I.6.1), we find that $p_1(G) = p_1(H) + p_1(K) - 1$. Hence F satisfies (i). It satisfies (ii) by construction.

Suppose some subfamily F' of F to sum to zero. Then its members in $(C_2, \ldots, C_h)$ sum to 0 or C, since $E(C)$ supports an elementary chain of $\Gamma(H, I_2)$. Similarly, for the members of F' in $(D_2, \ldots, D_k)$. Since F_H and F_K satisfy (iii), it follows that the intersections of F' with $(C_2, \ldots, C_h)$ and $(D_2, \ldots, D_k)$ are either both null or both exhaustive. Hence if F' is not null, it is F. □

Theorem XI.28. *Let (H, K) be a 2-separation of a 2-connected graph G, with hinges x and y. Let H_1 and K_1 be formed from H and K, respectively, by adjoining a new link A with ends x and y. Let H_1 and K_1 have planar meshes $(C_1, C_2, \ldots, C_h) = F_H$ and $(D_1, D_2, \ldots, D_k) = F_K$, respectively, where A belongs to C_1, C_2, D_1, and D_2. Then G is planar, having the planar mesh $F = (Q_1, Q_2, C_3, \ldots, C_h, D_3, \ldots, D_k)$, where $Q_1 = (C_1 \cup D_1)'_A$ and $Q_2 = (C_2 \cup D_2)'_A$.*

Proof. Deleting A from C_1, we obtain an arc $L(C_1)$, with ends x and y. Similarly, we have arcs $L(C_2)$, $L(D_1)$, and $L(D_2)$. The union of the two internally disjoint arcs $L(C_1)$ and $L(D_1)$ is a circuit, and it is Q_1. Similarly, Q_2 is a circuit, being the union of $L(C_2)$ and $L(D_2)$.

The construction of this theorem is illustrated in Fig. XI.4.1. Symbols for circuits are placed inside the corresponding faces.

Using Theorem IV.20 and Eq. (I.6.1) we find $p_1(G) = p_1(H_1) + p_1(K_1) - 1$. Hence F satisfies (i). It satisfies (ii) by construction.

Suppose that some nonnull proper subfamily F' of F sums to zero. Then its members in F_H, with the adjunction of C_i whenever $Q_i \in F'$, must sum either to $\varnothing$ or to $\{A\}$. Actually, they must sum to $\varnothing$, since $\{A\}$ is not the support of a cycle of G to I_2. Since F_H satisfies (iii), we infer that F' includes all the members of F_H except C_1 and C_2 and includes Q_1 and Q_2 as well, unless it includes none of these circuits. A similar rule holds for F_K. We thus have the contradiction that F' is either null or identical with F. So in fact F satisfies (iii). □

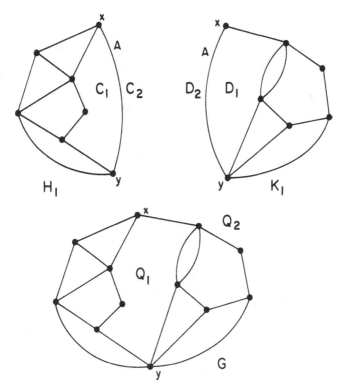

FIGURE XI.4.1

In view of Theorem XI.28, we can say that a test for planarity is now needed only for 3-connected graphs. The problem for a 2-connected graph G with a 2-separation (H, K) reduces to the problem for the two smaller 2-connected graphs H_1 and K_1. (See Theorem IV.21.) Using the theory of Secs. IV.3 and IV.4, we can say that a 2-connected graph is planar if and only if its 3-blocks are planar. Among these 3-blocks there may be some circuits and linkages. These are planar, being graphs of circuit-maps and their duals.

It is possible to have two nonisomorphic planar maps on the same graph. Consider, for example, the construction of Theorem XI.20, of which a particular case is shown in Fig. XI.4.2.

In this example we have complete freedom of choice as to how we open the cyclic sequence $(Y_1, Y_2, Y_3, Y_4, Y_5)$ into a linear sequence. We chose to break it at the turn (Y_5, Y_1). Had we broken it at (Y_2, Y_3), we would have got the map of Fig. XI.4.3, which is not isomorphic to that of Fig. XI.4.2. There is, of course, the same freedom of choice with the X's.

In another variation on the construction, we replace (X_1, X_2, X_3) by its conjugate sequence $(\theta X_3, \theta X_2, \theta X_1)$, to obtain the map of Fig. XI.4.4.

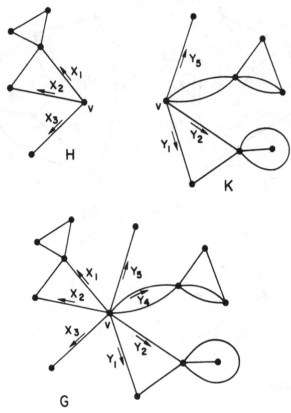

FIGURE XI.4.2

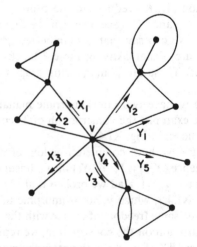

FIGURE XI.4.3

Connectivity in Planar Maps

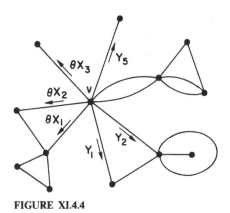

FIGURE XI.4.4

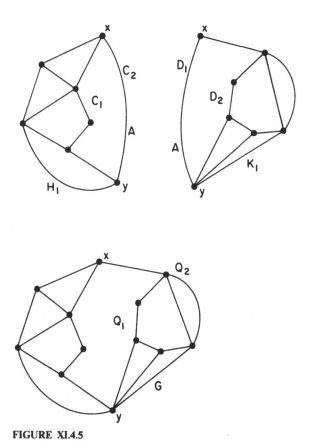

FIGURE XI.4.5

There is also an ambiguity in the construction of Theorem XI.28. There we could interchange the roles of D_1 and D_2, putting $Q_1 = (C_1 \cup D_2)'_A$ and $Q_2 = (C_2 \cup D_1)'_A$. The construction would then go as in Fig. XI.4.5. The final maps of Figs. XI.4.1 and XI.4.5 are not isomorphic, and their duals do not even have isomorphic graphs, for the map of Fig. XI.4.5 has a face of valency 8 and that of Fig. XI.4.1 does not. (The duals of the maps of Figs. XI.4.2, XI.4.3, and XI.4.4 do have isomorphic graphs.)

These ambiguities trouble us only for separable and 2-separable graphs. We shall prove a theorem of Hassler Whitney to the effect that a 3-connected planar graph has only one planar mesh.

XI.5. THE CROSS-CUT THEOREM

Let M be a planar map, C a circuit of $G(M)$, and H and K the residual graphs of C with respect to M.

A subgraph J of $G(M)$ is *enclosed* by H if each of its edges and vertices is enclosed by H. It is *marginally enclosed* by H if each of its edges and vertices either belongs to C or is enclosed by H. Referring to the definitions of cap-graph and cap-map in Sec. XI.3, we see that these definitions can be put equivalently as follows: J is *marginally enclosed* by H if it is a subgraph of $\text{Cap}(G(M), C, H)$. It is *enclosed* by H if it is such a subgraph not meeting C.

Theorem XI.29. *Let L be an arc in $G(M)$ avoiding C. Then L is marginally enclosed either by H or by K.*

Proof. Write the defining sequence of vertices of L as $(a_0, a_1, \ldots, a_n)$, and that of edges as $(A_1, A_2, \ldots, A_n)$. Since A_1 is not in $E(C)$, it is enclosed by either H or K, let us say H. If possible, let A_j be the first member of the edge-sequence not enclosed by H, that is enclosed by K. Then a_{j-1} is a vertex of C, by Eq. (XI.3.11), which is contrary to hypothesis. We conclude that each edge of L is enclosed by H, and therefore that L is marginally enclosed by H. □

An arc L in $G(M)$ avoiding C but having its two ends x and y in C is called a *cross-cut* of C from x to y. It is a cross-cut *through* whichever residual graph H or K of C marginally encloses it. Only one of them can do so since the edges of L are outside $E(C)$. By the definitions of arc and circuit x and y dissect C into two internally disjoint arcs L_1 and L_2 whose union is C and whose intersection is the edgeless subgraph U of C with vertices x and y. We can recognize L_1 and L_2 as the two bridges of U in C. We call them also the *residual arcs* in C of the cross-cut L, or of x and y.

Theorem XI.30 (The Cross-Cut Theorem). *Let L be a cross-cut of C from x to y through H, with residual arcs L_1 and L_2 in C. Let H' be the* subgraph of $G(M^*)$ obtained from H by deleting the edges in $E(L)$. Then H' has exactly two components, one being a residual graph of the circuit $L \cup L_1$ and the other of $L \cup L_2$.

Proof. One residual graph K_1 of $L \cup L_1$ must contain the connected subgraph K of $G(M^*):(E(M)-E(C))$. It therefore includes also the edges of L_2. So the other residual graph, H_1 say, of $L \cup L_1$ must be a subgraph of H'. Similarly, the circuit $L \cup L_2$ has a residual graph H_2 contained in H'.

H_1, and similarly H_2, is detached in H', for any edge of $G(M^*)$, not in $E(H_1)$ but incident with a vertex of H_1, is an edge of L or L_1. Hence H_1 and H_2 are components of H'. They are distinct components, being endgraphs of distinct bonds of $G(M^*)$.

Suppose H' to have a third component H_0. By the connection of $G(M^*)$, there is an edge A of that graph having one end a in $V(H_0)$ and one end b not in $V(H_0)$. If b is in $V(H)$, then A must be in $E(L)$. If b is in $V(K)$, then A is in $E(C)$. In either case, A belongs to one of the bonds $E(L) \cup E(L_1)$ and $E(L) \cup E(L_2)$ of $G(M^*)$, and its end a is a vertex of one of their end-graphs in H', that is, of H_1 or H_2. But this is contrary to the choice of A. We conclude that H_1 and H_2 are the only components of H'. □

There is a more symmetrical way of putting the Cross-Cut Theorem. Let us write $L = L_3$ and $K = H_3$. Then Theorem XI.30 can be restated as follows:

Theorem XI.31. *Let Q be a subgraph of $G(M)$ that is the union of three internally disjoint arcs L_1, L_2, and L_3 with common ends x and y. Then the subgraph $G(M^*):(E(M)-E(Q))$ of $G(M^*)$ has just three components H_1, H_2, and H_3, these being residual graphs of the circuits $L_2 \cup L_3$, $L_3 \cup L_1$, and $L_1 \cup L_2$, respectively.*

A graph such as Q in Theorem XI.31 which is the union of three internally disjoint arcs L_1, L_2, and L_3 with common ends, is called a *theta-graph* (θ-graph). Its three circuits $L_2 \cup L_3$, $L_3 \cup L_1$, and $L_1 \cup L_2$ constitute a planar mesh. Thus, by Theorem XI.24, we have the following:

Theorem XI.32. *Every theta-graph is planar.*

Theorem XI.31 has the following generalization.

Theorem XI.33. *Let M be a planar map. Let G be a 2-connected subgraph of $G(M)$, not a vertex-graph or a link-graph. Then the number of components of $G(M^*):(E(M)-E(G))$ is $p_1(G)+1$, $=n$, say. Suppose, moreover, that these components are enumerated as $H_1, H_2, \ldots, H_n$. Then G has a planar mesh $F = (C_1, C_2, \ldots, C_n)$ such that H_j is a residual graph of C_j for each suffix j.*

Proof. If G is a circuit, and in particular if G is a loop-graph, the theorem holds, by Theorem XI.9. So the theorem is true for $|E(G)| = 1$.

Assume it true whenever $|E(G)|$ is less than some integer $q \geq 2$, and consider the case $|E(G)| = q$.

By Theorem III.11, G is the union of a 2-connected graph G' and an arc L that avoids G' but has its two ends x and y in G'. We can apply Eq. (I.6.1) to show that $p_1(G') = p_1(G) - 1 = n - 2$.

Now G' is not a vertex-graph since it has at least two vertices. If it is a link-graph, then G is a circuit and the theorem holds. In the remaining case, we apply the theorem to G' as follows: The graph $G(M^*) : (E(M) - E(G'))$ has exactly $(n-1)$ components $H'_1, H'_2, \ldots, H'_{n-1}$, and G' has a planar mesh $F' = (C'_1, C'_2, \ldots, C'_{n-1})$ such that H'_j is a residual graph of C'_j for each suffix j.

Now L has an edge in common with some H'_j, let us say with H'_{n-1}. Then L is marginally enclosed by H'_{n-1}, by Theorem XI.29. Since it avoids G' it must be a cross-cut of C'_{n-1} from x to y through H'_{n-1}.

Let the residual arcs of L in C'_{n-1} be L_1 and L_2. Replacing C'_{n-1} in F' by the circuits $C_{n-1} = L \cup L_1$ and $C_n = L \cup L_2$, we obtain a planar mesh $F = (C'_1, \ldots, C'_{n-2}, C_{n-1}, C_n)$ of G, by Theorem XI.26. By the Cross-Cut Theorem, the graph $H'_{n-1} : (E(H'_{n-1}) - E(L))$ has exactly two components H_{n-1} and H_n, residual graphs of C_{n-1} and C_n, respectively. But the components of $G(M^*) : (E(M) - E(G))$ are those of $H'_{n-1} : (E(H'_{n-1}) - E(L))$, together with the components of $G(M^*) : (E(M) - E(G'))$ other than H'_{n-1}, since these other components involve no edges of L. So the theorem is verified for M and G. It follows in general by induction. □

The planar mesh F of Theorem XI.33 is uniquely determined by those bonds of $G(M^*)$ that have components of $G(M^*) : (E(M) - E(G))$ as end-graphs. We call it the planar mesh *imposed* by M on its subgraph G. Thus M imposes on $G(M)$ the planar mesh whose members are the trace-circuits of the faces of M. (See Theorem XI.23.)

Besides saying that M imposes on G the planar mesh $F = (C_1, \ldots, C_n)$, we can say that G decomposes M into the n cap-maps $\text{Cap}(M, C_j, H_j)$. Each face of M (vertex of M^*) is a face of just one of these. Moreover, all the faces of $\text{Cap}(M, C_j, H_j)$ are faces of M, except for the one "external" face, with trace-circuit C_j, associated with the second residual graph K_j of C_j. We can likewise say that G decomposes $G(M)$ into the n cap-graphs $\text{Cap}(G(M), C_j, H_j)$. For distinct suffixes j and k it is clear from the definitions that $\text{Cap}(G(M), C_j, H_j)$ is a subgraph of $\text{Cap}(G(M), C_k, K_k)$. Hence the intersection of $\text{Cap}(G(M), C_j, H_j)$ and $\text{Cap}(G(M), C_k, H_k)$ is $C_j \cap C_k$, by Eqs. (XI.3.10) and (XI.3.11).

We define a *Y-graph* as the union Y of three arcs, L_1, L_2, and L_3, which have one end y in common but are otherwise mutually disjoint. We call y the *center* and the arcs L_i the *arms* of Y. The end of L_i other than y is the *outer end* x_i of L_i. The three vertices x_1, x_2, x_3 are the *ends* of Y. Its other vertices, including y, are its *internal* vertices.

Suppose Y to be a subgraph of a graph G. It is said to *avoid* another subgraph H of G if it has no edge or internal vertex in H.

Theorem XI.34. *Let M be a planar map. Let C be a circuit of $G(M)$ with residual graphs H and K. Let Y, described above, be a subgraph of $G(M)$ and let it avoid C. Then the Y-graph Y is marginally enclosed by either H or K.*

Proof. The center y of Y is enclosed by H or K, let us say by H, by Theorem XI.11. Hence L_1, L_2, and L_3 are marginally enclosed by H, by Theorem XI.29. The theorem follows. □

Let us consider the special case of Theorem XI.34 in which the ends of Y are vertices of C. By the definitions of arc and circuit x_1, x_2, and x_3 dissect C into three internally disjoint arcs N_1 with ends x_2 and x_3, N_2 with ends x_3 and x_1, and N_3 with ends x_1 and x_2. The union of the N_i is C and the intersection of any two is a vertex-graph (Fig. XI.5.1).

Theorem XI.35. *Let M, C, and Y be as in Theorem XI.34. Let the ends of Y be in C, dissecting it into three arcs N_i as described above. Let Y be marginally enclosed by H. Then $C \cup Y$ is 2-connected. Moreover, M imposes on it the planar mesh whose members are C, $N_1 \cup L_2 \cup L_3$, $N_2 \cup L_3 \cup L_1$, and $N_3 \cup L_1 \cup L_2$. The corresponding residual graphs, components of $G(M^*):(E(M) - E(C \cup Y))$, are K, H_1, H_2, and H_3, respectively, where H_1, H_2, and H_3 are disjoint proper subgraphs of H.*

Proof. First we adjoin to C the arc $L_1 \cup L_2$. The resulting graph C' has the planar mesh $(C, N_3 \cup L_1 \cup L_2, N_1 \cup N_2 \cup L_1 \cup L_2)$, by Theorem XI.26. Moreover, by Theorem XI.30, this is the planar mesh imposed on C'

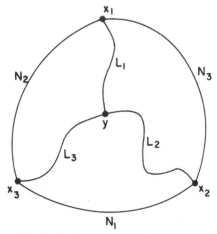

FIGURE XI.5.1

by M. The corresponding residual graphs are K for C and subgraphs H' and H_1 of H for $L_1 \cup L_2 \cup N_3$ and $N_1 \cup N_2 \cup L_1 \cup L_2$, respectively.

In adjoining L_3 we argue as for L in Theorem XI.33. Thus, L_3 is marginally enclosed by one of the three residual graphs, it is a subgraph of the corresponding cap-graph, and it has its two ends in the corresponding circuit of the planar mesh. This circuit can only be $N_1 \cup N_2 \cup L_1 \cup L_2$. An application of Theorem XI.30, as in the proof of Theorem XI.33, now gives us the planar mesh of the enunciation, imposed on $C \cup Y$ by M. For the 2-connection of $C \cup Y$ we can appeal to Theorem XI.22, or to Theorem III.10. □

Let L and L' be two cross-cuts of the circuit C in $G(M)$. Let the ends of L be x and y, and those of L' be x' and y'. We say that L and L' are *skew* if they are disjoint and the ends of L separate those of L' in the circuit C. For the second condition, we can say equivalently that L' has one end in each of the residual arcs of L in C.

Theorem XI.36. *Let M be a planar map, C a circuit of $G(M)$, and L and L' skew cross-cuts of C in $G(M)$. Then no residual graph of C marginally encloses both L and L'.*

Proof. Let the residual arcs of L in C be L_1 and L_2, L' having an end in each of them. Using Theorem XI.30 and taking L to be marginally enclosed by H (see Theorem XI.29), we find that M imposes on $C \cup L$ the planar mesh $(C, L \cup L_1, L \cup L_2)$. These circuits have residual graphs, components of $G(M^*):(E(M) - E(C \cup L))$, that are K, H_1, and H_2, respectively, where H_1 and H_2 are proper subgraphs of H. But now L' must be marginally enclosed by one of these three residual graphs, and it must have its two ends in the corresponding circuit of the planar mesh. This circuit can only be C: L' is marginally enclosed by K. Since L has no edge in $E(K)$, and L' no edge in $E(H)$, the theorem follows. □

Theorem XI.37. *Let M be planar map and C a circuit of $G(M)$. Let Y_1 and Y_2 be Y-graphs in $G(M)$ that avoid each other and C. Suppose, further, that they have their ends x_1, x_2, and x_3 in common and that these ends are vertices of C. Then no residual graph of C marginally encloses both Y_1 and Y_2.*

Proof. First we adjoin $Y = Y_1$ to C, obtaining the planar mesh described in Theorem XI.35. By Theorem XI.34, Y_2 must be marginally enclosed by one of the four corresponding residual graphs. It must be a subgraph of the corresponding cap-graph, and it must have its three ends in the corresponding member of the planar mesh. This member can only be C: Y_2 is marginally enclosed by K, whereas the description in Theorem XI.35 has Y_1 marginally enclosed by H. The theorem follows. □

XI.6. BRIDGES

Let C be a circuit in a connected graph G. If B is a bridge of C in G, we write $w(B)$ for the number of its vertices of attachment. Thus $w(B) \geq 1$ for each B. Moreover, if G is 2-connected, then $w(B) \geq 2$ for each B, by Theorem III.6.

If $w(B) \geq 2$, then the vertices of attachment of B decompose the circuit C into $w(B)$ internally disjoint arcs. If the vertices of attachment are enumerated, in their cyclic order in C, as $v_1, v_2, \ldots, v_{w(B)}$, then these *residual arcs of B* can be enumerated as $L_1, L_2, \ldots, L_{w(B)}$, so that the ends of L_j are v_{j-1} and v_j ($v_0 = v_{w(B)}$). The residual arcs of B in C are the bridges in C of the edgeless subgraph specified by the v_i.

Two distinct bridges B_1 and B_2 of C in G are said to *avoid* one another if either one of them has only one vertex of attachment or if all the vertices of attachment of one belong to a single residual arc of the other. This relation between two bridges is clearly symmetrical. Two bridges that do not avoid one another are said to *overlap*.

Two distinct bridges B_1 and B_2 of C in G are called *skew* if some two vertices of attachment of B_1 separate some two vertices of attachment of B_2 in C. They are *equivalent* if they have the same vertices of attachment. A bridge B with $w(B) = n$ is called an *n-bridge*.

Theorem XI.38. *Let B_1 and B_2 be distinct bridges of C in G that are either skew or equivalent 3-bridges. Then B_1 and B_2 overlap.*

Proof. Suppose B_1 and B_2 skew, with vertices of attachment a_1 and b_1 of B_1 separating two vertices of attachment a_2 and b_2 of B_2. Then a_1 and b_1 partition C into two internally disjoint arcs L and L', whereof one has a_2 and the other b_2 as an internal vertex. Each residual arc of B_1 in C is a subgraph either of L or of L'. Hence no such residual arc can include both a_2 and b_2; B_1 and B_2 must overlap.

If B_1 and B_2 are equivalent 3-bridges, then clearly each residual arc of B_1 in C includes only two vertices of attachment of B_2. □

Theorem XI.39. *Let B_1 and B_2 be distinct overlapping bridges of C in G. Then either they are skew or they are equivalent 3-bridges.*

Proof. Suppose first that some vertex of attachment v of B_1 is not a vertex of attachment of B_2. Then v is internal to some residual arc L of B_2 in C, with ends x and y say. Because of the overlap, B_2 must also have a vertex u of attachment that is internal to the complementary arc L' to L in C. Hence B_1 and B_2 are skew. Similarly, they are skew if some vertex of attachment of B_2 is not a vertex of attachment of B_1.

In the remaining case B_1 and B_2 are equivalent n-bridges for some integer $n \geq 2$. If $n \geq 4$, we can find two vertices of attachment separating two others in C, and therefore B_1 and B_2 are skew. If $n = 2$, then B_1 and B_2

avoid one another, contrary to hypothesis. Hence B_1 and B_2 are equivalent 3-bridges. □

Theorem XI.40. *Let x, y, and z be distinct vertices of attachment of a bridge B of C in G. Then B contains a Y-graph, avoiding C, with ends x, y, and z.*

Proof. There is an arc L in B that joins x and y and avoids C, by Theorem I.56. It has an internal vertex t, by Theorem I.49. There is an arc N in B that joins t and z and avoids C, by Theorem I.56. Choose t and N so that N has as few edges as possible. Then the intersection of N and L must be the vertex-graph at t, and $N \cup L$ is the required Y-graph. □

The next theorems show how bridge theory can be applied to problems of planarity.

Theorem XI.41. *Let G be the graph of a planar map M. Let B be a bridge of C in G. Then B is marginally enclosed by one and only one residual graph of C, with respect to M.*

Proof. Let the residual graphs of C be H and K. Choose an edge A of B. It belongs either to $E(H)$ or to $E(K)$, let us say to $E(H)$. If B is degenerate, having only the one edge A, the theorem is established.

If B is nondegenerate, then A has an end x in the nucleus $N(B)$ of B. But x is enclosed by H but not by K, by Theorem XI.11. If y is any other vertex of $N(B)$, then x and y are joined by an arc in B disjoint from C, by Theorem I.56. Hence y is enclosed by H but not by K, by Theorems XI.29 and XI.11. Each edge of B has an end in $N(B)$ and is therefore enclosed by H. Hence B is marginally enclosed by H, but not by K. □

Theorem XI.42. *Let G be the graph of a planar map M. Let B_1 and B_2 be overlapping bridges of C in G. Then they are marginally enclosed by distinct residual graphs of C, with respect to M.*

Proof. Suppose B_1 and B_2 skew. Then by Theorem I.56, we can find two skew cross-cuts of C in $G(M)$, one in B_1 and the other in B_2. Hence the residual graphs of C enclosing marginally B_1 and B_2 must be distinct, by Theorem XI.36.

In the remaining case B_1 and B_2 are equivalent 3-bridges, by Theorem XI.39, let us say with vertices of attachment x, y, and z. By Theorem XI.40 we can find Y-graphs Y_1 and Y_2, the first in B_1 and the second in B_2, such that each avoids C and has x, y, and z as its ends. Hence B_1 and B_2 are marginally enclosed by distinct residual graphs of C, by Theorem XI.37. □

We can often use Theorem XI.41 and XI.42 to demonstrate non-planarity. If, for example, we find that C has three mutually overlapping bridges in G, we can assert that G is nonplanar.

Bridges

Let us define the *bridge-graph* of C in G as the strict graph whose vertices are the bridges of C in G, and in which two vertices are joined by an edge if and only if they overlap. Then we have the following test for nonplanarity.

Theorem XI.43. *If the bridge-graph of C in G is nonbipartite, then G is nonplanar.*

Proof. Suppose G is planar, so that we can identify it with the graph of some planar map M. We can arrange the bridges of C in G in two classes, according as to which residual graph of C marginally encloses them (see Theorem XI.41). Two members of the same class do not overlap, by Theorem XI.42. Thus the bridge-graph of C in G is bipartite. □

Let a graph G_1, in which six vertices $a_1, a_2, a_3, b_1, b_2, b_3$ are to be specified as the *nodes*, be defined as the union of nine internally disjoint arcs L_{ij} ($i, j = 1, 2, 3$), the ends of L_{ij} being a_i and b_j. We call G_1 a subdivided $K_{3,3}$. It is indeed a graph obtainable from $K_{3,3}$ by repeated subdivision of edges. The arcs L_{ij} are its *branches*.

Similarly, we can define a graph G_2, in which five vertices a_1, a_2, a_3, a_4, and a_5 are specified as *nodes*, as the union of ten internally disjoint arcs L_{ij} ($1 \le i < j \le 5$) called the *branches*. The ends of L_{ij} are to be a_i and a_j. We call G_2 a subdivided 5-clique K_5.

Graphs G_1 or G_2 defined in these ways are called *Kuratowski graphs*. In each case the nodes are the nondivalent vertices and the branches are the bridges of the edgeless subgraph defined by the nodes.

Theorem XI.44. *The Kuratowski graphs are nonplanar.*

Proof. In the case of G_1 there is a circuit C made up of $L_{12}, L_{32}, L_{31}, L_{21}, L_{23}$, and L_{13}. (See Fig. XI.6.1.) Its three bridges L_{11}, L_{22}, and L_{33} are mutually overlapping. Hence G_1 is nonplanar, by Theorem XI.43.

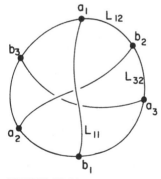

FIGURE XI.6.1

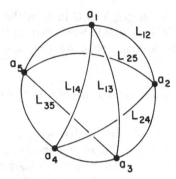

FIGURE XI.6.2

In the case of G_2, Fig. XI.6.2, there is a circuit C made up of L_{12}, L_{23}, L_{34}, L_{45}, and L_{15}. In the cyclic sequence $(L_{13}, L_{24}, L_{35}, L_{14}, L_{25})$ of its five bridges, each overlaps its two neighbors. Hence G_2 is nonplanar, by Theorem XI.43. □

XI.7. AN ALGORITHM FOR PLANARITY

We consider a circuit C in a 2-connected graph G. If B is a bridge of C in G, then its union with C is the corresponding *bridge-fragment* of G, with respect to C.

Theorem XI.45. *Each bridge-fragment of G is 2-connected.*

Proof. Let B be a bridge of C in G, and $B \cup C$ the corresponding bridge-fragment. If $B \cup C$ is not 2-connected, then it has a 1-separation (H, K) with a cut-vertex v. By Theorem III.4, we may suppose C to be a subgraph of K. But then G has the 1-separation (H, H^c), with cut-vertex v, contrary to the definition of G. □

If C has exactly one bridge in G, we call it a *peripheral circuit of G*.

Theorem XI.46. *A peripheral circuit of G is a member of every planar mesh of G.*

Proof. Let C be a peripheral circuit of G. Let $F = (C_1, C_2, \ldots, C_n)$ be a planar mesh of G. Then there is a planar map M whose graph can be identified with G and for which the C_j are the trace-circuits of the faces in $G = G(M)$, by Theorem XI.25. By Theorem XI.41, one of the residual graphs of C in M is edgeless, i.e., is a vertex-graph. Here $E(C)$ is the trace of the corresponding vertex of M^* and so C is the trace-graph of a face of M. It is thus a member of F. □

Theorem XI.47. *Let the bridge-fragments of G with respect to C be all planar, and let no two bridges of C in G overlap. Then G is planar and C is a member of one of its planar meshes. Moreover, if C has more than one bridge in G, then G has a 2-separation.*

Proof. Write m for the number of bridges of C in G. If $m = 0$, then $G = C$ and the theorem is trivial. If $m = 1$, then G is planar, by hypothesis, and the theorem follows from Theorem XI.46. Assume it true whenever m is less than some integer $q \geq 2$, and consider the case $m = q$.

Choose two bridges B_1 and B_2 of C. There is a residual arc L of B_1 in C, with ends x and y say, that includes all the vertices of attachment of B_2. Because no other bridge overlaps B_1, we can now separate the bridges of C in G into two complementary nonnull classes U_1 and U_2. Let L' be the complementary arc to L in C. Then U_1 consists of B_1 and all the remaining bridges of C, other than B_2, that have all their vertices of attachment in L'. The class U_2 consists of B_2 and those other bridges of C that have their vertices of attachment all in L but not all in L'. Some care is needed in this definition to take account of possible 2-bridges with vertices of attachment x and y, in both L and L'.

We can now recognize a 2-separation (H, K) of G. Here H is the union of L' and the members of U_1, and K is the union of L and the members of U_2. The hinges are x and y. Form H_1 and K_1 from H and K, respectively, by adjoining a new link A with ends x and y.

Consider the union G_i of C and the members of U_i ($i = 1, 2$). Clearly the members of U_i are the bridges of C in G_i. They have the same residual arcs and determine the same bridge-fragments as in G. So, by the inductive hypothesis, G_i is planar, and it has a planar mesh F_i having C as one of its members.

Now the two members of F_1 containing a given edge of L (one of which is C) must each contain the whole of L, for the internal vertices of L are divalent in G_1. Let us delete the edges and internal vertices of L from each of these circuits, replacing them in each by the single link A. We thus convert F_1 into a family F_1' of circuits of H_1 satisfying the conditions for a planar mesh of H_1. We note that one member of F_1' is formed from L' by adjoining A. Similarly, we can find a planar mesh of K_1 in which one member is formed from L by adjoining A. Applying Theorem XI.28, we find that G has a planar mesh F having $L \cup L'$, that is, C, as one of its members.

Thus the theorem holds when $m = q$, by Theorem XI.24. It follows in general by induction. □

Theorem XI.48. *Let the bridge-fragments of G with respect to C be all planar, and let the bridge-graph of C in G be bipartite. Then G is planar.*

Proof. We can arrange the bridges of C in G in two complementary classes U_1 and U_2 so that no two members of the same class overlap. Let G_i

denote the union of C and the members of U_j ($i = 1, 2$). The members of U_j are the bridges of C in G_j, having the same residual arcs and determining the same bridge-fragments as in G. Thus no two of them overlap as bridges of C in G_j.

By Theorem XI.47, G_j has a planar mesh F_j having C as a member, for each i. Hence G is planar, by Theorem XI.27.   □

Theorem XI.49. *Let G be any 2-connected graph that is not a vertex-graph, a link-graph, or a circuit. Then either G is a theta-graph or we can find in it a circuit C having at least two bridges in G.*

Proof. We can choose an edge of G, and this edge is not an isthmus. Hence we can find a circuit Q of G. But Q is not the whole of G and therefore Q has at least one bridge in G. If it has more than one, we have finished. Suppose therefore that Q is peripheral, with the single bridge B.

Let L be a residual arc of B in Q, with ends x and y. Let L' be its complementary arc in Q. By Theorem I.56, there is an arc N in B whose ends are x and y, and which avoids Q. Let C be the circuit $N \cup L'$.

Now L is a subgraph of G having x and y as its only vertices of attachment. It is thus a bridge of C in G, by Theorem I.55. If C has a second bridge, the theorem holds. If not, then G is a theta-graph, being the union of the three internally disjoint arcs L, L', and N.   □

From the foregoing theorems we can construct an algorithm for determining whether a given 2-connected graph G is planar. First we check whether G is a vertex-graph, a link-graph, a circuit, or a theta-graph. In all these cases it is planar. (See Theorems XI.3 and XI.32.) In the remaining case we find a circuit C in G with two or more bridges, by the method of Theorem XI.49. We examine its bridge-graph. If this is not bipartite, we can assert that G is nonplanar, by Theorem XI.43. If the bridge-graph is bipartite, then G is planar if and only if all its bridge-fragments with respect to C are planar, by Theorems XI.19 and XI.48. The problem for G is thus reduced to that for some smaller 2-connected graphs, by Theorem XI.45.

Consideration of this algorithm leads to the following necessary and sufficient condition for nonplanarity.

Theorem XI.50. *A 2-connected graph G is nonplanar if and only if it has a subgraph H in which there is a circuit C such that C has a nonbipartite bridge-graph in H.*

Theorem XI.51. *Let C be a circuit in a 2-connected planar graph G. Then C belongs to some planar mesh of G if and only if no two bridges of C in G overlap.*

Proof. Suppose C to belong to some planar mesh F of G. We can identify G with the graph of a planar map M so that the members of F are the trace-circuits of the faces of M, by Theorem XI.25. Since C is the trace-graph of a face of M, one of its residual graphs with respect to M is a vertex-graph. The other must marginally enclose each bridge of C in G, by Theorem XI.41. Hence no two bridges of C in G overlap, by Theorem XI.42.

Conversely, suppose that no two bridges of C in G overlap. Then the required planar mesh exists, by Theorems XI.19 and XI.47. □

Theorem XI.52. *Let G be a 3-connected planar graph, not a vertex-graph, a link-graph, or a circuit. Then G has one and only one planar mesh. The members of the planar mesh are the peripheral circuits of G.*

Proof. G has at least one planar mesh F, by Theorem XI.23. If C is a member of F, no two bridges of C in G overlap, by Theorem XI.51. Hence, by Theorem XI.47 and the 3-connection of G, C is a peripheral circuit of G. The theorem follows, by Theorem XI.46. □

Because of Theorem XI.25, we can identify this with Whitney's Theorem that a 3-connected planar graph can be drawn in the plane (i.e., on the sphere) in essentially only one way. Hence a 3-connected planar graph has, to within an isomorphism, only one dual graph [8, 9].

XI.8. PERIPHERAL CIRCUITS IN 3-CONNECTED GRAPHS

In this section we take G to be a 3-connected graph, not necessarily planar, with at least four edges. It is not a circuit, by Theorem IV.3. We describe an algorithm for finding peripheral circuits in G. It is based on the following theorem.

Theorem XI.53. *Let C be a circuit in G having two or more bridges. Let B be one of these bridges. Then some other bridge of C in G overlaps B.*

Proof. We can construct a 2-separation (H, K) of G if no other bridge overlaps B. We do it exactly as in the proof of Theorem XI.47, with B as B_1 and any other bridge of C as B_2. □

Let C be any circuit in G. Let B be one of its bridges in G. Suppose C to have a second bridge B' overlapping B. Consider two vertices of attachment x and y of B'. We can find an arc N in B' that joins x and y and avoids C. The vertices x and y partition C into two internally disjoint arcs L_1 and L_2. We describe the circuit $L_1 \cup N$ as a circuit *based on B'*. L_1 is its *common arc* with C, and L_2 is its *complementary arc* in C. It may happen that B has a vertex of attachment internal to L_2. In that case, we call $L_1 \cup N$ an *augmenting circuit* of B.

Theorem XI.54. *Let B and B' be overlapping bridges of a circuit C in G. Let D be a circuit based on B', and let it be an augmenting circuit of B. Then there is a bridge B_1 of D in G having both B and L_2 as proper subgraphs.*

Proof. The circuit D has a bridge with B as a subgraph, by Theorem I.57, and a bridge with L_2 as a subgraph, by Theorem I.55. It is the same bridge B_1 in each case, for B and L_2 have a common vertex not in D. Since B and L_2 have no common edge, they are proper subgraphs of B_1. □

Theorem XI.55. *Let A be any edge of G. Then G has a peripheral circuit through A.*

Proof. A is not an isthmus of G: We can find a circuit C of G through A. If it is peripheral, we have finished. If not, choose a bridge B of C and another bridge B' of C overlapping it. (See Theorem XI.53.) Retaining the above notation we take L_1 to be the residual arc of B' in C that includes A. Then, because of the overlap, B has at least one vertex of attachment internal to L_2. We construct a circuit D based on B' and having L_1 as its common arc with C. Then D is an augmenting circuit of B. By Theorem XI.54, it has a bridge B_1 with B as a proper subgraph. Moreover, it has A as an edge.

If D is peripheral, we have finished. If not, we repeat the process with D replacing C and B_1 replacing B. We then get a circuit through A with a bridge B_2 having B_1 as a proper subgraph. We repeat the process, making the chosen bridge bigger and bigger, until it terminates. We then have a peripheral circuit through A. □

Theorem XI.56. *Let A be any edge of G. Let P be a peripheral circuit of G through A. Then we can find a second peripheral circuit Q through A such that*

$$P \cap Q = G \cdot \{A\}.$$

Proof. Let the ends of A be b and c. Each is a vertex of attachment of the single bridge B_P of P, by Theorem IV.3. Hence they are joined by an arc in B that avoids P. Adjoining A to this arc, we obtain a circuit C such that $P \cap C = G \cdot \{A\}$.

Let S be the arc, with ends b and c, obtained from P by deleting A. It is a subgraph of a bridge B of C in G, by Theorem I.55.

If C is peripheral, we can take it as Q. If not we find a bridge B' of C overlapping B and proceed as in Theorem XI.55. The resulting bridge B_1 of D, like B, contains S as a subgraph, and the intersection of P with D is still $G \cdot \{A\}$. We infer that the procedure of Theorem XI.55 now leads to a peripheral circuit Q of G satisfying the stated condition. □

We now come to some theorems leading, in the planar case, to the construction of a planar mesh, and in the other case to a demonstration of nonplanarity. We use the same mod 2 addition as in Sec. XI.4. We identify a graph G with each of its orientations, and a member of $\Gamma(G, I_2)$ with its support and with the corresponding reduction of G. Thus we identify the elementary chains of $\Gamma(G, I_2)$ with the circuits of G (see Theorem VIII.34). We use linear algebra appropriate to a vector space over a field, in this case the field $GF(2)$. The number $p_1(G)$ is the rank of $\Gamma(G, I_2)$, the maximum number of linearly independent members of $\Gamma(G, I_2)$ and the maximum number of linearly independent circuits of G. (See Theorems VIII.12 and VIII.37.)

Theorem XI.57. *We can find $p_1(G)$ linearly independent peripheral circuits of G* (see [7]).

Proof. We can find two linearly independent peripheral circuits of G, by Theorem XI.56. Suppose we have found a set T of m linearly independent ones, where $2 \le m < p_1(G)$. We consider how to locate an $(m+1)$th peripheral circuit, independent of them all.

We can find a circuit C of G linearly independent of T. For the chain-base of $\Gamma(G, I_2)$ associated with any cell-base must include at least one such elementary chain. If C is peripheral, we have finished. If not, we choose a bridge B of C in G, and then locate another bridge B' overlapping it (see Theorem XI.53).

Suppose first that B' is skew to B. Keeping to the former notation, we choose our two vertices x and y of attachment of B' so that they separate two vertices of attachment of B in C. Then both $L_1 \cup N$ and $L_2 \cup N$ are augmenting circuits of B. But their mod 2 sum is C. Hence, one of them is independent of T.

In the remaining case B and B' are equivalent 3-bridges, by Theorem XI.39. Let their vertices of attachment be now x, y, and z. Then B' contains a Y-graph Y that avoids C and has x, y, and z as its ends, by Theorem XI.40. Let C_x denote the circuit made up of the residual arc of B and B' in C with ends y and z, and of the arms of Y to these two vertices. Let C_y and C_z be defined analogously. Clearly C_x, C_y, and C_z are circuits based on B', and they are augmenting circuits of B. But their mod 2 sum is C, and so at least one of them is linearly independent of T.

In each case, we have found an augmenting circuit D of B that is independent of T. By Theorem XI.54 it has a bridge B_1 containing B as a proper subgraph. If D is peripheral, we have finished. If not, we repeat the process with D replacing C, and with B_1 replacing B. Continuing in this way until the process terminates, we arrive at a peripheral circuit independent of T.

We can repeat the procedure for successively larger values of m until we have $p_1(G)$ linearly independent peripheral circuits. □

Our $p_1(G)$ linearly independent peripheral circuits constitute a basis for the vector space $\Gamma(G, I_2)$. So each circuit of G can be expressed as a sum of peripheral circuits.

Theorem XI.58. *Let x be a vertex of G. Let the set E_x of edges incident with x be partitioned into complementary nonnull subsets T and U. Then there is a peripheral circuit of G with one edge in T and one in U.*

Proof. Let A and B be edges of T and U, respectively. Let them join x to vertices a and b, respectively. Then a and b are distinct, by Theorem IV.1. Since x is not a cut-vertex of G, they are joined by an arc in G that does not include x. Hence there is a circuit C of G through A and B. It includes, of course, no other edge of E_x.

By Theorem XI.57, C is a mod 2 sum of peripheral circuits. So, for at least one peripheral circuit, the number of common edges with T is odd. Such a peripheral circuit has the required property. □

The next theorem provides another test for planarity.

Theorem XI.59. *A 3-connected graph is nonplanar if and only if it has three distinct peripheral circuits with a common edge.*

Proof. The application to the six graphs listed in Theorem IV.6, all planar, is trivial. So we can suppose $|E(G)| \geq 4$.

If G has three such peripheral circuits, it has no planar mesh, by Theorem XI.52. It is nonplanar, by Theorem XI.23.

In the remaining case, G has exactly two peripheral circuits through each edge, by Theorem XI.56. Let there be n peripheral circuits in all. They sum to zero mod 2. Hence, by Theorem XI.57, $n \geq p_1(G)+1$. But no nonnull proper subset of them sums to zero. Such a subset would define a partition of $E(G)$ into two complementary nonnull subsets E_1 and E_2 such that no peripheral circuit had an edge in each of them. By the connection of G, this would contradict Theorem XI.58. Hence, if we take all but one of the peripheral circuits of G, we have a set of linearly independent circuits. Since the maximum number of linearly independent circuits of G is $p_1(G)$, we conclude that $n = p_1(G)+1$.

We have now shown that the set of peripheral circuits of G satisfies the three conditions for a planar mesh. Hence G is planar, by Theorem XI.24. □

XI.9. KURATOWSKI'S THEOREM

The best-known test for planarity is Kuratowski's Theorem [4], which runs as follows.

Theorem XI.60. *A graph G is nonplanar if and only if it has a Kuratowski graph as a subgraph.*

The theorem can be derived from one or other of the tests for planarity already established. Perhaps it is fair to say that most of the published proofs, translated into the terminology of the present work, appear as developments of Theorem XI.50. Another approach is to work with 3-connected graphs and exploit Whitney's Theorem (see [6]). For variety we derive it here from Theorem XI.59. Kuratowski's Theorem has a generalization in the theory of matroids. There, tests for planarity are apt to reappear, perhaps in slightly more complicated from, as tests for graphicality.

All proofs of Kuratowski's Theorem seem to need the following lemma.

Lemma XI.61. *Let A be a link of a graph G. Then if G_A'' has a Kuratowski subgraph, so does G.*

Proof. Let the ends of A in G be x and y. Let K be a Kuratowski subgraph of G_A''. Let H be the reduction of G having the same edges as K, with the adjunction of A. The vertices x and y correspond to a single vertex v_A in G_A''. Otherwise we identify each vertex of G with its corresponding vertex in G_A''.

Suppose first that x is monovalent in H, incident only with A. Deleting A and x from H, we obtain a subgraph of G vertex-isomorphic with K, differing from K only in the replacement of v_A by y. So the theorem holds.

We may now suppose that each of x and y are at least divalent in H. Suppose x divalent in H. Then H is a Kuratowski graph, derived from K by lengthening one of the arcs L_{ij} by a single new edge.

We may now suppose each of x and y at least trivalent in H. This can happen only if K is a subdivided K_5 having v_A as a node. We may suppose v_A to be the node a_5, and that it is replaced in H by x as the end of L_{15} and L_{25}, and by y as an end of L_{35} and L_{45}. (See Fig. XI.9.1.) In this connection, a "subdivided K_5" may be simply a K_5.

We now have to verify that the graph H represented in Fig. XI.9.1 contains a subdivided $K_{3,3}$. One such subdivided $K_{3,3}$ has nodes x, a_3, a_4, y, a_1, and a_2, corresponding, respectively, to those denoted by a_1, a_2, a_3, b_1, b_2, and b_3 in the description of G_1 in Sec. XI.6. The branches are

$$H \cdot \{A\}, \quad L_{15}, \quad L_{25}, \quad L_{13}, \quad L_{23}, \quad L_{35}, \quad L_{14}, \quad L_{24}, \quad \text{and} \quad L_{45}.$$

This verification completes the proof. □

Theorem XI.62. *Let P and Q be distinct peripheral circuits in a 3-connected graph G, having at least two vertices in common. Then either $P \cap Q$ is a link-graph or G contains a subdivided $K_{3,3}$.*

Proof. Some common vertex x of P and Q is incident with an edge A of Q that does not belong to P, for otherwise P would include all the edges of Q, and P and Q would be identical, by Theorem I.27.

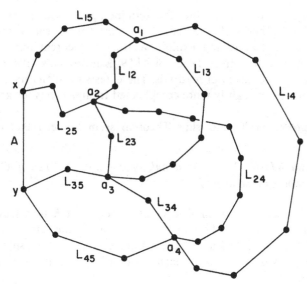

FIGURE XI.9.1

Now A must be part of an arc N_1 in Q that avoids P but has both its ends in P. One end is x. Let the other be y. The two ends of N_1 dissect P into two complementary arcs P_1 and P_2.

If A is the only edge of N_1, then $G \cdot \{A\}$ must be the single bridge of P in G. Hence, G is a 3-linkage, by Theorem IV.3, and the theorem is trivially true. A similar argument applies if either P_1 or P_2 is a link-graph, unless that link-graph is contained in Q as well as P. But if $P_i \subseteq Q$, then Q can only be the circuit $N_1 \cup P_i$. Then $P \cap Q$ is a link-graph and the theorem is still satisfied.

In the remaining case, each of N_1, P_1, and P_2 has an internal vertex, and so G has at least six edges. The bridges of P and Q in G each include all the vertices of G, by Theorem IV.3. By Theorem I.56, any internal vertex r of P_1 can be joined to any internal vertex s of P_2 by an arc N_2 in G that avoids Q. Choosing r, s, and N_2 so that N_2 is as short as possible, we can arrange that N_2 avoids P as well as Q. (See Fig. XI.9.2.)

In N_1 and N_2 we have two skew cross-cuts of P. As with N_1, we may suppose N_2 to have an internal vertex. Any internal vertex n_1 of N_1 can be joined to any internal vertex n_2 of N_2 by an arc L in G that avoids P, by Theorem I.56. Choosing n_1, n_2, and L so that L is as short as possible, we can arrange that L meets N_1 only in n_1, and N_2 only in n_2.

We have now constructed a subdivided $K_{3,3}$ in G. Its nodes are x, y, n_2, r, s, and n_1, corresponding to $a_1, a_2, a_3, b_1, b_2, b_3$ in the description of G_1 in Sec. XI.4. The branches are L, the arcs into which N_1 and N_2 are dissected by n_1 and n_2, respectively, and the four arcs into which P is dissected by x, y, r, and s. □

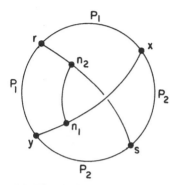

FIGURE XI.9.2

Corollary XI.63. *Let F be a planar mesh of a 3-connected graph G. Then the intersection of any two distinct members of F is null, a vertex-graph, or a link-graph.* (See Theorem XI.52.)

Proof of Theorem XI.60. First we note that, if G has a Kuratowski subgraph, it is nonplanar, by Theorems XI.19 and XI.44.

Conversely, suppose G to be nonplanar. We try to show that it has a Kuratowski subgraph.

Consider first the case in which G is 3-connected. By Theorem XI.59, it has three distinct peripheral circuits P_1, P_2, and P_3 with a common edge A. Let the ends of A be x and y. By Theorem XI.62 we may suppose that the intersection of any two of the P_i is $G \cdot \{A\}$. The single bridge of any P_i in G includes all the vertices of G, by Theorem IV.3.

Let the complementary arcs of $G \cdot \{A\}$ in P_1, P_2, and P_3 be L_1, L_2, and L_3, respectively. Each of these has an internal vertex, by Theorem I.49. Any internal vertex r_1 of L_1 is joined to any internal vertex s_2 of L_2 by an arc N_3 in G avoiding P_3, by Theorem I.56. We choose r_1, s_2, and N_3 so that N_3 is as short as possible. Then N_3 meets L_1 only at r_1, L_2 only at s_2, and L_3 not at all. Similarly, we construct an arc N_2 with ends r_3 in L_3 and s_1 in L_1, and an arc N_1 with ends r_2 in L_2 and s_3 in L_3. Figure XI.9.3 shows a case in which N_1, N_2, and N_3 are mutually disjoint.

Let us consider what happens if two of the N_i, say N_1 and N_2, have a common internal vertex. Then some internal vertex t of N_1 is joined to s_1 by an arc T in N_2. Choosing t and T so that T is as short as possible, we arrange that $N_1 \cup T$ is a Y-graph with center t and ends r_2, s_1, and s_3. The union of this Y-graph with the three arcs L_i is a subdivided $K_{3,3}$. (See Fig. XI.9.4.) The six nodes are x, y, t, s_1, s_3, and r_2.

We may now suppose N_1, N_2, and N_3 to be internally disjoint. The situation may not be quite so complicated as in Fig. XI.9.3. It may happen that $r_1 = s_1$, $r_2 = s_2$, and $r_3 = s_3$. In that "canonical" case, we have a subdivided K_5. Its nodes are x, y, r_1, r_2, and r_3. Its branches are

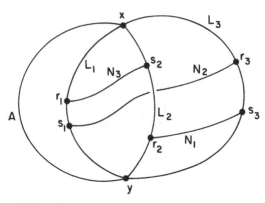

FIGURE XI.9.3

$G \cdot \{A\}$, N_1, N_2, N_3, and the six arcs into which L_1, L_2, and L_3 are subdivided by r_1, r_2, and r_3, respectively. Even if none, or not all, of these three coincidences occur we can transform to the canonical form by contracting the edges of the subarc of L_i, if any, that joins r_i and s_i, for each suffix i. The contracted G then has a subdivided K_5, and so the original G has a Kuratowski subgraph, by Lemma XI.61. This completes the argument for the 3-connected case.

Now let G be, if possible, a 2-connected graph which is nonplanar but contains no Kuratowski subgraph. Choose such a G with as few edges as possible. By the results already proved, G has a 2-separation (H, K), with hinges x and y, say. Form H_1 and K_1 from H and K, respectively, by adjoining a new edge A with ends x and y. By Theorem XI.28 we may suppose H_1 nonplanar. It has, by the choice of G, a Kuratowski subgraph J (see Theorem IV.21). But J remains a Kuratowski graph when we replace in it the edge A by any arc in the connected graph K joining x and y. (See Theorem IV.20.) We thus obtain a Kuratowski subgraph of G. This contradiction completes the argument in the 2-connected case.

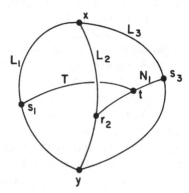

FIGURE XI.9.4

Exercises 325

For a general connected graph G, we observe that, if it is nonplanar, then one of its blocks is nonplanar, by Theorem XI.21, and that this block, a 2-connected subgraph of G, has a Kuratowski subgraph. If an arbitrary graph is nonplanar, then one of its components is nonplanar, and so has a Kuratowski subgraph. The proof of Kuratowski's Theorem is now complete.
□

XI.10. NOTES

XI.10.1. 3-Connected Planar Maps

The construction suggested in Examples 1 and 2 below has been carried through for graphs of up to 21 edges. This was in connection with the problem of dissecting a square into unequal squares, as few as possible. (See [2] and [3].)

XI.10.2. Cubic Planar Maps

Theorem XI.4, applied to a planar map with a cubic graph, shows that there must be at least one face with valency 5 or less. This result is exploited in proofs of the Five-Color Theorem.

EXERCISES

1. Let G be a 3-connected planar graph having at least six edges. If G is not a wheel, show that either G or its dual can be derived from another planar 3-connected graph H by link-adjunction.
2. Using the rule stated in Exercise 1, find all the 3-connected planar graphs of from 6 to 11 edges.
3. Discuss the maps on a given theta-graph.
4. For cubic maps on the torus, show that either every face has valency 6 or there is at least one face whose valency exceeds 6.
5. Consider the cube-map, whose graph G is that of a cube and whose faces correspond to those of the cube. Let C be a 6-circuit in G. Find its pinch-maps and cap-maps.
6. Show that the graph of a convex polyhedron has a planar mesh and is therefore planar.
7. The algebraic dual of a circuit is a bond. For 2-connected graphs, what is the algebraic dual of a bridge over the circuit?
8. Use Theorem XI.43 to show that the Petersen graph is nonplanar.
9. Verify Theorems XI.56 and XI.57 for the Petersen graph. How many peripheral circuits does this graph have through a given edge?

10. Find a Kuratowski subgraph in (i) the Petersen graph, and (ii) the graph obtained from the cube-graph by joining two opposite vertices by a new edge.

REFERENCES

[1] Brahana, H. R.., Systems of circuits on two-dimensional manifolds, *Ann. Math.*, (2) **23** (1921), 144–168.
[2] Duijvestijn, A, J, W, Thesis Eindthoven, *Philips Res. Reports*, **17**, (1962), 523–613.
[3] _____, Simple perfect squared square of lowest order, *J. Comb. Theory*, (B) **25** (1978), 240–243.
[4] Kuratowski, K., Sur le problème des courbes gauches en topologie, *Fund. Math.*, **15** (1930), 271–283.
[5] MacLane, S., A structural characterization of planar combinatorial graphs, *Duke Math. J.*, **3** (1937), 460–472.
[6] Thomassen, C., *Kuratowski's Theorem*, Aarhus Univ. Preprint series 1980/1981, No. 17.
[7] Tutte, W. T., How to draw a graph, *Proc. London Math. Soc.*, **52** (1963), 743–767.
[8] Whitney, H., Nonseparable and planar graphs, *Trans. Amer. Math. Soc.*, **34** (1932), 339–362.
[9] _____, 2-isomorphic graphs, *Amer. J. Math.*, **55** (1933), 245–254.

Index

Abstract graph, 6
Adjacent vertices, 2
  in graph of 3-blocks, 104
Algebraic duality, 206
Alternating barrier, 168-169
  accessible vertices of, 170
  and alternating paths, 169
  center of, 169
  components of, 168
  condition for, 169
  inner vertices of, 168
  irregular darts in, 169
  outer graph of, 168
  outer vertices of, 168
  Theorem of Alternating Connection, 170
Angle in map, 268
Arborescence, 126
  condition for, 138
  construction of spanning, 134-135
  converging, 127
  diverging, 126-127
  multiple diverging, 142
  paths in, 131-132
  residual, 135
Arc, 3-4
  adjunction of, 57
  in bridge, 29-30

characterization of, 17
condition for, by paths, 132
and connection, 16, 21-22
decomposition, by vertex, 66
directed, 131
edge-contraction in, 39
internally disjoint, 47
subdivision of, 77
as tree, 21
Arm, 64-66
  and blocks, 64-66
  connection of, 64
  and 1-separation, 65
Assignment Problem, 52
Automorphism, of graph, 6
  of premap, 263
Automorphism group, of graph, 8, 31
  of map, 263
  of premap, 263

f-Barrier, 171
  deficiency of, 172
  even and odd components of, 171-172, 173
  existence theorem for maximal, 176-177
  and f-factor, 172
  the f-Factor Theorem, 174-176

maximal, 176
outer graph of, 171
parity theoem for, 173
Basic permutation of preamp, 254
Beraha numbers, 250
BEST Theorem, 137
Bicursal unit, 167
  constituent of, 167-168
  edges of attachment of, 167-168
Binding number of complementary pair of
  subgraphs, 14
  and cutting pairs, 43-44
Binding set of pair of subgraphs, 14
Bipartite graph, 51
  complete, 79
  f-factors of, 180
Bipartition of graph, 51
Block, 60-64
  and circuits, 61
  and components, 61
  and 2-connected subgraphs, 60
  and cut-vertices, 62-64
  extremal, 64
  and loop or isthmus, 61
  of planar graph, 298-299
  of primitive chain-group, 208-209
  in Reconstruction Theory, 121-122
  and separable graphs, 61
3-Block, 83-104
  closed outflow from, 87
  edge-disjointness of, 92
  graph of 3-blocks, 93-95
  leading, 87
  outflow from, 87
  outlet of, 87
Block-graph of graph, 63-64
Bond of graph, 200-201
  end-graphs of, 200-201
  Hamiltonian, 236
Bond-vertex, 296
Boundary of 1-chain, 197
Bounding sequence of face, 265
Bridge, 27-30
  algorithm for, 118
  avoidance of, 311
  circuit with two bridges, 316
  degenerate, 28
  equivalent bridges of circuit, 311
  and Jordan's Theorem, 312
  nucleus of, 28
  outer vertex of, 28
  overlapping, 311
  overlap theorem of 3-connection, 317

skew, 311
  in subgraphs, 30
Bridge-fragment, 314
  2-connection of, 314
  and planarity, 315
Bridge-graph, 313
Brooks' Theorem, 233-235

Cap-graph, 294
Cap-map, 294
Capacity, of cut in digraph, 155
  of vertex in digraph, 154
Cell in map theory, 253
Cell-base of chain-group, 186
  condition for, 192
  and elementary chains, 187, 191
  existence theorem for, 186
  primitive, 189
Chain, 185
  0-chains and 1-chains, 197
  elementary, 186
  enclosure for, 186
  linear dependence of, 186
  nowhere-zero, 228-229
  primitive, 188-189
  support of, 186
  unit, 185
Chain-base, 189
Chain-group, 186
  primitive, 191
Characteristic polynomial, 249-250
Chromatic number of graph, 228
Chromatic polynomial of graph, 226-233
  of circuit, 229
  of clique, 230
  coefficients in, 231-232
  for positive integers, 228
  recursion formulae, 226-227
  sign, 233
  of tree, 229
  of union, 230-231
  and vertex-joins, 230
Circuit, as 3-block, 95-98
  characterization of, 17-18
  condition for, by paths, 132
  connection of, 16
  directed, 131
  edge-contraction in, 39
  Hamiltonian, 236, 243
Circuit-map, 287-288, 296
Cleavage, 95
  and virtual edges, 101
  relation to another cleavage, 102-103

Clique, 2-3
  connection of, 16
  3-connection of, 76
Coboundary of 0-chain, 200
  unoriented, 219
Coboundary group, 200
  cell-base of, 203
  as dual of cycle-group, 203
  elementary chains of, 201-202
  primitivity of, 202
  rank of, 204
Coboundary rank of graph, 204
  and planar duality, 288-289
Coloring of graph, 227
  and chromatic polynomial, 228
  condition for 2-coloring, 235
  and Hamiltonian bond, 236
Combinatorial sphere, 276
Complementary subgraph, 13
Component of graph, 14-18
  algorithm for, 118
  characterization by paths, 133
  connection of, 16
Component of premap, 256-257
Component of map, 260
Conductance of dart, 138
Conforming chains, 194-195
Conforming pair in map theory, 277
Connection of graph, 14-18
  by arcs, 21-22
  of bridge, 29
  list of graphs with, 22-27
2-Connection of graph, 54
  and bridges, 56
  and circuits, 56
  constructions for, 56-60
  and loops and isthmuses, 55
  in planar maps, 68, 297
  and recognizability, 117
  and separation numbers, 56
  and unions, 57
  and vertical 2-connection, 74
3-Connection of graph, constructions for, 74-83
  and edge-contraction, 105-111
  and edge-number, 71
  in planar maps, 113, 317-320
  and strictness, 71
m-Connection in graph, 70
  cyclic, 71
  vertical, 70
Connection-set of premap, 256
Connectivity of graph, 72

  and duality, 286
  and edge-contraction or edge-deletion, 73-74
  infinite, 72
Connectivity of chain-group, 210-215
  in planar maps, 296-306
Contraction of chain-group, 187
Contraction of graph, 32-36
  and components, 35
  repeated, 36
  and subgraphs, 35
Contraction of link in map, 272-273
  use of, 276-277
Cross, in map theory, 253, 283-284
Cross-cap, 277
Cross-cut of circuit in map, 306
  skew cross-cuts, 310
Cross-Cut Theorem, 306-307
Cubic graph, 5
Current in dart, 145
Cut-vertex, 54
Cutting pair of edge-sets, 43-44
Cutting pair of vertex-sets, 50
Cycle, 197
  bounding, 283
  elementary, 199
Cycle-group, 197
  cell-base of, 199
  as dual of coboundary-group, 203
  primitivity of, 199
  rank of, 200
Cycle-rank of graph, 288
Cyclomatic number of graph, 19
  of components, 40
  of forest, 19

Dart, 125
  cursal, 161
Deletion of edge in map, 273-275
Dichromate of graph, 243-248
  of 4-clique, 246
  coefficients, 246
  and duality, 289-290
  examples, 245-246
  recursion formulae, 244
  and spanning trees, 247-248
  and tree-number, 246-247
  Unimodal Conjecture, 246
Dichromatic polynomial, 225-226
Digraph, 125
  connected, 132
  duality for, 283
  equivalence with graph, 127

Eulerian, 125, 133–134
  graph-connected, 132
  strongly connected, 132
Disconnected graph, 121
Drawing of graph, 2
Drawing of map, 263–268, 286–287
Duality, of chain-groups, 188
  of cycles and coboundaries, 283
  intimations of, 36, 41
  of loop and isthmus, 290
  of premaps and maps, 259–261
  of primitive chain-groups, 193–194
  of spanning trees, 289

Edge of graph, 1
  of attachment, 166
  contraction of, 37–41, 67–68
  cursal, bicursal, unicursal, 161–162
  deletion of, 18–22, 66–68
Edge of premap, 255
Edge-reconstruction, 122–123
Edmonds' Construction, 255
End of arc, 5
End of edge, 1
End-graph of bond, 200–201
End-graph of isthmus, 18
  in arc, 21
  in tree, 20–21
Entrance of bicursal component, 163
Entry-dart, entry edge, 163, 168
Entry-path of bicursal component, 163
Erdös–Gallai Theorem, 181–182
Essential edge in 3-connected graph, 109–111
Euler characteristic of premap, 260
  additivity over components, 261
  maximal value for map, 276
Euler's Theorem, 137

1-Factor of graph, 51
  partial, 51
  theorem, 177
f-Factor, 170
  algorithm for, 183
  of bipartite graph, 180
  and parity, 171
  sufficient condition for, 171
  theorem, 177
Flow in digraph, 147
  5-Flow Conjecture, 240
  n-Flow, 240
  Flow polynomial, 237–240
Forest, 19–20
Four-Color Theorem, 237
Frame of graph, 199

Frame, in theory of 3-connection, 106
V-Function of graph, 221–226, 250

Girth of graph, 83
Graph, 1
  dual, 281
  of path, 132–133
  planar, 285
  of preamp, 255
  reconstructible, 116
Graph-function, 221
Graphic partitions, 181–182
Grinberg's Theorem, 236–237

Hadwiger's Conjecture, 52
Hall's Theorem, 50–52, 181
Hinge of 2-separation, 83
Hinge-graph of 2-separation, 95
Half-edge in premap, 255
Handle, in map theory, 277
Hinge of 2-separation, 83
Hub of wheel, 78

Identification of vertices, 149–152
Impedance, 143
Incidence number, 197
Incidence Matrix, 217–219
Intake of graph, in 3-block theory, 87
Interchange Theorem for cell-bases, 190
Internal vertices of arc, 5
Intersection of subgraphs, 10, 27, 63
Invalency of vertex in digraph, 125
Involution in map theory, 253
Isolated vertex, 5
Isomorphism of graphs, 5–9
Isomorphism of premaps, 261–263
Isomorphism class, 6, 261
2-Isomorphism, 114
Isthmus, 18–19

Jordan's Theorem, 290–296

Kelly's Lemma, 119–122
Kirchhoff matrix, 138, 141
Kirchhoff's Laws, 143
Klein bottle, 281
Kruskal's Conjecture, 52
Kuratowski graphs, 313
Kuratowski's Theorem, 320–325

Link, 1
  addition of, 24
  contraction of, 38–41
  of premap, 255

Index

Link-adjunction, 74–75
  and subdivision, 77
Link-graph, 2
Linkage, 26
  as 3-block, 96–98
Link-map, 262
  planarity of, 276
Loop, 1
  contraction of, 37–38
  of premap, 255
Loop-graph, 2
Loop-map, 262
  planarity of, 276

Map, alternating, 158–159
  canonical, 280
  canonical projective, 262, 266–267
  on circuit, 287
  dual, 260–261
  genus, 280
  on graph, 255, 298
  orientability-character, 272
  orientable, unorientable, 257–258
  planar, 276
  rooted, 263, 283
  self-dual, 262
  on tree, 287
  unitary, 275–276
Matrix-Tree Theorem, 140–142
Matroid, 71, 219
Max Flow Min Cut Theorem, 157–158
Menger's Theorem for graphs, 46–50, 53
Menger's Theorem for digraphs, 158
Minor of graph, 36
Minor of chain-group, 187
  of cycle and coboundary groups, 204–205
  and primitivity, 191
Multiple join, 2

Null graph, 2

Orbit of permutation, 254
  conjugate, 254
Orientability of map, 257–259
Orientation of graph, 128
  acyclic, 229
Orientation-class of map, 257
Oriented map, 258
  dual, 281
Orthogonality of chains, 187
Outflow, in 3-block theory, 87
Outlet, in 3-block theory, 87

Outlet of vertex-set in digraph, 155
Outvalency of vertex in digraph, 125

Path, alternating, 161
  circular, 130
  dart-simple, 130
  degenerate, 129
  in digraph, 129–133
  edge-simple, 132
  Eulerian, 133
  factorization of, 130
  in graph, 132
  head-simple, 130
  inverse of, 133
  length of, 129
  linear, 130–131
  origin and terminus of, 129
  product, 129–130
  simple, 130
  tail-simple, 130
Path-bundle, 152
  digraph of, 152
  f-limited, 154
  rotational and irrotational, 152–154
Peripheral circuit, 314
  and 3-connection, 317–322
  existence theorems, 318–320
  intersection of, 323
Petersen graph, 80
  3-connection of, 83
  and Tait colorings, 242
Petersen's Theorem, 177–178
Pfaffian, 183
Pinch-graph of circuit in planar map, 293
Pinch-map of circuit in planar map, 293
  planarity of, 294
Planar mesh, 299–302
  and circuit, 316
  imposition of, on subgraph, 308
  and peripheral circuit, 314
  uniqueness of, for 3-connection, 317
Planarity, 285–326
  algorithm for, 314–316
  condition for nonplanarity, 316
  MacLane's Test for, 301
  tests for, 299, 320
Pole in digraph, 145
Potential in digraph, 145
Premap, 254
  planar, 285
  self-dual, 262
Prism, 113
Projective plane, 266–267

Rank of primitive chain-group, 191
Realization of sequence of subgraphs, 120
Recognizability, 115–116
Reconstruction Problem, 115–122
  and characteristic polynomial, 250
  and dichromate, 248–249
  and Hamiltonian circuits, 249
  Generalizations of, 123
Reduction of chain-group, 187
Reduction of graph, 16
Regular chain-group, 194–196
  representative matrix of, 196
  residual chain-group of, 195–196
Regular element of ring, 188
Regular graph, 5
  recognizability of, 116
  reconstructibility of, 116
Residual graph of circuit in planar map, 291
  enclosure by, 291–292, 306
  marginal enclosure by, 306
  and partition of vertex, 292
Residual graph for Tait colorings, 241–242
Restriction of mapping, 11
Rim of wheel, 78
Rotor, 159, 250

Scalar product of chains, 187
Separable graph, 54
1-Separation of graph, 54–55
2-Separation of graph, 70
  A-major, 87
  A-maximal, 84–86
  hinge of, 84
n-Separation of graph, 70
  cyclic, 71
  vertical, 70
Separation number of edge-sets, 43–45
Separation number of vertex-sets, 50
Smith's Theorem, 243
Snark, 242
Splitting a vertex, in map, 270
Spoke of wheel, 78
Star, 23
Strict graph, 2, 7
String of blocks, 66
String of 3-blocks, 105
Structural property of graph, 11
Subdigraph, 125–126
Subdivision of edge, 77
Subgraph, 9–14
  bicursal, 163–167
  and darts, 163–166
  deficiency, 170

induced, 10
  of planar graph, 297
  primal, 115
  proper, 9
  spanning, 9
Submap, 296
Subpremap, 296
Superficial transformation, 272
Surface, 275–281
  genus of, 281
  orientable and unorientable, 281

Tait colorings, 240–243
  and bonds, 241
  and 4-flows, 241
Tait cycle, 242
Theta-graph, 307
  planarity of, 307
Thomsen graph, 79–80, 113
Tie between edge-sets, 47–50
Torus, 267–268, 281
Tour in digraph, 136
  Eulerian, 136
Trace of vertex or face in map, 291
Trace-graph of vertex or face in map, 291
Transpedance in digraph, 143
  as determinant, 146–147
Transpedance in graph, 148–149
Transportation theory, 152–158
Tree, 19–22
  and arborescence, 128
  recognizability of, 117
  spanning, 20
  as union of trees, 93
  uniqueness of arcs in, 22
Tree-number, 39–40
Triad in graph, 105
Triangle in graph, 104–105
Turn in map, 268
Twig, 22
Type of 2-connected graph, 84

Underlying graph, of digraph, 125
  of path, 131
Unimodal conjectures, 233, 246
Union of subgraphs, 10, 63
  and connection, 17
  for J-detached subgraphs, 27

Valency, in graph, 4, 20
  in map, 288, 325

Index   333

Vertex of attachment, 11–14
  and 3-blocks, 98
  and contraction, 41–42
Vertex of graph, 1
  bicursal and unicursal, 162
  and 3-connection, 71–72
  expansion of, into circuit, 80–83
Vertex of premap, 255
Vertex-bond, 103
Vertex-coboundary, 203
Vertex-graph, 2
Vertex-isomorphism, 33–34
Vertex-join, 230

Vertex-map, 276
Vertex-splitting in graph, 75–76
Virtual edge of graph, 91
  and cleavage, 101

Wheel, 78
  3-connection of, 78
  Theorem, 111–113

Y-graph, 308
  arms of, 308
  and circuit, 309–310
  existence theorem for, 312

ABOUT THE BOOK

Graph Theory makes a major contribution to the burgeoning literature on combinatorics. The central themes of graph theory, such as Menger's Theorem and network flows, the Reconstruction Problem, the Matrix-Tree Theorem, the theory (largely created by Professor Tutte) of factors (or matchings) in graphs, chromatic polynomials, Brooks's Theorem, Grinberg's Theorem, planar graphs and Kuratowski's Theorem, are covered in this reference book. Other topics include the theory of decomposition of graphs into 3-connected "3-blocks," electrical networks, and the classification theorem for closed surfaces. Professor Tutte unifies these topics into a coherent whole and imbues them with his highly individual approach to the subject and with his lively writing style.

Exercises, notes and exhaustive references follow each chapter, making this an outstanding textbook for advanced courses and an invaluable reference for anyone interested in graph theory.

FROM THE FOREWORD

"It is both fitting and fortunate that the volume on graph theory in the *Encyclopedia of Mathematics and Its Applications* has an author whose contributions to graph theory are—in the opinion of many—unequaled... The present book may be expected to play a considerable part in placing graph theory on a sound theoretical and technical footing."

Professor C. St. J. A. Nash-Williams
University of Reading

ABOUT THE AUTHOR

W. T. Tutte studied at the University of Cambridge, receiving the B.A. degree in 1938 and the Ph.D. in 1948, and was a Fellow of Trinity College from 1942 to 1949. He then taught at the University of Toronto from 1948 to 1962 and is presently teaching at the University of Waterloo. He is a Fellow of the Royal Society of Canada, which awarded him the Henry Marshall Tory Prize. In 1982, Professor Tutte was awarded the Izaak Walton Killam Memorial Prize by the Canada Council. He has published several papers and two previous books—one on graph theory and one on matroid theory—and has served for many years as Editor-in-Chief of the *Journal of Combinatorial Theory*.

Addison-Wesley Publishing Company, Inc.
Advanced Book Program
Menlo Park, California

Printed in the United States
By Bookmasters